AF690873

Darwin's Solution

The Impact of Evolutionary Thinking on
Medicine and Biotechnology

Darwin's Solution

The Impact of Evolutionary Thinking on Medicine and Biotechnology

Animesh Ray

Keck Graduate Institute, USA & California Institute of Technology, USA

World Scientific

EW JERSEY • LONDON • SINGAPORE • BEIJING • SHANGHAI • HONG KONG • TAIPEI • CHENNAI • TOKYO

Published by

World Scientific Publishing Co. Pte. Ltd.
5 Toh Tuck Link, Singapore 596224
USA office: 27 Warren Street, Suite 401-402, Hackensack, NJ 07601
UK office: 57 Shelton Street, Covent Garden, London WC2H 9HE

British Library Cataloguing-in-Publication Data
A catalogue record for this book is available from the British Library.

DARWIN'S SOLUTION
The Impact of Evolutionary Thinking on Medicine and Biotechnology

Copyright © 2026 by World Scientific Publishing Co. Pte. Ltd.

All rights reserved. This book, or parts thereof, may not be reproduced in any form or by any means, electronic or mechanical, including photocopying, recording or any information storage and retrieval system now known or to be invented, without written permission from the publisher.

For photocopying of material in this volume, please pay a copying fee through the Copyright Clearance Center, Inc., 222 Rosewood Drive, Danvers, MA 01923, USA. In this case permission to photocopy is not required from the publisher.

ISBN 978-981-98-2230-0 (hardcover)
ISBN 978-981-98-2231-7 (ebook for institutions)
ISBN 978-981-98-2232-4 (ebook for individuals)

For any available supplementary material, please visit
https://www.worldscientific.com/worldscibooks/10.1142/14559#t=suppl

Typeset by Stallion Press
Email: enquiries@stallionpress.com

To my mentors

Ethan, Frank, Ron
and
David Galas

Why and for Whom this Book

The genesis of this book was a conversation with a reputed engineer who questioned Darwinian evolution's ability to explain the intricate complexities of life. Similar doubts surfaced when discussing the enigma of human complexity with several physician friends.

Such skepticism resonates with many, as the truth of Darwinian evolution is often asserted with inadequate justification. The experience of practicing biologists is quite different. During my postgraduate years in India, my mentor suggested that I should find a lab that was doing cutting-edge molecular biology; that I should avoid people who "beat over their heads with evolution," because in his opinion evolution was largely a solved problem. In January 2016, I discussed this issue — that many non-biologists with highly technical knowledge are uncomfortable with accepting Darwinian evolution without questioning — with my friend Maurizio Chiurazzi at their cozy home in Naples over a delicious three-course Italian dinner that Maurizio and Aura graciously prepared. He suggested that I should write a book that explains how Darwin's ideas helped solve so many problems in biology. That appeared too daunting a task then, having completed a book barely three years before and promising myself never to write another. Alas, once the writing bug bites, the fever returns.

This book endeavors to bridge the gap between theoretical evolutionary biology and practical applications, highlighting how evolutionary theory has underpinned advancements in medicine and biotechnology. This is not a book on the biology of evolution. Anyone who has pondered the immense variety of life on earth must have had a spiritual experience. Darwinian evolution provides a sense of poetry that weaves into its fabric solutions to large problems of how we came to be what we are, and how we function in health and in sickness. And yet, to many, it is unclear why we should bother thinking about evolution if our goal is to treat diseases or to discover and develop drugs that treat the sick. I have chosen topics that, in my opinion, make this connection clearer.

There is no topic in applied biology that has not been influenced by Darwin. Therefore, the topics chosen here must reflect my personal bias. Inevitably, there are many omissions — some deliberate, others inadvertent. A glaring omission is that of plant biotechnology, which is another book.

The order in which I have organized the chapters has naturally to do with both impact and relevance as much as the logical progression of ideas. Part I of this book is a breezy introduction to why we care about evolution in the context of modern medicine and biotechnology. This part should be accessible to the general reader. Part II is an in-depth dive into various cases of medicine and biotechnology, which will be relevant for those aspiring to be physicians or those who would like to delve deeper into the molecular basis of medicine and biotechnology. Part II is not easy reading; it should be fortified with selected additional readings suggested at the end of the book for each chapter. Part III provides a more quantitative basis for many of the concepts touched upon in Parts I and II. More mathematically inclined readers and those who would like to do original research on these topics should find Part III useful. Parts II and III can be adopted as the background text for a senior undergraduate or a first-year graduate-level course, fortified with original papers referred to in the text. In several chapters of Part II, I have deliberately opted for a semi-historical narrative. This was done to emphasize the values that I consider important for science.

The book targets upper-level undergraduates and early-stage graduate students aspiring for careers in medicine or biotechnology, as well as those from physical or computational sciences seeking to expand their horizon. Card-carrying engineers and physicians may also enjoy the book. No more than a passing familiarity with the content of an advanced placement biology course in US high schools is needed to understand most of the materials in Parts I and II. Chapters 13–17 need a year of calculus; they are meant for those who would delve deeper. The book may be adopted as a textbook for a two-semester course, with further readings and exercises based on the main chapter topics, in a curriculum preparatory for medical school admission.

Acknowledgment

Prabir Pal and Siddhartha Ray Chaudhuri read an early draft, corrected errors, and encouraged me to continue. Invaluable suggestions from Ethan Signer and Pradip Mukerji enabled me to reorganize the book completely; comments from Frank Stahl and Ron Skurray helped correct numerous mistakes and provided a warning to be more careful with language. I have tried but not fully succeeded. To Ethan, Frank, Ron, and Pradip, I owe much gratitude. Chris Adami did a quick review of the first draft and assured me that I was not making a fool of myself. If I have, it is not Chris's fault.

Ravi Salgia and Julia Xu read parts of the first draft and patiently subjected themselves to many questions from a novice. To them, I owe sincere thanks. Ravi encouraged me to indulge in a more personal narrative. Upon much hesitation, I have done so.

Comments from Yun Liu helped me calibrate the text for my audience, and Shiva Abdolrahimi helped re-derive an important mathematical result. Insightful comments from Paul Sternberg, Shibu Yooseph, Jesse Frumkin, Atanu Sen, and Arup Sen helped me correct several errors. Comments from my students, especially Ashwin West, Princess Mae Valois, Ravneet Gill, and Jaylan Brown, were important. Invaluable corrections and critiques on the mathematical sections were provided by my friend Ken Alexander.

The journey of writing this book commenced amid the solitude of a rainy Ecuadorian December in 2021 and culminated during a sabbatical leave in 2023–2024. I extend my gratitude to my editor, Sandhya Devi, publisher of the book, and the Keck Graduate Institute for their support during this process. I was funded during this period by an NIH Director's Transformative Research Award (R01-AI169543) and a US Department of Energy award (DE-SC0025244). Most figures were constructed using the BioRender software licensed from www.biorender.com.

This book would not have been possible without the sacrifice of my wife, Sumita, and her constant encouragement. To her, I express my heartfelt gratitude.

Franklin W. Stahl, my mentor who profoundly influenced my life in science and as a human being, kindly wrote the Preface. He encouraged me to make this book a better read and wanted to see it published. On February 7, 2025, he wrote with characteristic brevity, "I'll stick around through 2025, and help when asked, if possible. Good luck!"

Frank passed away on April 2, 2025, without seeing this book printed.

Endorsements

"When I began my professional career nearly 70 years ago what I anticipated was a peaceful sojourn in academic science. At that time medical considerations were merely tangential and happily this is no longer the case. From that perspective this book with its range from molecular genetics all the way to therapy is quite remarkable, and will be particularly valuable for scientists just entering an expanding field with limitless prospects for human health."

—Ethan R. Signer
Professor Emeritus, Massachusetts Institute of Technology

"Animesh Ray has written a truly exciting book, a joy to read, that interweaves the science with the history behind some of the most significant advances in molecular biology in fields that include cancer and oncogenes, virology with an emphasis on coronavirus evolution, immunology, and aging. Uniquely, in some chapters, he employs imaginary conversations between key scientists to set the stage for the details of that chapter and two chapters are devoted to transcripts of insightful conversations between himself and two clinical scientists, an oncologist, and a hematologist, about their career paths, their day-to-day patient care, and their research. This is a book that I would love to have at hand to illustrate to my undergraduate and graduate students the influences of evolutionary

biology and molecular genetics on modern medicine and biotechnology, and hence on our everyday lives. Reading this fascinating book should also appeal to professionals in medical, biological, and wider fields."

—Ronald A. Skurray
Emeritus Professor, University of Sydney

Preface

*Franklin W. Stahl**

Animesh, my friend and former student, identifies three steps in his youthful search for the Meaning of Life. He quickly sees through Calcutta's three major religions — they cannot even explain themselves. He offers a sharp summary of the meaning of Darwinian life: God out, Nature in. It always works. But how?

A Darwinian adaptation by mutation of a gene encoding a relevant enzyme or substrate necessary for the adaptation would be ideal. We would then know that the meaning of Life is a planet, with the elements, temperature, and time that grant chemistry its needed elbow room.

Animesh finds Darwin to be such a reliable guide whenever called upon that he goes all out with Darwin when explaining disease progression at the molecular level. In the remaining pages of this remarkable book, Animesh offers his thoughts on recent advances in the molecular biology of disease processes. He has led us here with autobiographical passages outlining his road to understanding. I will use a similar autobiographical approach in this Preface. It has brought me to where I wanted to go.

In my first faculty position (1958), I was asked to teach Introductory Genetics in the Zoology Department of the College of Arts and Sciences of the University of Missouri. "What could be easier?" I thought than to follow that well-worn path of teaching Genetics, which starts with Mendel's analysis of the offspring from mating between inbred strains of garden peas. I soon found, to my chagrin, that I could not do it. With acknowledgements to Mendel's genius, I announced that, in the next lecture, we would start over. And we did — with exponential growth of E. coli. Molecule-based Genetics followed easily. Soon thereafter, I wrote the slim book, The Mechanics of Inheritance — on which Animesh bumped his toe.

The first review (by Dick Lewontin in Science) savaged my book. Dick thought that the book was inexcusably undignified (my words, more polite than his). On the beach a few days later, Barbara McClintock declared, "What a really bad book it is." I replied, "Barbra, what didn't you like about my book?" "Franklin," she sighed, "You've taken all the mystery out of it."

So, Dear Reader, take off your slippers, pull on your boots, and rejoice that Molecular Biology now holds a central place in the diagnostic laboratory. The mystery is out — technology in.

February 7, 2025, Eugene, Oregon

[*Dr. Franklin W. Stahl was a lifetime American Cancer Society Research Professor at the Institute of Molecular Biology, University of Oregon, Eugene, Oregon. He was a MacArthur "Genius" Fellowship awardee (1985), a member of the National Academy of Sciences (USA), a member of the American Academy of Arts and Sciences, and a winner of the Thomas Hunt Morgan medal awarded by the Genetics Society of America. In 1958, Matthew Meselson and Franklin Stahl published the unequivocal experimental evidence that DNA replicates by a "semi-conservative" mechanism, distinguishing among three possible modes of replication — a work that has been described as the "most beautiful experiment in the history of biology."]

Contents

A Word to Students

Imagine that you are stranded on a deserted island with this book, a box of chocolates, and a fishing rod.

After finding a nice cave to settle in, collect some coconuts and berries to eat for breakfast. Then, read Part I in one sitting.

Go catch some fish. But don't eat the chocolates yet.

From the following morning, read Chapters 1–5 of Part II in sequence, as leisurely as you wish. If you'd like to take a deeper look at how we make certain claims in these chapters, feel free to read Chapters 13–15. Then continue reading Chapters 6–12 of Part II in any sequence, and, if you like, you can also read Chapters 16–17 of Part III if you are curious. Upon reading each chapter of Part III, treat yourself to a chocolate or two. You deserve the dessert. You should, however, leave enough time every day to collect coconuts, birds' eggs, fish, and fresh fruits, to explore the island, and to enjoy the surf. If you are getting bored of the math in Chapters 13–17, then feel free to jump to Chapter 18. That last chapter is meant to make you feel uneasy.

By the time you have come to the end of the book, may a ship notice your smoke signal, and I hope you will be happily on board, sailing home. And you will have just emptied your box of chocolates.

Part 1

L'Aperitivo

I A Puzzle in the Galapagos

On December 7, 1907, a patient walked into the Chicago clinic of Dr. Arthur R. Elliott, MD. The patient, identified only as F. R., was a 17-year-old girl who exhibited symptoms of progressive ossification of her soft tissues. Dr. Elliott noted in his physical examination (Elliott 1911):

> *"The patient's stature was small, her face pale and round, her hair fine in texture and moderately abundant. The first point of interest to be remarked was the posture....(m)axillary movement was much restricted, so that it was possible to separate the teeth about an inch only...This restriction of movement and deformity was due to the presence of bony induration of the submaxillary tissue....When the patient was lying down, by placing a hand behind the head one could raise the trunk as if the whole spine were a rigid rod. A broad bony mass lying on either side of the lumber and lower dorsal spine could be felt as if the erector spine muscles were turned to bone....The pectoral muscles were hard and contracted, the arms being locked to the sides so that abduction was impossible...."*

Dr. Elliott's diagnosis was that of a rare disease termed *myositis ossificans progressiva*, now renamed *fibrodysplasia ossificans*. First described in the early 1740s, today this disease is known to be only one of several inherited genetic conditions, all leading to bodily malformations in babies, including an abnormally small eye on one side of the face, a cleft palate, and unusually small kidneys. In each of these inherited diseases, the function of one human gene, called by an enigmatic name *Bone*

Morphogenetic Factor 4 (*BMP4*)[1], is abnormal. As the child grows, other more debilitating conditions set in. The patient's muscles, tendons, and fibrous tissues are progressively replaced by abnormal bony tissues, confining the patient in a growing prison constructed of an ossified exoskeleton.

As it so happens, a close relative of the *BMP4* gene lays down the pattern of beaks in the embryos of birds, including finches that inhabit the Galapagos Islands about a thousand miles off the shores of South America. Observations on Galapagos finches contributed much to Darwin's thinking about evolution by natural selection.

***

January 7, 2022. From our cabin on Isla Santa Cruz, an island of the Galapagos, the sunrise was a light shade of pink, then a brighter hue turning into an orange splash. I stepped out on the wooden patio, a little damp from last night's drizzle. Five small ground finches hopped on the loose soil from which biting ants had swarmed my feet yesterday. The itch was a memory but for the red blotches on my ankles. Beyond, the Pacific glistened.

There was a lone vegetarian finch nibbling on bits of leaves on my left. Soon other finches came in waves — a dark-feathered medium ground finch, followed by a yellow warbler finch that caught a larva and furiously banged it against a twig (Figure 1), then a pair of small tree finches determined to claim a tree stump as their own. Then came a sharp-beaked ground finch higher up on a gum tree and a pair of Galapagos mockingbirds. The latter are not technically finches, though Darwin had described these as such. As the sun rose higher, a gaggle of large ground finches made their appearance.

Except for the mockingbirds and the yellow warbler finches,[2] none of these species that live on the Galapagos Islands are found on the mainland of continental South America, which lies nearly a thousand kilometers away.

Did these finches originate in the Galapagos Islands or did they come from South America's main continental landmass? If they did originate in

---

[1]Although this gene itself is not mutated, another gene, termed *ACVR1*, needed for the normal function of *BMP4*, is mutated. As a result, *BMP4* function is abnormal.

[2]The warbler finch is a late introduction to the Galapagos by man.

Figure 1 Some finches of the Galapagos islands. (a) vegetarian finch; (b) medium ground finch; (c) yellow warbler finch; (d) small tree finch; (e) sharp beaked ground finch; and (f) large ground finch.

the islands, why doesn't one find native finches in the Easter Islands, which are similarly located in the Pacific but further away from the coast of South America? Did the Almighty favor the Galapagos with finches but not the Easter Islands? If, on the other hand, the finches did come from the mainland, why are these specific species not found in continental South America? And ultimately, how is it that a gene that influences bone development in humans appears to lay down the bony beaks in finches?

Darwin provided one solution to these puzzles, which remains largely consistent with numerous observations and experimentations even today. Since Darwin's days, there have been many, many such puzzles uncovered in biology, and his theory of evolution provided the most intellectually pleasing and consistent solution to all of them. That's a staggering rate of success for a theory.

This book describes the paths to the solutions of a few of those puzzles, paths that have transformed medicine and biotechnology and promise an exciting future. Among those puzzles are the following. Why do genes that, when defective, cause diseases in humans appear to do similar things in birds or other animals? How do we get cancer? Why do we age?

Why are the elderly more susceptible to chronic diseases, like type II diabetes, Alzheimer's, or Parkinson's disease? Can we ever cure these diseases of the elderly? How does our immune system fight virtually endless types of foreign invaders? Can immunity defend us against totally new types of viruses that the world has not yet seen? How do viruses arise in nature? Can one evolve novel biological reagents in the test tube? This book tries to show that behind our dizzying progress in medicine and biotechnology there is a fundamental core of thinking that has powered us. In short, Darwinian evolutionary theory provides satisfactory and useful solutions to these puzzles and many more.

Let's take a step back and collect our thoughts. Darwinian evolutionary theory has two simple postulates, and everything follows from these.

First, variant organisms arise spontaneously and propagate their likenesses to later generations.

Second, those variants that happen to leave more descendants, relative to those of other variants, produce even more descendants that perpetuate the inherited variations. This is called "natural selection." The process is natural because evidently nature, i.e., a combination of environmental factors, does the selection. It is a selective process because a variant's descendants just happen to be favored in a particular environment to produce more progeny relative to those of other variants. No one is choosing — the organisms are just there, and together they sort into different classes by virtue of the efficiency with which each leaves its descendants to perpetuate the respective "races."

The selective process — abstract and unintentional — has been more vividly referred to as a struggle for existence. There might or might not be a psychological dimension associated with this selective process. A single mother in a dystopian city and a gold miner in the fields of *Serra Pelada* might be participants in natural selection as much as populations of bacteria swimming along a gradient of chemical attractants in the human lower gut or of coronaviruses infecting a vaccinated individual. The selection process could simply mean that a bird with a beak slightly more adept at breaking open hard yet plentiful seeds can eat better than other birds without a similar beak; eating better means laying better quality and/or more eggs. Or it could mean that a coronavirus spike protein's genetic variant that is slightly better at binding the cell surface receptor protein may enter human cells ever so slightly more efficiently than other coronaviruses; so, the former virus reproduces slightly more viral progeny that are themselves slightly better at infecting further human cells.

This simple theory is like a coiled spring; it has been enormously successful because much can be derived and/or deduced from these two elemental principles. This book tells stories of such uncoiling.

II A Skeptical Indian

I recall an encounter with a seasoned engineer, an alumnus of a renowned California institution, who said over coffee and donuts, "*Don't expect me to believe that the meticulously structured eye of a fruit fly, with its austere geometric symmetry, is merely a product of chance mutations*." Many engineers and even some physicians are perplexed by the proposition that mere chance — manifested through seemingly random mutations and the environment selecting competitive variants — can adequately account for the vast array and intricacy of terrestrial life. This variety includes the human body and mind, often perceived as remarkable, if not divine. Can we truly attribute the emergence of a Mozart, a Leonardo, or a Ramanujan solely to chance?

Two legitimate reasons contribute to this skepticism. First, it is challenging to accept that purely chance-induced variations in the genetic material can elucidate the beauty inherent in the unity and diversity of living nature, not to mention human creativity.

The second stems from the inherent difficulty of the human mind to intuitively grasp very large numbers. Wrap your head around the four billion years it took for evolution to shape humans. It's a tough one. We can't rewind life's tape to witness the evolution of current life forms. On the flip side, if you buy into this theory, explanations of all biological phenomena fall into place like a well-timed clock, exhibiting remarkable accuracy and predictive power. This book describes some of these puzzles and their solutions.

An important caution. In science, one can never state with certainty that an explanation of a natural phenomenon is true; one can only assert whether the explanation, the theory, is consistent or inconsistent with observations. In that sense, there has not been another theory in biology that has had more explanatory power than Darwinian evolutionary theory.

Ultimately, the willingness to admit Darwinian theory demands a prior intellectual agreement that we suspend our intuition provisionally until the evidence in support or against is weighed. Much has been written about the failure of human intuition ("naïve realism," in philosophical jargon) to grasp the truth of nature. Perhaps the most masterly articulation is the following (Bertrand Russell 1940):

> *"We all start from 'naïve realism', i.e., the doctrine that things are what they seem. We think that grass is green, that stones are hard, and that snow is cold. But physics assures us that the greenness of grass, the hardness of stones, and the coldness of snow are not the greenness of grass, the hardness of stones, and the coldness of snow that we know in our own experience, but something very different...Naïve realism leads to physics, and physics, if true, shows naïve realism to be false. Therefore naive realism, if true, is false; therefore it is false."*

It has not been an easy journey for me personally. I was born in a traditional Hindu family. By the time I was ten, I was devoutly religious. But within another year, I had begun to question small inconsistencies built into my belief structure.

I do not remember the exact triggers that might have led me along this path. There must have been multiple factors. Some of these might have had to do with being brought up in a Hindu family first by a Christian caretaker, whose daughter *Phulmoni* was my only childhood friend, and then by a widowed Muslim woman whose husband was murdered by a Hindu mob during the partition of India.

My fondest early memories are of attending Mass in the Catholic *Church of the Sacred Heart* a stone's throw from our rented home in a tiny northern suburb of Calcutta, whose bells would ring every hour of the day; of waking up in the morning to the sound of *Azan* floating in from the neighborhood Islamic mosque round the block, from where an irascible gaggle of geese would often chase me away.

Perhaps it was my chance encounter in the town library with ten volumes of a long-defunct children's encyclopedia that was published nearly forty years earlier. In utter neglect in a dark corner of the dusty library, they were slowly being turned to dust by bookworms. Among their crumbling pages, I discovered sections on astronomy, cosmology, and thoughts of ancient Greek philosophers. There, I discovered new heroes beyond those in the *Ramayana* and the *Mahabharata*. Socrates in his heroic defiance against authority; Archimedes, irascible and brilliant, dying under a javelin jab while solving a math problem, oblivious to the fall of his city; and Thales of Miletus measuring the earth's diameter by the seasonal shifts in the sun's shadow — they became my new obsession.

As I reflected on the superficial aspects of the religions I had encountered thus far — Hinduism, Christianity, and Islam — it struck me that their conflicting doctrines posed a challenge. Prohibition on beef in one, on pork in another; the varying importance of idolatry — these seemed irreconcilable. These and other paradoxes prompted me to question the divinity of Krishna, Muhammad, and Christ simultaneously. Despite the commendable Indian tradition of seeking a common core in various religious beliefs, my ignorance, relative immaturity, impatience, or perhaps sheer stupidity led me to reject all religions. At fifteen, upon discovering Bertrand Russell, I privately declared myself an agnostic. One more year would pass before I had read any Darwin.

During that period, typical of many teenagers, I grew wary of accepting textbook teachings that lacked ample evidence as unquestionable truths. This skepticism earned me notoriety among some of my teachers, a sentiment gleefully shared by classmates who reveled in the ensuing chaos. The concept of evolution, as presented in textbooks available to me at the time, seemed almost like a tautology. Often, biological explanation of "how" was a mere restatement of empirical facts accompanied by just-so stories resembling historical narratives. Peacock's feathers "are so" because they have been selected to be as such by evolution. The series of skeletal fossils of horse ancestors illustrated the progression of evolution because natural selection led to their adaptation[3] or extinction.

It was hard for me to believe that such an old bird as living nature could be explained away by two seemingly unrelated and apparently trivial postulates of Darwinian theory. The analogy of biological evolution

[3]Adaptation is a measure of how well an individual or a species can survive in a particular environment or a range of environments — a measure of long-term evolutionary success.

to history was not much help either, because my mother had a higher degree in history — familiarity breeds contempt. I needed convincing from another source.

That source turned out to be the emerging understanding of how biology works at the level of molecules. Perhaps the clue to how living things came about is hidden within the smallest things that make up life — just like the story of how a radio works is about electrons flowing through valves, resistors, and capacitors. Only the former story promised to be mightily more interesting than radio engineering because there's unlikely to be an ultimate engineer. The first of these promises was a slim book that my tenth-grade biology teacher had lent me, *The Mechanics of Inheritance*, by Franklin W. Stahl. It was also my first book on biology, which had posed problems at the end of each chapter, the solutions to many of which had eluded me.

But do we really need all these special pleadings by molecular-level biology to take Darwin seriously? The British-Indian polymath J. B. S. Haldane (1892–1964) wrote in his book *The Causes of Evolution* with characteristic pithiness (Haldane 1941):

> *"Darwinism has been a subject of embittered controversy ever since its inception. The period up till Darwin's death saw a vast mass of criticism. This was mostly an attack on the doctrine of evolution and was almost entirely devoid of scientific value. The few really pertinent attacks were lost amid a jabber of ecclesiastical bombinations. The criticism was largely dictated by disgust or fear of this doctrine, and it was natural that the majority of scientific men rallied to Darwin's support. By the time of Darwin's death in 1882, Darwinism had become orthodox in biological circles. The next generation saw the beginnings of a more critical attitude among biologists. It was possible to criticise Darwin without being supposed to be supporting the literal authenticity of the Book of Genesis. The criticism came from all sides. Palaeontologists, geneticists, embryologists, psychologists, and others found flaws of a more or less serious character in Darwin's statements. But they almost universally accepted evolution as a fact."*

I cannot match Haldane's temerity or the depth of his intellect. I therefore take a view that scientists have found relatively useful and epistemologically humble — I urge you to suspend your belief for the time being on the question of whether the crux of Darwinian evolutionary theory is true. I invite you to tentatively assume for the purposes of this book the two main tenets of Darwinian evolutionary theory mentioned above and then see how the consequences of these two rather simple assumptions unwind like a coiled spring to provide numerous predictions, some mathematically precise, which turned out to be observably true. I also hope to show how such an unwinding of the coiled spring has provided, over the past five decades, a gateway to biotechnological feats worth trillion-dollar yields.

The mechanism of evolution is a living and growing science. Science feeds on inconsistencies to correct itself. Inevitably, our current understanding of the mechanisms of evolution is filled with inconsistencies, gaps, and controversies. These controversies are not the stuff of this book. This book, instead, deals with the consequences and the impacts that evolutionary theories have had on matters not directly concerned with the study of evolution as a science. Here, we examine how thinking in explicitly Darwinian terms has been enabling groundbreaking strides in medicine and biotechnology.

III The Molecular Connections

In the modern synthesis of Darwinian evolutionary theory, old notions of similarity and diversity of anatomical characteristics — think of a bird's beak or a horse's jaw — unite in quantitative rigor with DNA and protein sequence diversity and similarity. Darwin's thoughts find new life by entwining evolutionary notions, enriched in quantitative rigor, with concepts of modern genetics, about which Darwin in his time had no direct knowledge — Darwin did not read Mendel's experiments on the inheritance pattern in sweet peas, although Mendel did read Darwin. Mendel had no idea of genes or chromosomes, let alone of DNA or protein. Mendel's discoveries were algebraic abstractions that only in the early 20th century found physical meaning in genes arranged on strings of chromosomes.

The crux of Darwin's theory of evolution in its modern synthesis can be stated in just two points: inherited genetic variations are produced because of errors in DNA/RNA replication or damage to these molecules that persist; the environment then selects for the biological trait (i.e., the phenotype) of individuals impacted by gene variants.

Thus, genetics forms the backbone of this new narrative, with mutation as the spark of change. Yet, the environment wields its sway, sculpting the fate of species. It is a delicate waltz — genes and environment — crafting the evolving tree of life. In this synthesis, we see Darwin's vision through the lens of microscopes and DNA sequencers. It reveals the intricate mechanics of evolution, a symphony of genes and selection, fashioning the rich tapestry of life's diversity. It is an unfinished journey, reminding us that life's story never truly concludes.

Darwin's collection of finches was important for him to make the crucial hypothesis that organisms continue to evolve when they are geographically isolated. This eureka moment came to him when he compared finches from the South American mainland with finches from various Galapagos Islands. Some of these islands were too far from one another for these small finches to fly across the ocean. These islands formed natural geographic niches in which the birds, isolated geographically, could evolve away from the original species that colonized the islands.

It is uncertain how the Galapagos Islands were first colonized by the finches from the mainland — probably a rare storm had carried a few little birds from the mainland to these volcanic isles. A few must have survived, and they colonized the islands because there was no competition from other birds. Easter Island just happens to be too far away from the shores of South America for finches to have survived the trip buoyed on a storm, which is probably why there are no finches there.

Each Galapagos island has multiple different habitats — some have slender and tall cactuses, others have short, squat ones; some have trees with nuts that required sturdier beaks to crack the nuts open; there are insects also, different insect species at different levels above ground, which are food to the birds. Trees had their different heights — some were short, others tall. The finches evolved to inhabit, to the best of their advantage, these various resource-rich layers of natural habitats.

Recent work on Darwin's finches has shown that a relative of the human *BMP4* gene, whose abnormal function is associated with a range of human diseases described at the beginning of this chapter (see Section I), is responsible for the variation in beak sizes in these bird species. In some species of finches, this gene is more active in distinct patterns of cells in developing embryos than in the equivalent regions of embryos of other species of finches. These patterns of gene activity correlate well with their respective adult beak shapes (Abzhanov et al. 2004; Sinervo 2005).

Here's an explanation in a nutshell for these findings: subtle genetic changes, evidently due to DNA sequence variation or mutations in or around the *BMP4* gene, had occurred at random. If such a change had occurred in the cells that gave rise to an egg or a sperm cell, then the chick could inherit this variant. Most of these DNA sequence variations either do not change anything at all for the next generation of birds, or if a few of these changes do affect the beak shape, most of these variations were likely detrimental to the birds, so these birds died out. Rarely, very rarely, a chance sequence variant provided a slight, perhaps very slight, advantage to the bird that had inherited that gene variant from its parent bird, by allowing it to eat better than most of its neighbors. This was perhaps because its variant beak structure allowed it to break slightly harder nut shells than its neighbors could. This obviously led to slightly better nutrition for the possessors of this gene variant, which could have resulted in a slightly higher number of eggs having been laid by the variant birds. Note that the gene sequence variant that the parent birds had inherited from their ancestors will be passed on to the subsequent progeny. Further variants similarly would arise in subsequent generations, and a few very rare further variants could have produced an even better ability to crack these hard nuts, which would have been similarly passed to subsequent generations of birds, leading gradually to the selection of a beak shape difference. Ultimately, these birds came to populate a different ecological niche characterized by trees with harder nutshells that were not previously edible by the ancestral finches.

This, in principle, is the same mechanism of evolution that has given us the alpha, beta, delta, or various omicron subvariants of the SARS-CoV-2 coronavirus.

Every species of extant organisms on earth must have had its origin from a corresponding ancestral organism that had a slightly different gene variation. In principle, therefore, one can trace back the ancestry of every organism by following the likenesses of their genes through a family tree representing their descent, ultimately from a single organism. This tree is more like a web because sometimes DNA is transferred between species — so-called horizontal transfer of DNA.

All organisms, including us, are therefore related by a grand common ancestry — the web of life.

IV How We Got to Think This Way

The picture sketched above was not always so clear. Al-Jahiz (c. 776–868/869) was a 9th-century Arab scholar from Basra, in present-day Iraq. In his book *Kitab al-Hayawan* (*The Book of Animals*), Al-Jahiz discussed perhaps for the first time in recorded history the concept of natural selection. He wrote about how the environment influences the characteristics of animals and how species can change over time in response to their surroundings:

> *"Animals struggle for survival, for resources, to avoid predation and to reproduce. Environmental factors influence organisms to develop new characteristics to ensure survival, and thus transform into new species. Animals that survive to reproduce can pass on their successful characteristics to their offspring."*
>
> — *from The Book of Animals, by Al-Jahiz*

Al-Jahiz's speculations, however, were mostly unknown in Europe until the late 19th century. To my understanding, he did not explicitly propose that the generation of variation and the selection by the environment were independent processes. Enter Carl Linnaeus (1707–1778), a botanist with a knack for organization. With a methodical approach, he started naming and classifying every living thing he encountered, from majestic oaks to tiny insects. It was as if he was giving order to God's whimsical creations. Linnaeus, however, did not attempt to find a reasonable explanation for these resemblances and differences among organisms, relegating these to the unquestioned workings of the original creator.

Fast forward to the Galapagos Islands, where Darwin observed diverse species and their unique adaptations, in which different species flourish in different ecological sectors (niches). His observations led to the groundbreaking theory of evolution by natural selection. It seemed God's perfect plan was a bit more dynamic than previously thought. Darwin, however, searched in vain for a mechanism for producing variations and their inheritance.

Then came Gregor Mendel, a monk with a passion for gardening. Through his experiments with pea plants to attempt to quantify methods for generating hybrid seeds, he uncovered the fundamental laws of heredity, revealing how traits are passed from one generation to the next. His work, refined by Bateson (1861–1926), provided a missing piece in Darwin's theory, explaining the mechanism of trait inheritance.

In the early 20th century, working with fruit flies, Thomas Hunt Morgan (1866–1945) found that he could generate heritable variants (mutants) using physical agents, such as X-rays. Around the same time, using the humble bread mold in their experiments, George Beadle (1903–1989) and Edward Tatum (1909–1975) showed that genes direct the production of enzymes. This discovery expanded with Edward Lewis's (1918–2004) work, which showed that genes also control the synthesis of other proteins that are not necessarily enzymes.

Bacteria and viruses soon joined the story; they, too, have genes that can mutate and recombine. This set the stage for studying life at the molecular level. Key experiments with bacteria and bacterial viruses identified DNA as the carrier of genetic information. Starting in 1953, the double-helical structure of DNA with complementary pairing of four nucleotide bases explained how information could be duplicated, propagated, and even mutated.

A new problem arose: DNA, the code of life, resided in the cell's nucleus, but proteins — the agents of biological work — were synthesized in the cytoplasm in higher organisms. How does the information flow from the DNA in the nucleus to proteins in the cytoplasm? The solution came in 1961 from Sydney Brenner (1927–2019), Francis Crick (1916–2004), Matthew Meselson (1930–), and François Jacob (1920–2013), who discovered mRNA, the messenger that carries genetic information from DNA to ribosomes, where proteins are made. This discovery established the central dogma of molecular biology: DNA directs RNA synthesis, which in turn directs protein synthesis. Shortly thereafter, it was found that the three-letter genetic code was universal — all terrestrial living beings use the same triplets of RNA nucleotide bases to encode proteins — a remarkable capstone of unity in the immense diversity of life.

Thus, from Linnaeus's early taxonomy to the molecular discoveries of the 20th century, the story of life's complexity unfolded. Every discovery built upon the last, creating a cohesive understanding of how life's intricate processes are orchestrated. This narrative of curiosity and scientific inquiry revealed the profound interconnectedness of all living things in the web of life.

V Genes Behind Human Diseases

A consequence of this Darwinian view of life is that all extant life on earth inexorably leads to a single primordial ancestor. This is not an easily intuitive inference but requires special mathematical pleading, which we will pursue later in the book (Chapter 16). For the moment, we will assume this to be true. If we do so, it also leads to the idea that genes, or, more appropriately, DNA sequences, of various organisms ought to be related because we are all related by descent, and it is only the genes that are passed down in ancestry. This latter idea is commonsense because we often inherit the shape of our uncle's nose, grandmother's eyes, or our father's bald patch.

So also does the gene that shapes the beak of a finch, or a similar gene, exist in humans.

From the observed defects caused by mutations that alter the function of a gene termed *ACVR1*, which is needed for the function of BMP4 gene in human patients, we deduce that the function of the normal *ACVR1* gene in humans must be to shepherd the development of a well-formed pair of eyeballs, suture the lips and palates, make a good plumbing system, and keep the bones where they should be. Remarkably, in finches, *BMP4* gene's function is to lay down the bones of the beak in the embryo according to a specific pattern depending on the finch species, such that a species-specific beak shape emerges.

The finding that a gene like the one that controls beak shape in finches can affect human anatomy is staggering to contemplate. The last common ancestors of birds and humans existed about 310 million years ago. A gene that patterns a finch's beak has evidently been maintained through inheritance over 310 million years to shape human organs. There are many, many such examples, some even more spectacular. A relative of a flowering plant gene known as *AGAMOUS* that controls the wonderful whorls of flowers is of importance in sustaining the division of human blood cells (where the gene relative is known as the *SERUM RESPONSE FACTOR*). The last common ancestors of plants and humans had existed about 1.6 billion years ago — here is a gene that attests to over a billion years of evolutionary conservation.

Even more interestingly, the product of the mammalian version of the *BMP4* gene, a member of the so-called Transforming Growth Factor β (TGF-β) family of signaling proteins, is an important constituent needed throughout life. Besides controlling the development of bone, eyes, and brain, it also is needed for the appropriate control of sugar utilization by humans. When TGF-β is in short supply, several disorders of metabolism occur. One of these disorders is a form of diabetes (gestational diabetes) that often strikes expectant mothers. Intriguingly, abnormalities in TGF-β function are involved in certain types of cancer. Thus, this ancient gene's activity is enmeshed within the orchestra of human cellular physiology, providing essential ingredients of a complex life.

What allows for the perpetuation of such ancient genes over more than a billion years? Is it the chance evolution of these genes in the remote past that had succeeded in contributing certain traits and therefore a biological advantage to living organisms, which kept being propagated in succeeding generations?

VI Uncomfortable Questions

Two questions arise. The first is methodological: How can we confirm that the gene governing beak shape in finches is indeed a relative of the gene dictating organ shape in humans? We've developed methods to make such inferences and even measure the genetic divergence on a kind of family tree — a visualization akin to a genealogical tree linking us to our ancestors and so-called blood relatives. We discuss some of these principles later in the book (Chapters 13–16). Following this well-trodden path enables us to make surprisingly useful connections to medicine and biotechnology.

The second question is more fundamental and philosophical. To avoid error in science, we examine a natural phenomenon from multiple perspectives. Up to this point, we have presented a singular viewpoint — gene variations arise spontaneously and are refined by natural selection to increase the number of descendants left by the variant-bearing individual. We have not yet explored the nature of the entity encoded by, for instance, the *BMP4* gene. What if this molecule, the TGF-β protein, is a necessary contingency for performing its biochemical task? In this alternate perspective, the product encoded by the *BMP4* gene could stand alone in the universe, uniquely capable of fulfilling its function, much like the essential role a fulcrum plays for levers in displacing mass: *Sine fulcro nulla displacio* — without a fulcrum there is no displacement!

It's conceivable that the structure of protein molecules, such as that of the TGF-β protein, could undergo fine-tuning, much like an engineer designs a fulcrum of a particular geometry for leverage or a carpenter adjusts the shape of a chisel for smoothing wood. In an analogous manner, a particular version of the TGF-β protein might be necessary for shaping a beak, a bone, or a jaw. From this perspective, the relatedness of genes across organisms may not solely arise due to a common descent but rather from the chemical or physical necessity of performing a specific molecular function. Like how a fulcrum has a unique function, a TGF-β-like protein could be a necessity for shaping organs, with no other vastly different substitute capable of performing that task.

As is true of most philosophical problems, there is no clear solution to this question: whether the relatedness of organisms is fundamentally due to inheritance or due to chemical necessity. Therefore, is the notion of relatedness by inheritance the most reasonable answer?

If we do accept this assertion even provisionally, it immediately leads to the consequence that alternate solutions to the same adaptive problem should exist, which is not necessarily the case if we assume chemical or physical contingency.

For if it were contingent that one biochemical process or step is obligatorily dependent on this or that kind of molecule, then there must not exist drastically different alternate ways to make the same process occur. A chisel might have a wooden or a plastic handle, but for the task of scraping wood a beveled blade is a necessity for a chisel; however, can the task conducted by a TGF-β-like protein be done by a protein that is vastly different, bearing no structural resemblance to TGF-β?

To generalize, are the precise biological solutions for an organism to survive and flourish in the natural world contingent, i.e., physical necessities? Or are they optional, in the sense that there are other alternatives to a particular biochemical entity but are just the way things have turned out among a myriad of different possibilities?

In biotechnology, we are beginning to engineer alternate solutions to the same biochemical problem by synthesizing numerous alternate molecular forms and reagents that, so far as we can tell, had never existed on earth. In fact, generative artificial intelligence is beginning to contribute importantly to this effort. This may seem to support our notion of chance as an important facet of nature, as Darwinian evolution would have it, rather than the theory of physical or chemical contingency. However, so far most of these alternate solutions are reconstructions built on scaffolds that nature had already elaborated. Therefore, this question remains an unresolved, perhaps an insoluble, philosophical question.

VII The Landscape of Evolution

Here's a related problem. If we assume that organisms adapting to a new environment do not have genetic engineers to fashion their genes to make these more suitable to the new conditions, then we must entertain the possibility that random chance occurrences (mutations) would enable them to do so, albeit over many generations. A well-known problem in doing so is this: since mutations occur at random, some mutations — probably most of them — would degrade biological function rather than improve it. An organism adapted to one environment where it is at one measure of healthy or successful livelihood (i.e., fitness) must therefore go through a series of mutations — a "mutational pathway" — to go from one optimum of adaptation to another. Let us look at this problem more closely.

Imagine a graph in which we plot on the x-axis a series of gene variants a, b, c, d...etc., and on the y-axis another series of gene variants m, n, o, p...etc., (Fig. 2). Note that an organism has many genes — humans have at least 20,000 genes. For every variant combination, for example, *am*, *an*, *ao*... or *bm*, *bn*, *bo*...etc., the organism has a certain adaptive value or

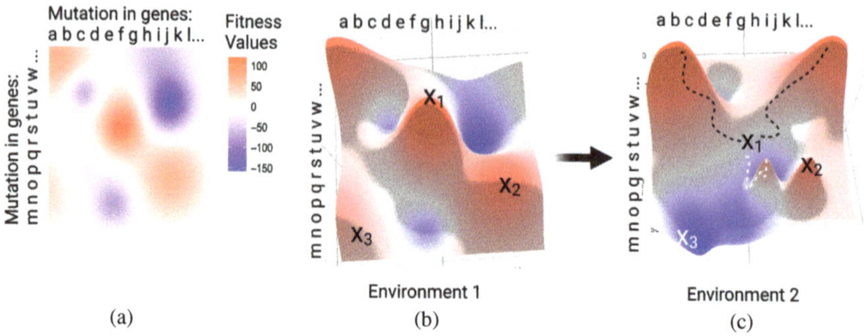

Figure 2 A fitness landscape. (a) Gene variant combinations represented arbitrarily on the x and y axes provide different levels of fitness in a scale of –100 to +100, color coded on the right. (b) In environment 1, the variant combination X_1 has the highest fitness, but X_2 and X_3 are also reasonably fit. (c) The environment changes to state 2. Now X_1 and X_3 have lost fitness in this environment, but X_2 is still doing well. However, X_1 may find its way out of the low fitness value by following one or more of the black dashed lines by steadily improving its fitness through accumulating further mutations. However, it is unlikely to reach the local fitness peak (white dashed path) because that involves a selection against its fitness. The variant combination X_3 will likely go extinct because it has few paths for steady climbing.

fitness. The variants are ordered such that each variant's neighbors are its closest related gene variants. Let us arbitrarily define the fitness scale as a continuous variable in the range of –150 to +150 units and encode these values in shades of blue and red color, respectively. We can expect to get a diagram such as in Fig. 2a, where patches of red define high fitness values for a particular gene variant combination, whereas patches of blue define low fitness for other variant combinations. One can see this better in a pseudo-3D view in Fig. 2b. Imagine that this diagram represents a particular environmental condition, here labeled arbitrarily as Environment 1. Clearly, the variant combination X_1 has a high fitness as it is on a steep red peak in this environment. The variant X_2 has a reasonable fitness, though not a very high one. The variant X_3 is also on a somewhat shallow fitness peak.

Now assume that the environment changes to a different condition (Environment 2). As it so happens (Figure 2c), X_2 is still on a local fitness peak, but both X_1 and X_3 find themselves in rather unfit states. To climb back up to a higher fitness state, X_1 must follow one or more of the steady "fitness hill-climbing" paths (in dashed black) to a higher fitness value by exploring other variant combinations (by mutation, for example). It is not likely, however, that it will ever find a smaller fitness peak close to it, because it must first sacrifice fitness further to reach it (white dashed path). X_1 is unlikely to take that latter path, because being less fit implies being out-competed by other variants. The variant X_3, however, is unlikely to ever find a higher fitness state because it has no path leading steadily up the hill, and therefore it will likely go extinct.

This diagram, an evolutionary fitness landscape, explains why organisms need genetic diversity to survive in a changing environment. The fitness landscape is multidimensional, because there are many genes and potentially many variants of every gene, so it is not possible to capture the complexity using a two-dimensional image, but it can be analyzed mathematically. The pseudo-three-D representation in Plate 2 merely captures the essence.

Darwinian evolutionary theory, coupled with genetics, leads to this essential idea of evolutionary landscape, and opens many interesting questions as to how organisms traverse this potentially rugged fitness landscape. Must individual organisms with gene mutations make a random walk such that only those few lucky enough to be at or near a fitness peak survive? Or has evolution already selected the gene combinations in a way that organisms generally dwell upon smooth slopes of fitness valleys, such that their fitness-hill climbing is possible due to small effects of additional mutations built upon the existing combination of genes? These questions are central to an understanding of evolution and thus perhaps one day to engineering novel synthetic organisms to perform precise tasks in a complex environmental scale, like reducing the atmospheric carbon dioxide burden or terraforming Mars for human habitation.

VIII From Molecules to Men

When I first heard of evolution in a test tube in the mid-1980s, the idea had appeared too risky to bank one's career on trying to accomplish it. The principles were clear enough: generate lots of variants by mutagenesis, then gradually increase the "stringency of selection"[4] for a biological property. Intuitively, however, it didn't seem likely that one could achieve in a test tube a feat that normally takes thousands to millions of generations. Would it work?

Then several advances occurred nearly simultaneously. First, three separate groups of researchers realized that instead of attempting to evolve whole organisms, one could apply evolutionary principles to evolving molecules, such as RNA, DNA, and proteins. One could easily start with a trillion molecules in a small test tube — one mole of a protein or RNA or DNA contains Avogadro's number ($\sim 6.022 \times 10^{23}$) of each molecule, so even a hundred-billionth of a mole of such a substance in a test tube represents over a trillion molecules. If there was a clever way to generate numerous variants of the molecules, then one of the conditions of evolution would be satisfied. If one now selected for a function of the molecule and had devised a way to keep only those that perform the function better than other variants, and a way to copy these molecules, then all the conditions of evolution would be satisfied. Albeit for molecules made by organisms, not for the living organisms *per se*.

This was attempted by three different groups of researchers nearly simultaneously in the late 1980s. All three groups were successful within a remarkably short time in evolving molecules of the stuff of life — RNA, DNA, and proteins — with unnatural biochemical activities that, insofar as we can tell, had never existed on earth.

If molecules could be evolved artificially, then why not whole organisms? It took another ten years to demonstrate that cultures of bacteria grown and maintained over thousands of generations accumulated enough natural variants upon which artificial selection in the laboratory could be imposed. Thus, we are now able to successfully evolve bacteria in the laboratory (Tenaillon et al. 2016). These successes with evolving enzymes and bacteria in the test tube have been rather useful for practical applications, because novel enzymes — agents that could catalyze valuable chemical reactions, such as removing dirt efficiently from clothes at high temperatures in the presence of caustic laundry detergents — can now be evolved.

[4]This concept, "stringency" or "strength" or "intensity" of selection has precise mathematical definition, provided in Chapter 14. Here we use it in a commonsense term.

IX Evolution of Variation

As the thread-like DNA (or RNA, for example, in RNA virus genomes) is duplicated in a cell, the newly made complementary chains[5] with the string of bases become polymerized successively. Adenine (A) on the parent chain brings in thymine (T) (or uracil, U, in RNA) on the new chain, guanosine (G) on the parent brings cytosine (C), and likewise T (or U) brings A, and C brings G. As the new chain grows in a stepwise fashion, each of the two new double-chained molecules gets a fresh new complementary chain but keeps one old chain. This occurs relatively faithfully along the length of the DNA or RNA molecule with rare mistakes.

An error occurs on average about once in 10,000 bases as the polymerization of the new chain grows on the parental DNA chains.[6] Such an error is often the incorporation of an incorrect base. For example, an A on the parental chain might bring in a C in the new chain by mistake, or a G could bring in A, etc. When such mismatches occur, the normally smooth DNA double helices get bulges or kinks. These structural irregularities are the targets for molecular surveillance machines. These machines attack one of the two chains near the deformed spots, remove a stretch of the newly synthesized DNA, and reduplicate it, allowing for a second attempt at a correct DNA replication over the previously erroneous site. If the error rate is constant for all DNA replication, then the probability that a second error will occur at that very spot is the square of the error rate, or one in a hundred million bases. These rare errors — mutations — may persist and will be inherited through subsequent generations.

The diploid (2 sets of chromosomes) human genome has about 6.8 billion base pairs (~3.4 billion per chromosome set). We would expect to get about 60 to 80 errors of DNA base incorporation per diploid genome per round of DNA replication in human cells. Add to that the mutations that occur due to other, rarer kinds of DNA replication errors

[5]Or 'strands', as they are commonly called, but we will generally Chapter 12 use the word 'chain' because it is more descriptive of the fact that it is composed of repeating units. We will use the word 'strand' of a DNA chain only in the context of a helical state, for example, 'DNA double strand break'.

[6]This error rate considers the proofreading activity of the respective DNA or RNA polymerases.

that cause the removal or addition of one or multiple DNA bases simultaneously (termed deletion and insertion mutations, respectively). In addition, the collision of DNA molecules with high-energy cosmic rays, exposure to ultraviolet rays, damaging chemicals, and errors of chromosome movement during cell division all produce mutations. These are the sources of genetic variation.

Genetic variations are the products of natural processes, apparently unrelated to the subsequent selection of gene properties (phenotypes). The variant genes are the stuff on which natural selection works.

In the above story told in most textbooks of undergraduate biology in excruciating detail, there is seldom any mention of selection, of environmental selection, *directly affecting the rate* at which mutations or variations occur. We implicitly assume that the mutation rate is approximately uniform.

In principle, however, natural selection may favor a higher or lower mutation rate under certain conditions. Potentially, every environmental condition may 'have to work with an' optimum rate of variant production. Below this rate, populations might collapse because there are not enough variants that can explore the extremes of environmental conditions, so they die off; above this optimal rate, populations might be unstable because too many defective genes or gene combinations are produced. Importantly, these selective events should affect populations of cells just as they do individual organisms. Seen in this light, the generation of variation is itself a biological property (phenotype, if you will), which therefore could be selected by the environment. Therefore, the rate of production of genetic variation itself can evolve.

Enter cancer. We now know cancer to be a disease of competition between normal body cells kept disciplined by many checks and balances, and a few cells that become undisciplined. In a sense, cancerous tumor cells become more evolutionarily adept than normal body cells, by increasing their rates of evolution through an increased rate of mutation and proliferation. Numerous variants thus produced consume the available resources like parasites, starving the host, ultimately going extinct as the host itself dies. We begin Part II of this book with the story of the discovery of the molecular causes of cancer.

Part 2

Il Piatto

Chapter 1

Of Emus and Oncogenes

The central problem we consider in this chapter is what causes cancer. We show that scientists used the framework of evolutionary relationships among viruses of birds and mammals to work out the genetic causes of cancer. Now we understand that oncogenes are cancer-causing genes in mammals and other vertebrates. These genes are integral parts of vertebrate genomes, normally necessary to perform numerous essential biological functions. Mutation or abnormal regulation of these genes can cause cancer. This chapter tells the story of how these genes were discovered, especially with the help of birds (emus, ducks, chickens, etc.). The presence of these genes in birds could be related by evolutionary theory to cancer-causing genes in humans as normal and likely essential constituents of multicellular life. We try to capture the intellectual excitement in the early days of these discoveries, consider alternate theories, and how a few important protagonists inspired and influenced progress. Chapter 1 ends by describing how these ideas led to a rationale for genetic testing for cancer susceptibility.

Important terms:

A *genome* is the sum total of an organism's genetic endowment. It is generally organized into chromosomes, each of which is a discrete molecule of DNA wrapped by various proteins. Humans have 23 pairs of chromosomes, including two sex chromosomes (either a pair of X chromosomes in females or an X and a Y chromosome in males). Every human cell (except sperm and egg cells) contains the same number of chromosomes. Sperms and eggs have half as many. The total DNA content of the human genome is about 3.4 billion base pairs, unequally distributed among the chromosomes.

A *Gene* is a unit of inherited biological property. It can be a stretch of DNA or RNA depending on the organism. It may or may not encode a protein. *Mutation* is a stable change in the sequence of DNA or RNA.

RNA polymerase is an *enzyme* (or catalyst) that allows the making of RNA from a DNA template (transcription); *DNA polymerase* replicates DNA; *reverse transcriptase* makes a DNA copy of RNA; and *RNA-dependent RNA polymerase* replicates RNA.

Retroviruses are viruses with RNA genomes, but upon infection the viral RNA is copied into DNA and is integrated into the host cell's genome. This viral DNA then replicates along with the rest of the genomic DNA at every cell division. The HIV-AIDS virus is an example of retroviruses.

Bacteriophages, or simply *phages*, are viruses that infect bacteria. *Virulent* phages are those that destroy the host bacteria following infection and multiple rounds of replication of their genomes within the host cells (the *lytic* replicative cycle). *Temperate* phages under certain conditions can integrate their DNA genomes into the host chromosome such that the viral genome is replicated along with the host's chromosomal DNA about once per cell division. In this state, the host cells are not harmed; in such an integrated state, the phage DNA is called a *prophage*. However, if the host's life is threatened, say, if it encounters UV rays, the prophage excises out of the host chromosome and undergoes a lytic replicative cycle to make further infectious phages.

An *operon* is a set of prokaryotic genes, generally encoding multiple proteins that together function in a related biological process, the corresponding coding sequences of which are transcribed into a single long messenger RNA (mRNA) under the control of a common regulatory mechanism. An example is the *lac* operon, which encodes three proteins necessary for metabolizing lactose as the sole carbon source by *Escherichia coli*. All three protein-coding genes are co-transcribed as a single mRNA. Regulation of transcription of this mRNA therefore affects all three proteins.

Cancer is a disease that is recognized as due to abnormal and continued proliferation of certain cells. In advanced stages, cancer can spread (also called *metastasis*) to various bodily organs. The spreading cancerous cells parasitize the human body, consume abnormal amounts of nutrition at the expense of normal cells, and exhaust the organs. Death is usually accompanied by organ failure.

Oncogenes, more appropriately, *proto-oncogenes*, are normal human or other animal genes that, if mutated or abnormally expressed, can directly cause cancer. *Oncogenic* implies the ability to induce cancer; oncogenes are potentially oncogenic, but so are certain chemicals, such as components of coal tar. *Tumor suppressor* genes normally prevent the cancerous transition of a cell. A *tumor suppressor* gene may act by reducing the mutation rate, by inhibiting the activity of an *oncogene*, by suppressing abnormal cell proliferation, or by killing a cell that is fully or partially cancerous.

Homology is a concept of structural relatedness. An elephant's forelimbs are homologs of human hands even though the former function as legs. Similarly, an elephant's trunk is not a homolog of human hands even though their functions overlap. Likewise, two genes are *homologs* of each other if their structures, namely, the sequences of DNA/RNA or of their corresponding encoded proteins are similar. *Paralogs* are homologous genes that had potentially evolved by duplication of an ancestral gene copy followed by mutational diversification. Paralogous genes, always with sequence relatedness, may or may not have similar or related functions.

Melting or *dissociation* of a nucleic acid generally refers to the coming apart of two chains of the nucleic acid, which were held together by weak hydrogen bonds between complementary bases (base pairing). This can be accomplished by simply heating a solution of the nucleic acid. *Annealing* is the reverse process, in which the two complementary chains become bonded together, which can be accomplished by slowly cooling the solution. DNA–DNA or DNA–RNA *hybridization* refers to the annealing between two DNA or RNA chain(s). Annealing can be imperfect, meaning that the two mostly complementary chains are not perfectly complementary — they have some base-pairing *mismatches*. Obviously, such imperfect hybrids are less stable (held together less tightly) than are the perfectly complementary duplex hybrids.

In this book, we follow the following conventions. A gene's name given by scientists (e.g., *MSH2*) is always written in italics; its normal or wild-type copy (or *allele*) is written in block letters, whereas its mutant allele is written in small letters (e.g., *msh2*). The gene's RNA or protein product is written in block letters without italics (e.g., MSH2 mRNA and pMSH2 protein). An exception, for example, is p53, which is the protein product of the gene *TP53*. We will specify the exceptions as they occur.

A *ligand* is a molecule that binds to a *receptor* (often a protein) to modulate the function of the receptor. A ligand can be either a small molecule, even an atom, or a macromolecule, such as another protein. A ligand to a receptor is like a finger to a fingerprint detector that unlocks your smartphone.

Somatic cells constitute the body cells of a multicellular organism derived by divisions and specialization of the primordial *zygote* (the fertilized egg). Mutations arising in somatic cell genomes are not inherited by the next generation of the organism but are destined to be confined to the body and to perish at death. *Somatic mutations* are therefore mutations that are only inherited by the somatic cells; *germline mutations* are those we inherit from our ancestors and may pass on; a few new mutations might occur in the lineage of cells that give rise to sperms or eggs, the so-called *germline cells*.

***

In a parallel universe, it is a crisp English morning. A long table covered with white linen is lined by four filigreed lawn chairs. Delicate China cups and saucers, sugar cubes, a tin of Britannia biscuits are on the table. There is a small aviary across, where colorful canaries were hopping around. The characters are Charles Darwin the British naturalist who was the "father of evolutionary theory," Gregor Mendel, the Czech clergy who proposed the eponymous laws of heredity; James Clerk Maxwell, the English theoretical physicist who had unified electricity, magnetism, and light, and Louis Pasteur, the French biologist par excellence who pioneered the science of modern microbiology.

Darwin: Padre Mendel, what do you think of cancer?

Mendel: I told you a long time ago that genes are the stuff that make a character trait. Young folks have now shown that traits are formed by genes, and cells evolve from normal cells by acquiring new genetic properties that give them evolutionary fitness. Cancer can sometimes be considered a trait, though not a desirable one.

Darwin: Now, now . . . don't confuse things. I said the species evolves, not genes.

Mendel: Yes, I understand that. How could you have said anything about genes? You didn't even know they existed.

Maxwell: Father Mendel, if I be allowed to speak in such matters as are not electricity or magnetism . . . even you didn't know what genes are nor that they mutate.

Mendel: True, but as a mathematician you understand that those are mere details. Darwin here had proposed that a species accumulates heritable changes at random, which affect the properties of individuals of the species, and nature selects for variants that are the fittest. Nature changes; those variant individuals that had existed before are no longer fit; new heritable changes appear that now impart higher fitness. So, the population begins to change in terms of its collective variant properties. I showed that these properties are due to genes — well, I didn't know what they exactly were, but I didn't need to. Whatever they are, they are very stable but occasionally they change. Now everyone calls them DNA. It is the same situation with cancer. Cancer cells accumulate heritable variations in their DNA. Nowadays they call these variations mutations. So, mutations happen within your genes.

Maxwell: I am not a mathematician but a physicist. There's a difference.

Darwin: Padre, your tea is getting cold.

Pasteur: Friends, I am not a doctor but I am interested in how cancer patients can be cured. In my time, patients used to suffer a lot because every patient seemed to be different in their cancer symptoms. Some would respond to the treatment but most would not. I now understand, because young fellows have shown, that every patient's cancer cells have a unique genetic signature. But there are some common or similar parts in these signatures, like the alphabets within the names of people — like a barcode that you scan before going on a flight. I wish I could fly too but, in my time we only had steam locomotives, which I took in the summer for holidays to L'Espiguette.

Darwin: Professor Pasteur, you have too many holidays in France. I don't know when you do your work! In any event, what you both are saying can make sense only if someone first shows that these cancer-causing trait variants occur because the genes are already there in all normal cells in the first place.

Pasteur: Not quite. There could sometimes be a virus that brings in a cancer gene to a human — so you should always pasteurize your milk. But I say that in jest. You can't get cancer from infected milk though you can get very sick, even die from tarnished milk. But some rare cancers are due to infection by viruses, especially if your immunity is not strong. Most cancers however are due to mutations in genes that already exist in all cells. We all carry the genes for cancer in every cell, although in forms that don't normally produce cancer. It is just a chance that one of these genes might mutate to cause the cancerous trait. Then you have the disease.

Maxwell: Does this mean if you knew the genes by looking at the DNA, you could tell exactly what cancer signature is there, and so you can vary your treatments for individual patients?

Pasteur: Precisely. Now they call this, personalized medicine.

Darwin: How did they find all of this out?

Mendel: This pair of young west-coast doctors, who like you, Darwin, disliked seeing patients and thought like you did about the Galapagos finches. They thought that if some cancer-causing genes are present in chickens as well as in flightless birds like Emus, these genes must be very ancient, because emus and chickens had diverged many million years ago, and maybe even humans have them. They then found these genes in humans — normal genes that all vertebrates need for their normal traits. These can become the cause of cancer by mutation. Evolution takes care of the rest.

Darwin: Padre, I disliked seeing the innards of humans so much that I dropped out of medical school. Friends, this is so interesting but complicated that my head spins. Let's drink tea and be merry. Here's to personalized medicine.

He raises a Britannia biscuit to toast, dips into his tea and misses it as the softened piece drops into the cup.

***

The decade of the 1970s was a heady time. The Vietnam War had just crossed its peak amid global anti-war protests and fears of nuclear conflict. Despite the intense Cold War, there was newfound optimism due to NATO-Soviet détente. The first moon landing in 1969 and advances in space and telecommunication technology had signaled a new era, while computers began revolutionizing business and industry. Meanwhile, a series of Nobel Prizes went to scientists who were shaping a revolution in molecular biology. This new kind of biology had electrified the atmosphere with the promise of new cures for thus far intractable diseases, such as cancer.

There was a rising fear too. Advances in biology prompted debates over societal control of research, mirroring fears about unchecked military research and the arms race. The advent of genetic engineering in the early 1970s, such as gene cloning, sparked fears of weaponization of biology and of eugenics, evoking memories of Nazi experiments. Against this backdrop in 1973, the great geneticist Theodosius Dobzhansky (1900–1975) wrote:

> *"Nothing in biology makes sense except in the light of evolution."* (Dobzhansky 1973)

Despite near universal acceptance of the theory of evolution among practicing biologists, a debate on the very legitimacy of Darwinian evolution was reaching a new crescendo due to its apparent conflict with deeply held religious beliefs in the center of America: the rise of the so-called "Creation Science" following the 1968 Supreme Court decision in *Epperson v. Arkansas.*[1]

1.1 A War on Cancer

In the charged atmosphere of the 1970s were born a few of the most significant conceptual advances that revolutionized the treatment of cancer. On December 23, 1971, US President Richard Nixon had declared a "war on cancer" by providing unprecedented funding and autonomy to the National Cancer Institute (NCI) and the National Institutes of Health. Facilities for biological warfare at Fort Detrick were repurposed for cancer research. James D. Watson (1928–2025), co-discoverer of DNA's structure, had believed cancer, thought at the time to be of viral origin, could be cured within a decade.

Before the 1970s, cancer was often a death sentence. Fifty years later, at least in the developed countries, most types of cancer could be reduced to a manageable chronic disease status, with death delayed by about 3 to 5 years if not cured outright. This has largely been possible because of those very advances made in the 1970s. This chapter will explore how Darwinian thinking revolutionized our understanding of human cancer.

1.2 Retroviruses Cause Cancer in Birds and Rodents

Cancer is a dreadful word associated with a literary aphorism implying insidious and capricious decay. Francis Peyton Rous (1879–1970), the discoverer of tumor viruses, opened his 1966 Nobel lecture with these words:

> *"Tumors destroy man in a unique and appalling way, as flesh of his own flesh, which has somehow been rendered proliferative, rampant, predatory, and ungovernable* (Rous 1967)."

---

[1] The state of Arkansas had passed a law prohibiting the teaching of human evolution; Epperson, a teacher, sued. In 1968, the state's Supreme Court decided in favor of Epperson, holding that the state violated the First Amendment of the US Constitution.

Around 1911, Rous took fluids filtered from chicken tumors. The filter pores were too fine for chicken cells or bacteria to pass through. He showed that this filtered fluid could cause tumors in healthy chickens. Later studies revealed that it contained a tiny virus, now called the Rous sarcoma virus (RSV).

Similar viruses were later found in mice. The murine sarcoma virus (MSV-Harvey) and mouse mammary tumor virus (MMTV) could cause tumors in mice. The mouse leukemia virus (MLV) could lead to leukemia. In monkeys, a virus called Simian Virus 40 (SV40) caused tumors when infected monkey cells were implanted in rodents.

But what about humans? Could human cancer also be caused by viruses? Could scientists find a human cancer-causing virus?

In 1964, a human tumor-causing virus, the Epstein–Barr Virus (EBV), was discovered. This raised the exciting idea that viruses might cause cancer in humans, leading many researchers to investigate. If they could eliminate these viruses, by vaccination, for example, they might cure cancer.

Another big question was what part of these viruses caused cancer. Could one find a specific "gene" responsible?

Renato Dulbecco (1914–2012), an Italian virologist working at Caltech in Pasadena, focused on a simpler model. He chose SV40 and the newly discovered mouse polyomavirus to study. He aimed to identify the tumor-causing gene within these viruses. This approach inspired others to join the effort.

A few words on these early workers in tumor virology are necessary to appreciate how thinking in terms of Darwinian evolution had influenced the mental framework of these early cancer researchers.

Dulbecco had studied medicine in Italy along with Salvador Luria (1912–1991). Luria had escaped from Italy into the United States of America in the early phase of World War II. By 1943, he, along with Max Delbrück (1906–1981), had discovered that bacteria and their viruses had genes that spontaneously mutate, as do those in pea plants and fruit flies, and that bacteria can be subjected to natural selection. In other words, these organisms should also evolve, just as Darwin had speculated for plants and animals. Importantly, the appearance of new variant bacteria was independent of selection — an important postulate of the Darwinian evolutionary theory, which was demonstrated for the first time by this duo.

Delbrück moved to Caltech in the late 1940s. He soon recruited Dulbecco, fostering a critical intellectual atmosphere focused on simple hypotheses to explain genetic underpinnings of complex processes like evolution and cancer. This novel reductionist approach, inspired by theoretical physicists, was disfavored by most biologists at the time.

Delbrück's experience with reductionism began in Europe in the 1930s, working with Niels Bohr (1885–1962) (Fig. 1.1), who had proposed the atomic model with electrons in discrete orbits, and Lise Meitner (1878–1962), who would go on to discover nuclear fission. Delbrück himself discovered how gamma rays scatter in a static electric field, known as Delbrück scattering. By the mid-1930s, he shifted his interest to biology, popularizing the reductionist approach with Luria.

In the early 1950s, this approach led to the search for a simple biological model of cancer, akin to the bacteriophages studied by Luria and Delbrück, which had influenced James Watson, a former PhD student of Luria. A tractable model of cancer could potentially be understood in simple molecular terms. Could tumor viruses present such a model?

1.2.1 *Are viruses or chemicals the cause of cancer?*

One of the hardest things in science is telling the difference between a true cause and a coincidence. Sometimes, two things happen together, but that doesn't mean one caused the other. To infer causation, several criteria must be met. First, the two things must happen repeatedly, not just once. Second, the cause must come before the effect. Third, there needs to be a clear explanation of how the cause leads to the effect. Finally, this is the most difficult yet rigorous, all other reasonable explanations must be ruled out to ensure something else isn't the real cause. This careful process helps avoid mistaking coincidences for causes.

While viruses were known to cause tumors, other things like radiation and chemicals from coal tar also caused tumors. Cancer sometimes runs in families, suggesting gene mutations without any link to viruses.

Figure 1.1 Niels Bohr on a motorbike, with (from left to right) Johann Kudar, George Gamow, Margrethe Bohr (Niels Bohr's wife), Max Delbrück, and Leon Rosenfeld. The photo was taken in 1931 at Bohr's summer house, Tisvilde, North Zealand, Denmark. Niels Bohr's model of atomic structure unified decades of experimental observations on atoms and spectroscopy, and later he catalyzed the foundation of quantum mechanics. Kudar contributed to atomic theory. Gamow later elaborated and extended the "Big Bang Theory" of the origin of the universe and was also the first person to propose that the genetic code comes in triplets of DNA bases. He wrote several highly popular books, including *One, Two, Three, Infinity* and the *Mr. Tomkins* series. Delbrück influenced a generation of early molecular biologists. Leon Rosenfeld did groundbreaking work on condensed matter physics and quantum mechanics. Photo courtesy of Niels Bohr Institute, Copenhagen. (Reproduced with permission from the Niels Bohr Institute).

This created uncertainty about whether viruses really caused cancer. It wasn't clear if cancer was caused by chemicals, viruses, or something entirely different.

Rous suggested that tumor viruses acted as triggers, starting the process of cancer but not being needed to keep it going. This idea also applied to chemicals or radiation, which could cause cancer long after exposure. These triggers didn't need to stay in the body for cancer to continue, unlike infectious diseases where the disease-causing agent remains in the infected area as the disease progresses.

Two important steps in understanding how these viruses could induce tumors occurred in the 1960s. The first was the demonstration that RSV contained an RNA genome (Crawford and Crawford 1961). The so-called C-type of viruses of which RSV was a member, all appeared to contain RNA. This led the famous geneticist J. B. S. Haldane (1892–1964) to write in a letter in July 1963 (Dronamraju 1987):

> *I disagree that 'DNA is the basis of life and evolution'. There are viruses with only RNA. We don't know how it is copied but since some RNA viruses will grow in a variety of hosts, it is improbable that it needs DNA to be copied.*

The second discovery, reported in 1970, solved a part of Haldane's problem: the identification of a virally encoded enzyme that uses RNA as a template to make a DNA chain that is complementary to the RNA chain — the reverse transcriptase enzyme (Baltimore 1970; Temin and Mizutani 1970). Integral to the life cycle of this group of tumor-inducing RNA viruses is the making of a copy of the viral RNA genome to DNA (Fig. 1.2). Since then, this class of RNA viruses has been called retroviruses because they could transfer information from RNA to DNA, the reverse (or *retro*) of the more common direction of genetic information transfer.

1.3 Two Hypotheses on How Retroviruses May Cause Cancer

1.3.1 *The provirus theory of cancer*

There were two competing hypotheses in the field at the time. The first one, the cancer provirus theory, was favored by Howard Temin (1934–1994). A PhD student of Dulbecco and a co-discoverer of the reverse transcriptase enzyme, Temin proposed the idea that DNA copies of retroviruses occasionally integrate into the host cell's chromosome, which causes a mutation, some of which may lead to cancer.

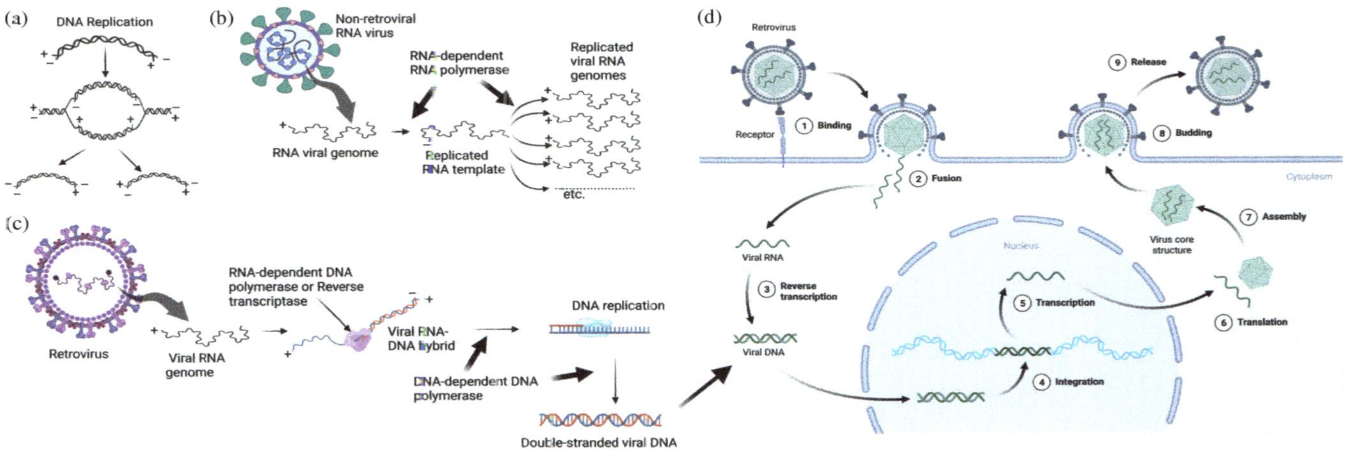

Figure 1.2 Replication of DNA and of the genomic RNA of two different classes of animal viruses. (a) Replication of DNA proceeds by way of copying each of the two complementary chains into two daughter DNA chains, producing two progeny double-chained DNA molecules, each of which retains one of the two parental DNA chains. (b) A non-retroviral RNA virus (such as the coronavirus) releases its RNA genome from the virus particle, which is copied to a complementary chain of RNA by an RNA-dependent RNA polymerase (RdRP), which subsequently is used as a template RNA for making further copies of the viral genomic RNA by RdRP. Newly synthesized genomic RNA molecules are subsequently packaged into fresh virus particles. (c) For retroviruses, such as certain tumor viruses, the single-chained genomic RNA is first used as a template to produce a DNA copy of the RNA (a complementary DNA chain) by the virally encoded enzyme called RNA-dependent DNA polymerase or reverse transcriptase. This DNA chain then serves as a template for at least one further round of DNA replication to make its complementary DNA chain by the action of a DNA-dependent DNA polymerase as in A. The double-chained DNA thus produced contains all information of the original RNA genome of the retrovirus. This piece of DNA integrates into the host genome (d). As the host cell's chromosomes replicate during the host cell division, the viral genome (now called the pro-virus) replicates as if it is a part of the host's chromosomal DNA. [This figure was drawn using Biorender.com under an academic license]

The above hypothesis was based on several observations chief among which was that RSV infects yet remains quiescent for generations only to reemerge as infectious virus particles at a later time (Hill and Hillova 1972). This is also a rather neat hypothesis because viruses that integrate into the host cell's chromosome were already known in the bacterium *Escherichia coli* (or *E. coli*).

This class of bacterial viruses, temperate bacteriophages, was being intensely studied by molecular biologists at the time. These bacteriophages can sometimes integrate into a host's chromosome in a lysogenic cycle, remaining dormant for many generations before replicating and destroying the cell (Fig. 1.3). The famous temperate bacteriophage lambda, which infects *E. coli*, significantly contributed to the revolution in molecular biology and gene cloning. It was found that lambda and its relatives can cause mutations at their integration sites in the host chromosome. Evidently, this integration must cause a change of the DNA

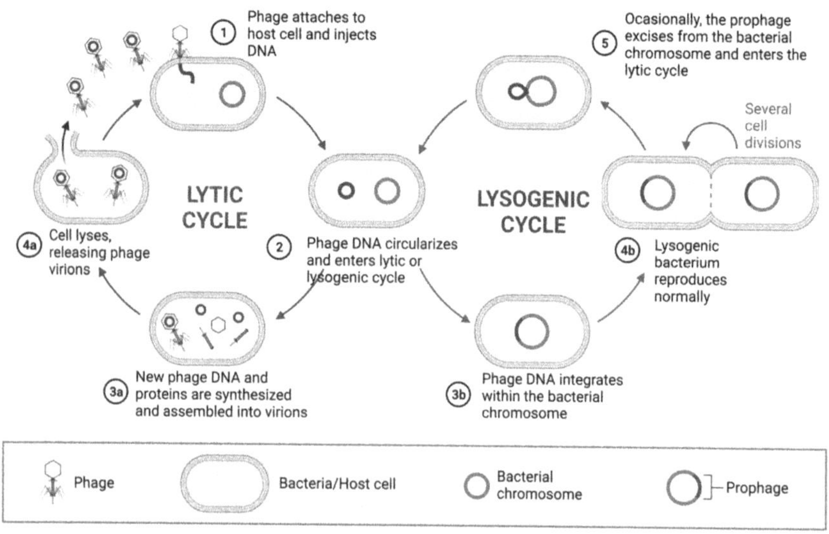

Figure 1.3 Lytic and lysogenic life cycles of bacterial viruses (bacteriophages), such as that of the temperate bacteriophage lambda. In a lysogenic cycle, the pro-viral (prophage) DNA integrates within the host's chromosome and replicates as part of the host DNA replicative cycles. When conditions are suitable, the integrated phage DNA excises out of the genomic DNA and replicates freely as an independent entity. [This figure was drawn using Biorender.com under an academic license]

sequence at the insertion site; therefore, integration must cause a mutation, where mutation is defined by any changed DNA sequence.

Jonathan Beckwith (1935–) and Ethan Signer (1937–) had utilized this property of lambda phage to clone a bacterial gene, showing how to make many identical copies of a gene for biochemical analysis (Beckwith and Signer 1966). They got the phage to integrate near a gene in the host bacterium's chromosome. Obviously, when the bacteriophage's DNA is inserted into that location on the chromosome, one would call that a mutation due to the insertion of the viral DNA. When the viral DNA excised itself, it also picked up the adjacent host gene, replicating it along with the phage DNA.

This method of gene cloning and mutation induction by temperate phages became a standard practice of that time. Might a similar mechanism explain cancer caused by oncogenic RNA viruses? Temin speculated that viral DNA integration into host chromosomes could be an evolutionarily conserved process from bacteria to humans. Perhaps retroviral infection would cause critical mutations by inserting into host cell chromosomes (Fig. 1.4a), potentially knocking out nearby genes and leading to cancer. Alternatively, an imprecise excision of the provirus from the genome could cause a loss of important host genes, triggering cancer (Fig. 1.4c).

1.3.2 *The "Virogene" theory of cancer*

The second hypothesis (Huebner and Todaro 1969), the oncogenic "Virogene" model, was more subtle and yet grounded on a deep evolutionary basis. It was postulated that some of these retroviruses had infected ancestral animals in the remote past and had integrated into the genomes of germline cells. These viral genes, including those that cause cancer — the oncogenes — are then passed on to succeeding generations (Fig. 1.5).

Just like the lambda bacteriophage carrying a bit of the host chromosomal DNA from one cell to another, as Beckwith and Signer had utilized, these retroviruses could potentially ferry around bits of the host genetic material from one species to another. One imagines that these ferried genes, perhaps gained from a different species altogether, would normally remain silent or even provide an evolutionary advantage by providing

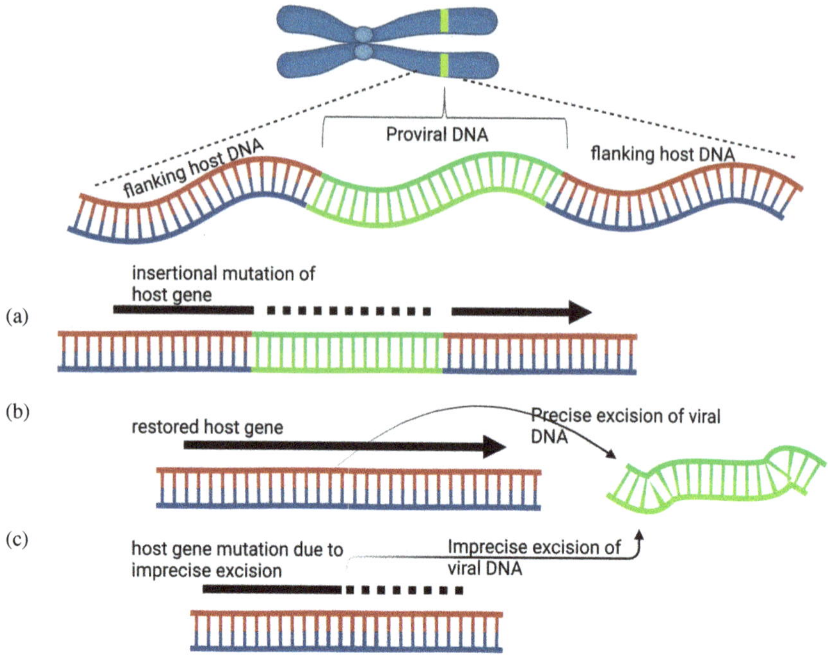

Figure 1.4 Temin's model of viral oncogenesis by a mutation directly induced by a retrovirus. A retroviral DNA copy inserts somewhere into a host cell's genome. (a) The locus where the provirus genome is inserted, a mutation is created, thus causing cancer. (b) The provirus DNA might be excised cleanly, without causing any mutation or cancer. (c) Imprecise excision might leave a mutational scar on the host DNA, which could cause cancer, perhaps depending on the presence of other factors that enable a phenotypic activation of the cancer-causing genetic lesion. Induction of cancer therefore occurs in the same individual that was originally infected by the retrovirus, though later in time. [This figure was drawn using Biorender.com under an academic license]

some important functions and yet could potentially cause cancer when subjected to influences by chemical or physical agents (Fig. 1.5).

There is no *a priori* reason to believe that which is true for mice or cats is necessarily true of humans. The Darwinian evolutionary paradigm, however, makes this quite a reasonable argument. That which worked for an ancestral organism of both mice and men might indeed persist as an inheritance and shared by both.

The main distinction between the provirus and the "Virogene" hypotheses was that the former was a somatic model of cancer in which

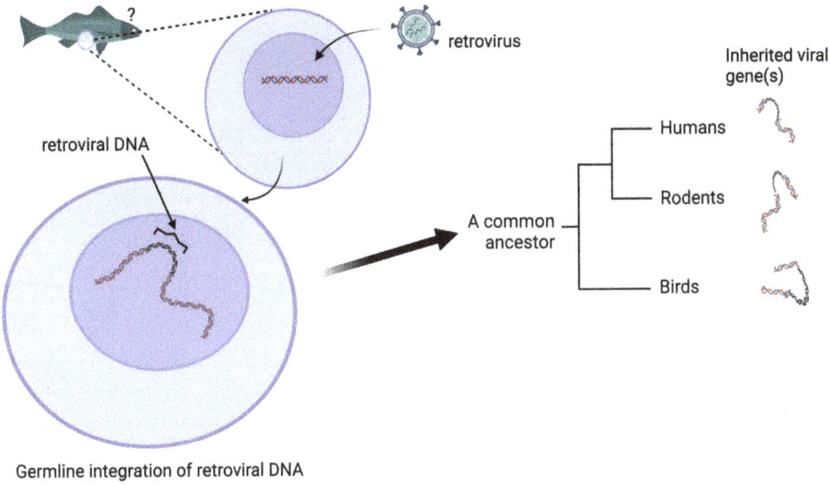

Figure 1.5 The "Virogene" model. In an ancestral organism (here arbitrarily designated as a fish), a retroviral infection had occurred, which had led to germline transmission of the virus-encoded gene. Through subsequent evolutionary steps mammals inherited this gene. Abnormal function of this gene in some cells leads to cancer. Alternately, a retrovirus might have picked a copy of an ancestral gene from one organism and transmitted that to another ancestral organism, which is then inherited through the germline. The gene thus introduced into the host genome induces cancer when an appropriate condition favors the gene's phenotypic activation. Cancer induction, therefore, occurs in an individual that is far removed, perhaps by millions of years, from the individual originally infected by the retrovirus that had contributed the oncogenic gene. [This figure was drawn using Biorender.com under an academic license]

viruses had infected an individual's body cells and thus caused cancer; the latter posited that the infection had occurred in some remote previous generation (or in an ancestor) of the animal species (Fig. 1.5).

What would constitute an experimental test to distinguish between these two vastly different yet competing hypotheses: Temin's provirus model or the "Virogene" model?

1.4 From Poetry to RNA Viruses

I first saw Harold Varmus (1939–) in the summer of 1989 when he was on sabbatical leave from the University of California San Francisco at the Massachusetts Institute of Technology. Varmus was a frequent attendee at afternoon seminars at the Whitehead Institute. I particularly remember a summer evening when he walked into the seminar hall in jeans and a light green T-shirt, slim and youthful — he was yet to turn fifty. In another 4 months or so, he would win the Nobel Prize in Physiology or Medicine with his long-time collaborator J. Michael Bishop (1936–) for the discovery of oncogenes.

While in college, Varmus might have been more interested in poetry and political affairs than in discovering mechanisms of cancer. He went to graduate school at Harvard to study metaphysical poetry of the 17th century but might have soon realized that a professional career as a poetics professor might be less fulfilling than that of his friends who were training to be physicians. He went on to study medicine in Columbia, and while doing so, he did a volunteer internship in Bareilly, about 250 km east of New Delhi, India. One might speculate, though I don't know, that that internship experience in a developing nation might have been enough for Varmus to look for another more intellectually satisfying preoccupation. In 1966, after his graduation from medical school, Varmus took a basic-science research internship at the National Institutes of Health. In Varmus' own words (Varmus 1990):

> *"Perhaps because his wife was a poet, Ira Pastan agreed to take me into his laboratory, despite my lack of scientific credentials."*

Ira Pastan (1931–) was intrigued by how a simple chemical regulated the expression of a gene in *E. coli* bacteria. He and Max Gottesman (1935–) were doing an intricate series of experiments on the regulation of a set of genes, called the *lac* operon, by the chemical, cyclic adenosine monophosphate (cAMP). These studies would eventually become classic by defining a model for gene regulation (Fig. 1.6). If cAMP could turn on the expression of a gene in *E. coli*, why can't another molecule turn on the expression of a cancer gene in human cells?

Varmus joined Pastan's lab, where he learned how to design experiments that could give clear and precise results to test different ideas. He also learned how to use nucleic acid hybridization, a method to compare the similarities between two genes or their RNA products in exact terms. This technique would soon help him settle the debate between two ideas about how cancer develops.

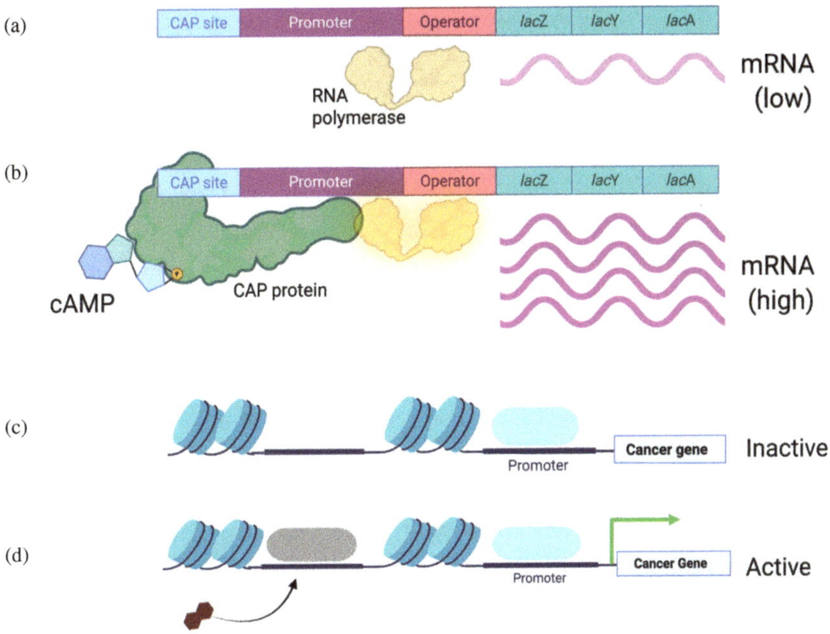

Figure 1.6 Regulation of gene expression by a small molecule. (a) In *E. coli* a group of genes named *lacZ*, *lacY*, and *lacA* are normally transcribed from a DNA sequence (termed, promoter-operator site) where the RNA polymerase molecule binds. However, the transcription into mRNA of this gene set occurs at a low rate if a nearby site, termed the CAP site, is unoccupied. (b) If there is sufficient cyclic AMP (cAMP), a small molecule, is present then it augments the binding of a protein termed the CAP protein at the CAP site. Binding of the CAP protein in turn allows a high rate of mRNA transcription by RNA polymerase. (c) In a similar manner, it was thought possible that a cancer-producing gene might be inserted into the mammalian cell's chromosome, but the gene is normally not active for transcription. Chromosomal DNA is shown here as spooled into folded structures (blue ovals). The light blue oval is an RNA polymerase. (d) If a carcinogenic molecule is present (purple double hexagon), it allows another protein (the grey oval) to bind to a DNA site (which was otherwise vacant), which leads to active transcription of the cancer-causing gene (green arrow). Note that (c) and (d) were purely hypothetical and were conceived by analogy to (a) and (b). There is little current evidence in favor of such a process being important for cancer, although we cannot eliminate its existence. CAP site is the DNA sequence where the CAP protein binds; promoter is a DNA sequence necessary for transcription of the genes into mRNA through the binding of the enzyme RNA polymerase; operator is a DNA sequence where regulatory proteins may bind to regulate transcription. [This figure was drawn using Biorender.com under an academic license]

1.4.1 *Nucleic acid hybridization kinetics*

If two genes are identical, then their respective DNA sequences are identical — the so-called Watson chain of one duplex would be identical to the Watson (or +) chain of the other duplex. Similarly, the complementary Crick (or –) chain would be identical to the Crick chain of the other (Fig. 1.7a).

In a liquid solution, the two chains of a DNA helix can be separated by heating. This is because the chains are held together by weak hydrogen bonds, which can break when heated as water molecules collide with them (Fig. 1.7a). The temperature at which the two chains of a DNA or RNA–DNA hybrid separate (or "melt") depends on two things: the GC content and the similarity between the sequences.

GC base pairs have three hydrogen bonds, while AT base pairs have only two. Therefore, molecules with more GC base pairs will have a higher melting temperature. Chains with identical sequences will hold together more tightly and melt at higher temperatures. Chains with partial similarities, such as a mutant gene and its normal version or genes related through evolution, will melt at lower temperatures.

Since DNA carries genetic information, the similarity between sequences may reflect their evolutionary relationship. These partially similar sequences are often called "homologous." This term comes from evolutionary biologist Richard Owen (1804–1892), who used it to describe traits inherited from a common ancestor that have similar structures (see Chapter 16 for more details).

But how would one make a partially complementary double-helical DNA molecule? Molecules in which the two chains are not fully complementary but have mismatched DNA bases?

If one heats duplex DNA molecules from one organism, their two complementary single chains separate ("melt"). If, likewise, one heats another collection of DNA molecules from another organism, having a slightly different DNA sequence, then one gets the latter's two complementary chains melted apart too. Now, if one mixes the two DNA solutions together and slowly cools the resulting mixture, the DNA chains will collide with one another, and if they find complementary chains, they will anneal.

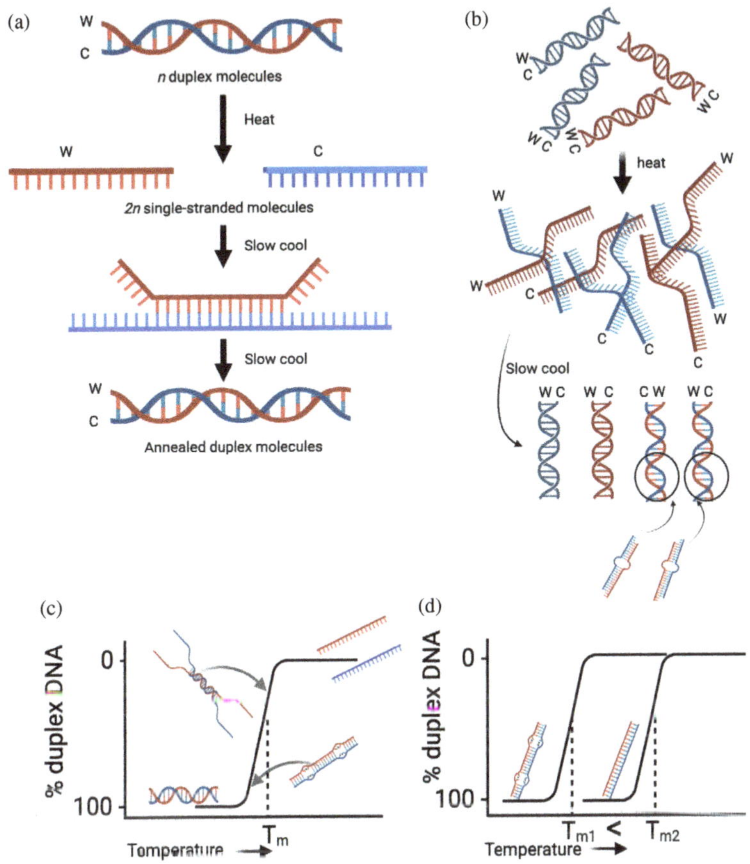

Figure 1.7 Melting and annealing of DNA duplexes. (a) DNA double chains dissociate because the hydrogen bonds that hold the double helix together are weak and cannot withstand higher energy collisions with solvent molecules as the temperature is raised. At a certain temperature sufficient to break the hydrogen bonds (and overcome the so-called stacking interaction between pairs of nucleotides along the axis), the two complementary chains (W and C for Watson and Crick, respectively) fall apart. If they are cooled slowly, the two complementary chains come together and reform the duplex. (b) If two DNA duplexes with largely similar (with some differences) sequences are mixed and heated, followed by slow cooling, some not-perfectly complementary DNA chains would still pair, forming heteroduplexes, producing some mismatched pairs. (c) When a perfectly complementary duplex is heated, the duplex melts at a defined temperature (T_m), which corresponds to the breaking of the hydrogen bonds that hold the duplex together. (d) In the case of a slightly mismatched duplex, however, there are fewer hydrogen bonds per unit chain of the duplex molecules. If such molecules are heated, they will melt at a lower temperature than a perfect duplex. In such cases T_m will be dependent on, among other factors, the extent to which the DNA sequences are dissimilar (i.e., divergent) between the two chains. The more the divergence, the lower the T_m. [This figure was drawn using Biorender.com under an academic license]

However, when there are partially complementary chains available (as between the two slightly different DNA molecules), then some of the time a chain will find a wrong complementary partner and pair with it partially — this will generate an odd hybrid DNA molecule with little bulges in its structure where the base pairings are not perfect (Fig. 1.7b). Once formed, such "hybrid" duplexes containing mismatched bulges can be heated once again to find the temperature at which the two imperfectly matched chains melt apart.

Intuitively it should be clear that a duplex with perfectly matched complementary chain pairs would be more stable than an imperfectly matched chain pair. Thus, the temperature at which the chains fall apart — the melting temperature — should be a measure of the extent of mismatches (Fig. 1.7c,d). In other words, the melting temperature of various DNA hybrids could be a sensitive measure of the degree of DNA sequence homology. The more the similarity, the more stable, and the higher the melting temperature. In this fashion, the melting temperature becomes a yardstick for measuring the likenesses (or homology) of DNA sequences.

Varmus became adept at this laboratory technique. Importantly, this technique allowed the qualitative ideas of relatedness between any pair of organisms — including those of different viruses (or of their genes) — to be framed as a precise number — the melting temperature of the hybrid DNA molecules.

While at Pastan's laboratory, Varmus read about oncogenic RNA viruses and realized that he had just the tools to distinguish between the two competing hypotheses. At the end of his internship at NIH, he obtained a postdoctoral position at the University of California San Francisco, in the laboratory of J. Michael Bishop. Soon afterwards Varmus became a faculty member in the same department, and the Bishop–Varmus team collaborated for over a decade in nailing down the basis of cancer induction in mammals by retroviruses.

1.5 As Flesh of His Own Flesh, Gene of His Own

By 1970, Peter Duesberg (1936–2026), working across the Bay Bridge at the University of California, Berkeley, and his collaborator, Peter Vogt

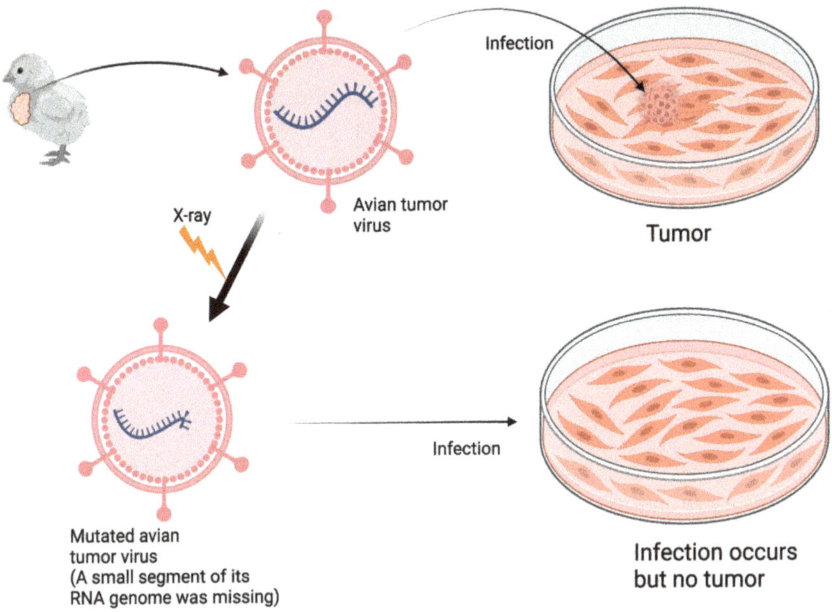

Figure 1.8 A defective avian retrovirus is infectious but fails to make tumors. The mutated virus produced through X-ray treatment of a tumor-competent virus had a shorter RNA genome. Does the missing segment of the RNA contain at least part of the tumor-causing gene of the virus? [This figure was drawn using Biorender.com under an academic license]

(1932–), then at the University of Washington, Seattle, had shown that cancer-associated retroviruses carried pieces of extra genetic material in addition to the viral genome (Duesberg and Vogt 1970).

Duesberg and Vogt took a tumor-inducing avian RNA virus, RSV, that was mutated by X-ray to make it unable to induce tumors (Fig. 1.8). This defective virus, however, was still infectious. This mutant virus lacked an RNA piece sufficient to encode a protein of about 500 amino acid residues long. Evidently, bombardment with X-ray had knocked off the coding region for this fragment of RNA from the virus genome. Interestingly, this missing RNA piece was present in all avian tumor viruses they examined but not in avian leukemia viruses. Leukemia viruses never produce tumors. Provocatively, Varmus and his colleagues wondered whether this missing RNA fragment might have contained the Virogene, the hypothetical cancer-causing agent of tumor viruses.

Note that a biochemical identification of a putative cancer-causing gene, now renamed an "oncogene," although supportive of the notion of a retroviral gene as an inducer of cancer, did not distinguish between the provirus and the Virogene hypotheses. This is because both hypotheses predicted that the cancer-causing genes could be ferried around by RNA viruses from cell to cell.

To reiterate, the provirus hypothesis proposed that cancer is caused by a virus by directly inducing tumorigenic mutations in the host chromosome, while the Virogene/oncogene hypothesis proposed that an ancestral infection by a retrovirus had already left behind a copy of its tumor-inducing gene presumably many generations ago; the latter sometimes gets activated in a few cells of an individual. By means of pure Darwinian thinking Varmus, Bishop, Vogt, and Dominique Stehelin (1943–2019) came to a critical test for distinguishing between these two hypotheses.

1.5.1 *Ancient footprints of oncogene relatives*

Darwin wrote in his "*On the Origin of Species*":

> *"I cannot doubt that during millions of generations individuals of a species will occasionally be born with some slight variation, profitable to some part of their economy."*

Some 50 years later, Haldane, Sewall Wright (1889–1988), and Ronald Fisher (1890–1962), each working independently, had solidified the idea that variants of characteristics within individuals in an evolving population emerge spontaneously and independently of any other factors related to natural selection. Natural selection then proceeds to refine and select only those spontaneous variants in the population that increases fitness in a changing environment. A vast majority of these variants, however, must be maladjusted to the environment because it is far easier to destroy a character than to build one! The rare ones that do contribute to fitness, persist due to natural selection and contribute to the genetic divergence among species.

Extending this idea to genes, we can anticipate that rare DNA or RNA sequence changes due to mutation, having been fixed over many generations, will accumulate slowly and should have the effect of causing greater and greater divergences between the ancestral nucleic acid sequence and those that are subsequently derived thereof. In a later chapter we will see how to measure evolutionary time by the extent of this sequence divergence, assuming such changes occur randomly in genes and proteins.

Varmus and his collaborators realized that if viruses had contributed genes to an ancestral organism from which modern birds had originated, then all or most birds should inherit some form of these same genes. These inherited forms should exhibit a certain degree of sequence divergence. Importantly, Varmus and his colleagues realized that if they measured the degree of this divergence among various species of birds, then the extent of the divergence should correlate with their evolutionary distances.

Varmus and his colleagues (Stehelin et al. 1976) managed to isolate chicken chromosomal DNA and asked whether it hybridized to the tumor-inducing viral RNA fragment missing in leukemia viruses.

Their first observation was that parts of chicken chromosomal DNA hybridized to the 'tumor-inducing' RNA piece (to be precise, to the DNA copied from that RNA piece using reverse transcriptase). In other words, some DNA existed in the chromosomes of uninfected chickens that had sequence relatedness with the tumor-inducing piece of a retrovirus. The second observation was that the endogenous chicken chromosomal DNA was related but not identical to the tumor-inducing RNA.

Third, they collected chromosomal DNA from different birds — emu, quail, turkey, duck, and chicken — which were thought to be evolutionarily distant from each other by fossil evidence. By hybridizing these bird DNA samples to tumorigenic DNA from chicken and then measuring their melting temperatures, Varmus and his colleagues showed that the melting temperatures (i.e., the DNA sequence correspondence or homology) varied among the species of birds. Strikingly, the degree to which the DNA sequences varied among any two species of birds appeared to correlate well with the approximate evolutionary distance between the pairs as determined from the occurrence of fossils in chronologically timed geographic strata (Fig. 1.9).

In this sense, Varmus and his associates showed that there existed endogenous copies of a tumor-inducing "principle" — the Virogene, now renamed *oncogene* — in chicken chromosomes, which are remarkably conserved over at least 100 million years of evolution.

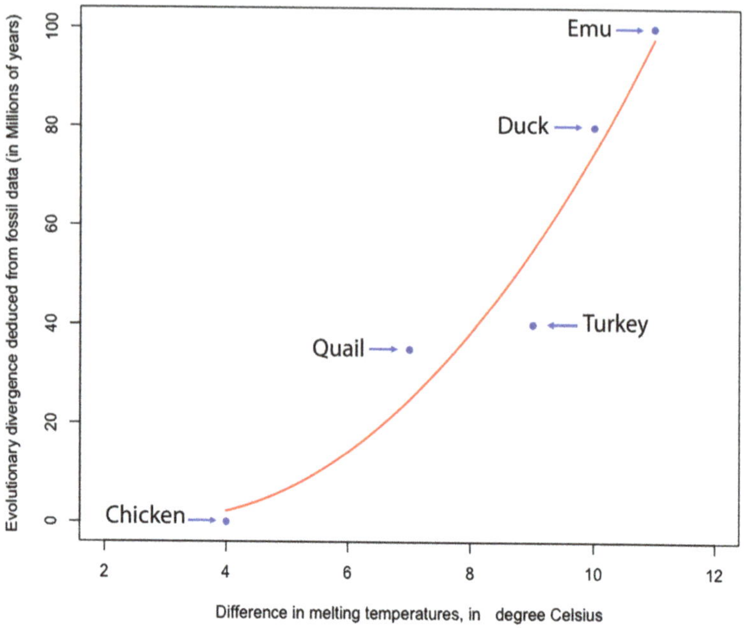

Figure 1.9 DNA melting temperature differences of endogenous viral oncogenes reflect evolutionary distance among birds. See text for explanation. The *x*-axis is the difference in melting temperature (ΔT_m) between the perfect complementary duplex of the integrated viral oncogene and the heteroduplexes formed between the viral oncogene and the endogenous copies of the oncogenes in various birds (in °C). The hybrids were: viral oncogene: chicken endogenous genes (ΔT_m = 4°C), viral oncogene: quail endogenous genes (ΔT_m = 7°C), viral oncogene: turkey endogenous genes (ΔT_m = 9°C), viral oncogene: duck endogenous genes (ΔT_m = 10°C), viral oncogene: emu endogenous genes (ΔT_m = 11°C). [Plot constructed by curve fitting of data from Stehelin et al. (1976)]

Indeed, subsequent studies in the Varmus–Bishop laboratory demonstrated that this gene is highly conserved not only among birds but also among humans, bovine, and even fish DNA (Spector, Varmus, and Bishop 1978).

A separate, perhaps equally important question was how could this gene be maintained through millions of years of evolution with relatively minor changes in their sequences unless these ancient genes also had some properties that provided fitness to the animals?

To reconcile their observations, Varmus and Bishop proposed the existence of "proto-oncogenes" — genes that are derived ancestrally from retroviruses, incorporated into the vertebrate host chromosomes generations ago, and most likely were recruited to perform some important functions in the ancestral organism. Only then one could imagine how these genes might have contributed to the fitness and survival of the descendant organisms. Cancer could in principle occur when these proto-oncogenes become dysregulated, in the sense that mutations caused by physical or chemical agents could turn on these gene activities in cells at the wrong place at the wrong time. And thus, cancer is formed.

1.6 Oncogenes Galore

The gene discovered by Varmus and his team, which was earlier identified by Duesberg and Vogt as the tumor-causing factor in RSV (Duesberg and Vogt 1970), was named *src* (short for "sarcoma"). It turned out that the *src* gene family is found in all animals — from worms like *Caenorhabditis elegans* to insects and mammals, including humans.

We now know that this gene family members code for enzymes called tyrosine kinases. These kinase enzymes chemically modify other proteins by adding a phosphate group to the latter. Such chemical changes may either activate or inactivate the function of these enzymes. Some of these chemically modified proteins are also kinases themselves, which go on to modify even more proteins. This creates a chain reaction that amplifies the original signal many times. The result is a change in the cell's behavior. This process is called intracellular signaling, and many oncogenes are part of this signaling system (Abram and Courtneidge 2000; Brown and Cooper 1996) (Fig. 1.10).

The human version of the gene, written in big letters as *SRC*, encodes the soluble (non-receptor) tyrosine kinase (Sicheri, Moarefi, and Kuriyan 1997). Mutations SRC are frequently associated with malignant colorectal cancer. However, the normal function of the *SRC* gene is widespread in both embryonic and adult tissues, functioning in multiple signaling pathways, including immune response pathways, cell cycle progression, cell migration, and adhesion.

Is there only one *SRC* gene in humans? As it so happens, there are at least 32 relatives of the *SRC* gene in humans.

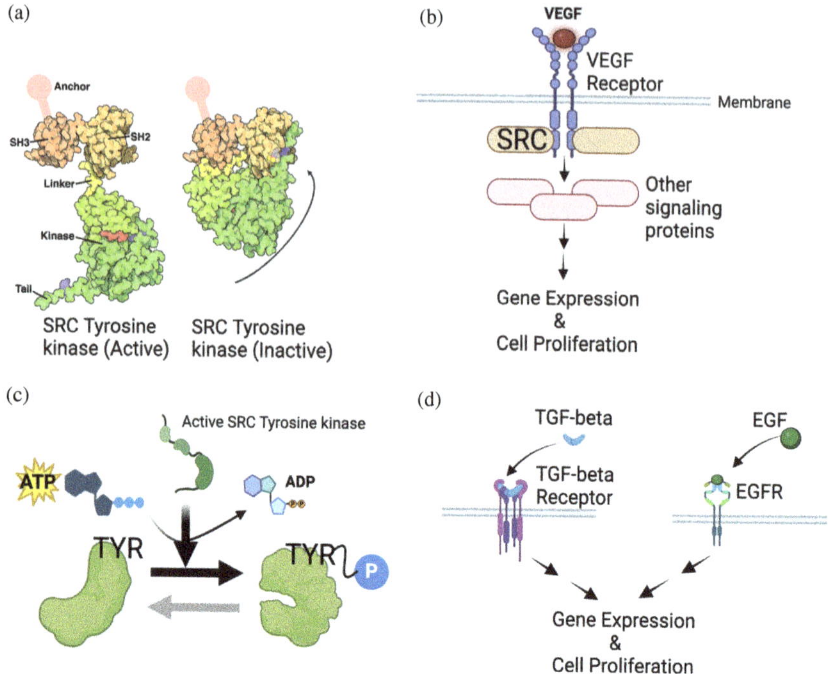

Figure 1.10 The SRC family of protein kinases catalyzes a chemical reaction that signals the cellular proteins to perform altered functions. (a) Proteins encoded by the *SRC* gene family exist as either a compact structure (inactive) in which its three structural domains are closely folded or an extended structure (active). (b) The structural transition of SRC protein is controlled by other proteins, such as the membrane-bound "receptor-protein kinases," which are activated the binding of a ligand molecule (e.g., VEGF). (c) Following activation, the SRC protein, which is an enzyme, catalyzes hydrolysis of ATP to ADP and simultaneously attaches a phosphate group to another (substrate) protein molecule (green). The recipient of the phosphate group is a tyrosine, an amino acid in the substrate protein, with a free hydroxyl group. Upon phosphorylation of the tyrosine residue, the substrate protein changes its conformation and thus its activity. This latter protein subsequently participates in further chemistry, which culminates in the altered activity of many additional signaling proteins and thus alters cellular behavior. The endogenous copy of *SRC* therefore encodes a regulatory protein that signals the cell to perform a variety of cellular functions for which the gene's function is under evolutionary selection. (d) Similarly, there are other signalling pathways, for example, TGF-β/BMP4, EGFR, and many others. [This figure was drawn using Biorender.com under an academic license]

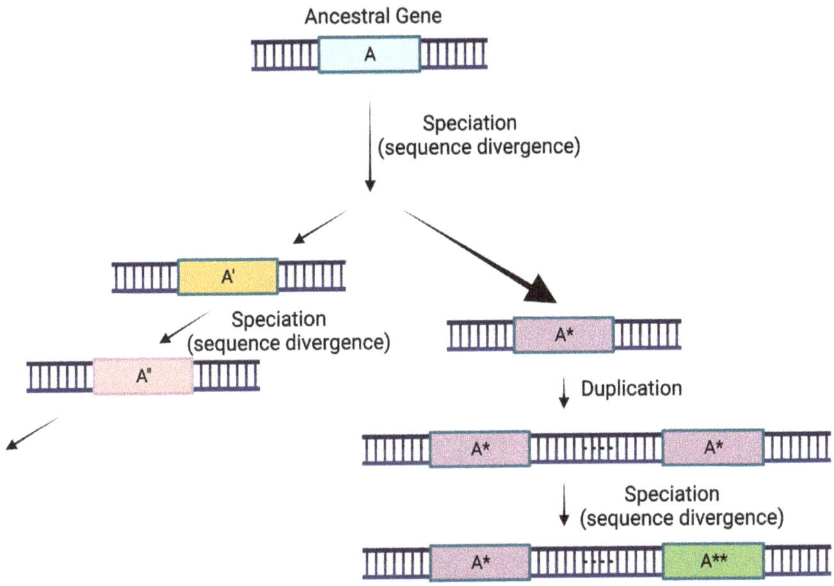

Figure 1.11 Homologous, orthologous, and paralogous genes. Gene homologs are genes in different species of organisms related by their DNA sequence similarity, whereas orthologous genes are related by direct ancestry through speciation events. Paralogous genes are genes thought to have evolved following a gene duplication event followed by sequence divergence. Note that all paralogous genes are homologous to one another. In this diagram, all genes are homologous to one another because they have largely similar sequences. A, A**, A″, and A* are orthologous to one another. A′ and A* are paralogous to each other. [This figure was drawn using Biorender.com under an academic license]

Gene relatives have been defined by two words: homologs and paralogs. Homologs are pairs of genes that are demonstrably related by sequence conservation (see Chapter 16 for a description of how to measure sequence conservation). A paralog (Fig. 1.11) is a homologous gene that most likely arose by the duplication of a common ancestral gene — likely in an ancestral species. In other words, homologs are genes related by ancestry among different species, and paralogs are multiple divergent copies of homologous genes within the same organism. Thus the 32 copies of the *SRC* gene relatives in humans are paralogs of one another. One important paralog of *SRC* is the gene *YES1*, mutations in which are also associated with various malignancies.

The identification of the first proto-oncogene opened the door to the discovery of many other proto-oncogenes, such as *MYC, RAS*, and at least 38 others. Besides these, there are nearly 500 additional candidates that are suspected to be oncogenes. Modern diagnosis and prognosis of cancer are increasingly based on deciphering the status of these genes (see Chapters 2 and 3). For example, in non-small cell lung cancer, *SRC*-activating mutations are associated with the dissociation of tumor cells from the extracellular matrix and subsequent migration of these cells to distant metastatic sites (Haura 2006; Zhang et al. 2007). If a lung cancer patient's tumor cells exhibit the presence of such a mutation, the treatment recommended is different than otherwise. Moreover, drugs that can inactivate this mutant form of the SRC protein have been developed, thus providing a targeted therapy for these patients.

1.7 Oncogenic Proteins Are Often Drug Targets

If a gene, when abnormally activated (to make RNA → protein), causes cancer, one could potentially find a chemical that inhibits the activity of that oncogenic protein. This is the logic behind targeted cancer chemotherapy of today.

However, just because *SRC* is abnormally activated in a particular cancer does not imply that only this gene's product can be targeted therapeutically. For example, lung cancer tumors that exhibit *SRC* activation are now successfully treated with the drug Navitoclax (also known as ABT-263) (Mohamad Anuar et al. 2020), which triggers the death of the cancer cells that are dissociated from the extracellular matrix. Thus, Navitoclax does not target the SRC protein directly. It instead targets the product of another oncogene, *BCL2*, which is synergistically activated when *SRC* is activated. Thus, this drug targets another component of the cancer cell machinery, which is turned on by the *SRC* oncogene.

Several other oncogenic proteins have been the direct or indirect therapeutic targets against cancer. These include *HER2/ERBB2*, *BRAF*, and *EGFR* (Leiphrakpam et al. 2023; Siena et al. 2018; Uribe, Marrocco, and Yarden 2021). Specialized treatment options for cancer patients who harbor mutational signatures in their tumor genomes now exist. For example, patients who exhibit an overproduction of the oncoprotein (the protein encoded by an oncogene) HER2 in certain breast cancer cells (i.e., HER$^+$ breast cancer) are often successfully treated with a drug called Herceptin. Herceptin blocks the activity of this oncogenic protein (Gray, Cameron, and Early Breast Cancer Trialists' Collaborative Group 2021; Leiphrakpam et al. 2023). Metastatic skin cancer or melanoma tumor cells that contain mutations in the oncogene *BRAF* are often treated with the drug vemurafenib; this drug inactivates the product of this oncogene (Gutzmer et al. 2020). Likewise, a certain type of mutations in the oncogene *EGFR* is often seen in metastatic non-small cell lung cancer tumors, which can be treated by several targeted anti-cancer drugs, including Erlotinib and Gefitinib (Xu, Johnson, and Grandis 2017). EGFR is a receptor protein that binds to an extracellular molecule (a general term for such molecules is "ligand") that activates the EGFR's enzyme activity (also a tyrosine kinase, like SRC). In some cancer cells, the *EGFR* gene is mutated such that the mutant enzyme is active even without the ligand being present. Erlotinib and Gefitinib inhibit this enzyme, thus reversing the effects of that cancer-causing mutation.

1.8 Tumor Suppressor Genes Are Defenders Against Cancer

Classical ideas in Darwinian evolution propose that natural selection acts on the traits of individuals existing within a population. Reproductive success of the individual — the frequency with which the individual passes on her or his genes to the next generation — is what determines the evolutionary fitness. Thus, a high rate of childhood or pre-adulthood cancer incidence in the population causing pre-reproduction lethality or morbidity would diminish the reproductive success of individuals carrying any such gene. Such traits, therefore, should be disfavored in the population; individuals carrying these traits would be out-competed over generations.

What molecular genetic mechanisms might account for disfavoring the spontaneous production of cancer cells?

We have implied above that an oncogenic transformation in a cell is to the advantage of that cell in relation to its noncancerous neighboring cells.

Due to such mutations, a tumor (or cancer) cell multiplies, and its clonal descendants propagate at the expense of other cells. It is as if the cancerous cells parasitize the host, ultimately killing the latter by usurping available resources and by producing waste products, providing little benefits to the host. This line of thinking also implies that we should anticipate the existence of an opposing class of genes that have evolved to prevent or at least reduce the occurrence of cancer. Evolution is a game of tug-of-war.

Indeed, multicellular organisms have evolved intricate mechanisms to both reduce the frequency of spontaneous mutations and to eliminate cells with oncogenic mutations. Genes that are central to these defensive mechanisms against cancer are collectively termed "tumor suppressor" genes.

One of the earliest tumor suppressor genes to be discovered is called *TP53* (Levine 2020). Nearly 80% of solid tumor cells contain inactivating mutations in the *TP53* gene, suggesting that once *TP53*'s protective activity is lost, cancer might occur more easily. Besides *TP53*, at least 30 additional tumor suppressor genes have been discovered (Kontomanolis et al. 2020; Sherr 2004).

As the reader can anticipate, some of these tumor suppressor genes may normally function to reduce the rate of spontaneous mutation. The fewer the number of mutations that occur, the lower is the chance of a tumorigenic mutation. A few examples of this class of genes are *MLH1, MSH2, MSH6, BRCA1, BRCA2*, and *APC1*. Other tumor suppressors, such as *RB1* and *CDKN2A*, are direct or indirect inhibitors of cell division. Yet others, such as *NF1* and *NF2*, enhance cell differentiation processes; the latter act by way of removing the reproductive potential of presumptive cancerous cells, enabling these cells to enter cellular stages that prevent further cell division. There are also a few tumor suppressor genes, including *TP53*, *CDKN2A*, and *VHL*, that can turn on a program of biochemical events leading to cellular death. A cell that might by accident have encumbered an oncogenic mutation could be committed to death by the action of such a tumor suppressor gene.

The activity of certain tumor-suppressor genes is involved in the immune recognition of cancer cells and the latter's removal by immune cells in the blood. Such immune cells are known as cytotoxic T-lymphocytes (Blagih, Buck, and Vousden 2020; Mergener and Peña-Llopis 2022). Manipulating the tumor-killing ability of cytotoxic lymphocytes, the so-called CAR T-cell immunotherapy (see Chapter 5, Section 5.4), has recently emerged as a highly effective tool for treating cancer (Akhoundi et al. 2021; Teoh and Chng 2021). Exciting developments are anticipated within the next five years.

1.9 Pervasive Genetic Testing for Cancer Susceptibility Gene Mutations Is Important

Discoveries of oncogenes and tumor suppressor genes brought major changes to how physicians diagnose and treat cancer. For example, certain precancerous growths, like adenomas, are more likely to turn into cancer if their cells have specific mutations in the *TP53* or *RB1* tumor suppressor genes. Discovering and specially treating such adenomas can save future cancer occurrences.

Family members of breast cancer patients, and even those with little information on breast cancer in their own families, are now often advised to have their *BRCA1* and *BRCA2* genes checked (Buchanan et al. 2018). If these individuals carry inherited mutations in these tumor suppressor genes, they can start early and have regular medical checks, even if they don't yet have cancer. Similarly screening the relatives of certain colon cancer patients to determine whether they harbor mutations in *MLH1*, *MSH2*, and *MSH6* tumor suppressor genes is often advised (Foulkes, Knoppers, and Turnbull 2016; Lerman and Shields 2004; Mai et al. 2014).

Cancer patients with diagnosed cancer in one organ (say, pancreatic cancer) may return with an unrelated cancer in a second organ later if they harbor one or more of the cancer susceptibility gene mutations. Relatives of such patients appear to have a higher than basal level of risk for developing a host of different cancer types. Therefore, early genetic screening for these mutations can favorably impact many potential cancer patients.

Programs for early screening or screening of the relatives of cancer patients based on gene sequencing have increased the frequency of early detection of several cancer types and have significantly reduced cancer mortality rates in affluent countries. Lamentably, as of this writing, even in the United States, over 90% of cancer patients are not genetically tested for the presence of cancer-susceptibility mutations (Kurian et al. 2023), thus missing a very large number of their relatives at risk of cancer. In countries where most people are unable to access these diagnostic or screening options, the national burden of nearly all cancer types continues to increase (Sung et al. 2021).

As the reader can appreciate, fundamental to modern approaches to diagnosis and treatment of cancer is the concept of oncogenes as a normal endowment of the human genome, which may occasionally go awry and cause cancer. This critical insight was molded fundamentally by the Darwinian concept of the relatedness of different species on earth, without which neither the observations described in this chapter can make sense nor much progress could have been made in understanding the molecular basis of cancer. Similarly, arguing along Darwinian views, we saw how tumor suppressor genes prevent cancer and therefore allow us to decipher the signature of mutations in known tumor suppressor genes. These signatures serve as prognostic markers for cancer treatment and inform public health policy.

Most of our knowledge of cancer susceptibility gene variants is based on observations of people of European ancestry — a result of research done in western countries, conducted upon relatively affluent ethnic groups. There is evidence of additional susceptibility variants important for cancer in relatively less studied ethnic groups, but our knowledge of such is quite incomplete (Park, Cheng, and Haiman 2018). Conversely, certain variants that constitute a risk factor for the European ancestry may not constitute a similar risk for other ethnicities. This has resulted in increasingly asymmetric health outcomes along the so-called north-south global axis. Urgent effort is needed to mitigate this asymmetry.

In summary, cancer, if diagnosed early, can often be reduced to the status of chronic diseases. Individualized therapies are essential for treating cancer effectively. This is due to the fact that both inherited natural variants and somatically produced natural variants (mutants) arise within the patient. These undergo natural selection, giving rise to the cancerous cell property unique within a patient's body, as if each person's tumor, even individual tumors, are akin to individual "organisms" or "species" in a natural ecosystem. To kill these unique cancer cells within the patient's body obviously requires that we tailor the therapy targeted to these unique tumors. These profound advances in understanding the biological basis of cancer, its diagnosis, and treatment have been made possible through the insights accorded by Darwin's evolutionary theories.

Chapter 2

An Oncologist Speaks

Dr. Ravi Salgia is a real-life oncologist at the City of Hope's Comprehensive Cancer Center, one of the premier cancer treatment hospitals in the United States. He is also the Arthur & Rosalie Kaplan Chair in Medical Oncology, leading over 100 medical oncology faculty members in his department. His undergraduate degree was in mathematics, biology, and chemistry, following which he studied medicine and simultaneously obtained a PhD in biochemistry. He was a faculty member at Harvard Medical School and at the University of Chicago before moving to the City of Hope Comprehensive Cancer Center near Los Angeles. On a sunny November morning in Southern California, we sat down to chat.

Author: Ravi, tell me about your work . . . what do you do?

Dr. Salgia: I am a clinician scientist. I see patients in the clinic but I also am a scientist in the lab, trying to come up with a better, less toxic therapy. . . . If one kind of treatment worked with one patient but didn't work or not so well with another patient, then I investigate these in the laboratory. I try to see what was important for these differences.

Author: What motivates you?

Dr. Salgia: At the end of my journey, it's always the patient. What is it that the patient needs the most? What are the patient's family's needs or their caregiver's needs? And then how can we utilize the tools and the science that can affect these patients' outcomes and then go beyond that.

Author: Tell me a bit about your specialty.

Dr. Salgia: I am a lung cancer specialist. I take great pride in caring for lung cancer patients. Despite my administrative responsibilities, I still have a large clinic within my hospital. People ask me, "in your position, why don't you go down on your clinic size?," but when I was growing up, I didn't tell my parents that I would be an administrator. I wanted to be a physician or a scientist or a mathematician. But then, administration is crucial, for it helps you advance your colleagues, your institution, yourself, and to make it better for our patients. Ultimately, it is always the patient.

Author: How did you get here, in this role?

Dr. Salgia: You know that initially I started out in mathematics. As a child I liked math, and as an undergraduate I treasured the mathematics. I loved every minute of calculus and complex analysis. I lectured around the country. I wrote some papers on a new integration technique. I loved all of that. It is the same as a math problem now — we have a problem before us [a patient's condition] and we will have to solve it. There are known solutions to certain problems and there are certain problems with solutions to arrive at. I always think like a mathematician that way. But when the time came during my undergraduate days to decide what to do in my life, I had two choices. Either I would become a mathematician or I would apply what I learned to something else. As a junior in college, I decided to apply to medicine what I was learning. Again, that wasn't enough for me. Should I only treat patients or should I not go beyond and try to learn something about the disease to make the treatment better.

Author: How did you decide to do that?

Dr. Salgia: I also loved science, every minute of it was exciting. I became an MD-PhD student. The first two years of medical school is basically memorization — but it is an important aspect of being a physician because you must have the basics, the fundamental knowledge. You can't get that by ChatGPT; you can't get that by somebody else telling you what it should be. You must learn yourself. Then I went into the laboratory. I loved it, you know, every moment of it.

Author: Laboratory?

Dr. Salgia: It's like this . . . for MD-PhD you take your first two years of preclinical medicine courses, then you spend the next few years depending on how your work goes doing research on the laboratory bench or doing informatics in the computer, even do a political science PhD, whatever interests you, really honing your interest and your skill, your passion and your vision. I went into a biochemistry lab. I worked on an inherited metabolic disease, Gaucher disease. I used some math as well . . . doing enzyme kinetics.

[**Author's note:** Dr. Salgia's PhD research in the 1980s discovered that in hereditary Gaucher disease, where the patient has reduced activity of an enzyme in the lysosomes — a cell organelle required for the appropriate degradation of proteins and other substances — the fluidity of cell membranes is affected. A patient who has Gaucher disease displays an enlarged spleen, liver, and yellow spots in the eye; has abnormal eye movement; suffers from easy bruising, lung problems, extreme fatigue, anemia, fragile bones, and bone pain; and often has brain damage. Dr. Salgia used a spectroscopic technique to determine that the membranes of fibroblast cells taken from Gaucher disease patients have different degrees of fluidity relative to normal cell membranes, helping to demonstrate that the reduction of activity of one enzyme can have profound effects in many parts of the body.

Interestingly, recent evidence indicates that the prevalence of this inherited disease in certain ethnic groups might be associated with an evolutionary advantage (Beutler 1993; Slatkin 2004). Patients having a mild form of Gaucher disease are less frequently affected with type II diabetes or a liver disease called nonalcoholic steatohepatitis. The available data cannot be explained by a chance occurrence of this mutation in a small group of individuals reproductively isolated from the larger population, leading to an increase in the frequency of this mutation — a phenomenon known as the "founder effect" (Slatkin 2004). This suggests that there is a different explanation for the prevalence of this mutation within this small ethnic group; the most likely explanation is a selective advantage for the Gaucher disease mutation. When Gaucher disease patients are treated with a therapy to replace, in high doses, the missing enzyme, some of the patients become more prone to developing two other diseases (type II diabetes and/or liver disease), underscoring the need to carefully regulate the dose of this enzyme (Ilan, Elstein, and Zimran 2009).]

Author: You did your first two years of preclinical studies in medicine, then you did PhD research in the laboratory. What did you do after that?

Dr. Salgia: In the MD-PhD program, after getting PhD you return to medical school to complete your clinical studies for the next two years. My motivation for doing PhD was, okay, we will take great care of patients but we will also try to understand why so and so has this disease. So, in the future you can perhaps prevent the occurrence of the disease in the first place.

Medicine is incredibly complex . . . it has the science part and then it has the art part of medicine. And you must integrate both parts together. I am thankful to all my mentors and my colleagues who taught me. After graduating from the medical school at Loyola University in Chicago, I went off to Baltimore at the Johns Hopkins for my "residency" training. But I also did a little bit of research there. I did a deep dive into internal medicine. How to take good care of patients, how to get great patient outcomes, and if I don't get great outcomes then how to deal with that as well. Can you make that experience into a betterment? I learned a lot. Then I went to the Dana Farber Cancer Institute in Boston to do my fellowship at the Harvard Medical School.

Author: Tell me about those days of yours.

Dr. Salgia: I was a fellow in medical oncology at Harvard Medical School in the early 1990s. That's where this basis of molecular oncology was instilled in me. At that time, we didn't have a lot of understanding of molecular oncology. But we knew there must be a molecular basis to how cancer occurs, the diagnosis of cancer, the treatment of cancer, and, as you say in your book, the evolution of cancer. I learned how to do structure-function analysis of molecules (involved in cancer), how to use this for the diagnosis and treatment of cancer, how to put it all together. Those were my formative years. I didn't forget the mathematics, but truly became much more the experimentalist in the lab, exploring structure-function analysis of therapeutics, outcomes, and of course my lung cancer patients . . . I truly love taking care of them.

Author: Were you always treating lung cancer patients?

Dr. Salgia: Initially I was treating hematological cancer patients . . . many had leukemias. Then I started treating lung cancer patients.

Author: Ravi, tell me about some of the major challenges you face in the clinic.

Dr. Salgia: The challenges are many. They start from the very diagnosis of cancer, what stage is it . . . meaning, has it spread to the lymph nodes, has it spread to other organs. I am also a molecular oncologist — so I ask, what are the molecular abnormalities that can happen, whether they have happened or not, in this individual patient. How can we piece together these various clues to find the best course of treatment?

Author: What factors contribute to solving some of these problems?

Dr. Salgia: You know, it always comes down to the institutions that allow me to work on my patients and then give me the protected time to work on the science of the solution. When I first interviewed at Harvard, at Dana Farber, they asked me, "What do you want to do?" I gave an answer that might have been naïve or it might have been a deep answer, I still don't know. I said, "I want to cure cancer. Period." Then, as I started taking care of lung cancer patients, my goal was, "I want to cure lung cancer." I still hold to that truth. If you ask me, what is my passion and vision, I will say, I want to cure lung cancer. Period.

Author: How did your career develop after Dana Farber?

Dr. Salgia: I became a junior faculty at Harvard, then I became a tenured faculty at the University of Chicago. What I like about the field that I am in is that you get mentored, but then you also mentor others, and this kind of relationship never goes away. Now to fast forward to Chicago, we had made several seminal discoveries at Harvard and brought them to Chicago. Maybe there was an even higher impact during my time in Chicago in terms of the therapies and the clinical trials we did there . . . there were a lot of clinical trials. Then this opportunity came up, to come to the City of Hope.

Author: Why the City of Hope?

Dr. Salgia: In this department at City of Hope we are basically asking, how do we impact our patients not just from the ivory towers but from where the patients are. Does Mohammed come to the mountain or the does the mountain go to Mohammed? In this case the mountain would be the patient. Many times, in many institutions, the mountain goes to Mohammed. But we really need to make Mohammed go to the mountain. So, how do we deliver all that care to the patient effectively? How do we energize the next generation of physicians? We train new physicians, and they take over. At the end of the day, I hope we are out of our job — that we can prevent cancer and that's about it.

Author: What do you do with lung cancer patient now when the patient comes through your clinic's door? Can you give some examples?

Dr. Salgia: Let's say they have a newly diagnosed lung cancer, they come to the clinic . . . truly you'll have to be a physician in the clinic. So, you get their history. Generally, my patients have a diagnosis of mesothelioma.

[**Author's note:** Mesothelioma is a type of rather vicious lung cancer, which has been linked to exposures to environmental factors, such as asbestos, smoking, and radioactive gases like radon.]

Have they been exposed to radon gas or to smoking; talk to them about their family history; do they have other diseases . . . diabetes or hypertension, because that also helps us for decision-making. Then examine the patients.

I can't overemphasize how important it is to examine the patient. In my practice, I have now listened to about 350,000 lungs. With experience comes pattern building. You get to learn patterns. Something is skewed, I immediately know it by listening to the lungs, listening to the heart, feeling the liver, the lymph nodes. I really can't emphasize it enough to young people who will be going into medicine, the importance of physical exam of the patients and knowing the patterns. Then, you look at their data . . .

Author: Let me stop you there for a moment. Before you look at their data, do you, during your physical exam, use your intuition developed over the decades or do you look for certain objective measures? This is important for young physicians for they have not yet developed those intuitions from experience.

Dr. Salgia: Both. You look for abnormal heart and breathing sounds, you look at their feet to see if they are swollen. Those are objectives. But then you feel their pulse, to see if they are tachycardic or bradycardic — heart running fast or slow — those are more subjective, and then you put these all together . . . It's not just the heart sound or the lung sound but the conglomerate of all these factors that you learn to see through experience.

What can you learn from these observations for deciding on therapy, for you will have to decide on a therapy. The other day there was this patient whom I asked to walk but (s)he was not walking properly. The patient had had brain metastasis that was treated with radiation, but I also looked at the eye movement and based on that I recommended another MRI even though the patient had already had a brain MRI. Lo and behold, the patient had another metastasis in the brain. You'll have to put things together. You get information from all the things you did but then you put them together and come back to say, what do they mean? Only then you look at the laboratory data on tests. I will look at their chemistry panels, their sodium-potassium levels, their kidney functions, liver functions, look at how their white blood cells are, hemoglobin, platelets, all of that. Sometimes you'll have to order special tests based on what you see. Let's say somebody has rheumatoid arthritis so that you might have to order the detection of some antibodies. You also look at their skins. Very crucial, where is the tumor? Is it just a single, isolated tumor in the lung, at multiple sites within lungs? Is there a tumor in the liver? If the brain is affected, then obviously you look at that. Many people do it the opposite way, but I am a little different from the traditional oncologist in that I like to see the patient first then see the data. The data might be provided to me ahead of time, that's ok. But it is crucially important to examine the patient.

One example, I see a CT scan of the lungs and I see a diffuse opacity, and I say, oh I can see lung cancer. But I talk to the patient and he says, "I had Covid a few months ago." So now I think, oh it is a complication from Covid infection! I am a very systematic person. You do what is necessary but sufficient. When I talk to the patient and examine the patient, I get many more clues than just from the laboratory test numbers . . . and then you look at the tumor biopsy! Does the pathology make sense? Someone was diagnosed with small cell lung cancer but he never smoked . . . well, you say, how is that possible? The two don't come together . . . then you ask for molecular testing.

Author: Would you do molecular testing routinely for all lung cancer patients who come to you?

Dr. Salgia: Yes. That is our policy now.

Author: To what extent? Are these tests looking at selected gene mutation signatures? Or are they looking at their whole genomes?

Dr. Salgia: You can do all of that. You can do whole genome sequencing, whole transcriptome sequencing, you can do deep RNA-seq.

[**Author's note:** This allows assessment of gene expression states.]

Author: Would you do these routinely?

Dr. Salgia: Yes, we like our internal platform, with which we look at City of Hope's chosen gene sets first. But let's say, we have a 23-year-old patient with metastatic lung cancer, and we can't identify a molecular cause using our internal platform, then I think you'd have to look deeper. We have an "algorithm" for this: when the biopsy sample arrives, we first use our internal platform but depending on the findings we then may have to look at whole exome sequencing, RNA-seq and all that.

Author: Based upon what you find there, do you then decide on the course of the therapy?

Dr. Salgia: Exactly. If the tumor is resectable, we surgically remove it. But even it is early stage resectable, should you do chemotherapy or immune-therapy first, and then surgery? Or only immunotherapy first, then surgery, followed by more immunotherapy or targeted small molecule drugs? There are so many nuances we go through . . . no textbook helps. And all that data that are generated in the process, you'll have to be a super-specialist to understand all that data and decide what you'll do. I am not saying our job is difficult or challenging, I am just saying that you'll have to keep up with the sciences.

Author: Do you do this alone or is there a team?

Dr. Salgia: We have multiple layers of teams. Today at three o'clock I will be meeting with my clinical team. We will go over all the patients we will see this week. There will be myself, our nurse-practitioner, our nurse, pharmacist, two research nurses, two of our clinical research associates, and our scribe.

Author: What do the clinical research associates do?

Dr. Salgia: They help me go through the patient's history, the clinical records charts, they help collect blood, or we have them follow up with the research staff. The clinical research associates (CRAs) I work with, sometimes I ask them, hey, let's do a journal club and discuss these findings. As an example, we saw a patient this last Friday, who has been treated for the past five years. And they missed it! I saw the patient on a consultation for the first time. And we found that the patient had a unique chromosomal fusion that was in on page 2 of the patient's report, and nobody had paid attention to it. Suddenly it opened so many opportunities! Because I already knew the literature around that chromosomal fusion. I told the clinical research associate, let's pass this paper around today.

So, it's a group effort . . . it's also a shared decision-making effort. Our nurse-practitioner has great clinical knowledge. We have clinical team meeting on Tuesdays and Fridays, and by the time we see the patient again basically 80–85% of the background work is done. We might not have seen the patient yet, but at least we have done the detailed background work with a fine-toothed comb. So that's one layer of the team.

The other one that we have is what's called our "tumor board." We present our cases to our colleagues, who are surgeons, radiologists, radiation oncologists, pulmonologists . . .. We go over our patients, not everyone, but those for whom we need multi-disciplinary decision making.

Author: You prioritize some cases, whose cases you will present.

Dr. Salgia: Yes, when there is a routine standard of care for some patients, we may decide not to present those cases to the tumor board. But even for some routine cases, we prioritize decision-making on whether we need a surgical opinion, and I will present them. Do I need a radiology opinion, so I present those cases to the radiologists.

We also have built here what is called the "Complex Case Series," in our Medical Oncology department. I wanted to have each one of us be empowered by the thought leaders of the world. People can present in our department, for example, very tough lung cancer cases, and I will go over with my colleagues in a focused setting. We do that once a month. If we have a sarcoma patient, we bring in a sarcoma expert. If we have a head-neck cancer patient, we bring in a specialist. In this way we go beyond the other ways we just discussed, by coming together as a group we find out, "this is what the leaders of the world would do." A whole different ballgame . . . . We should think very similarly in the sense of how our patients deserve standard operating procedures, quality assurances, quality control but they also deserve the thought leaders of the world looking after them. Of course, it comes with a price because everybody must join forces, everyone is busy but imagine the end product! So, we do this shared decision-making process, the care modeling that we do. I probably have the most experience within our group on molecular oncology, so I provide that expertise. Other members of our team have clinical research expertise, others have outreach experience, they provide that. That's what I like in team building — not to have each other as clones but to synergize. You are a big believer in team science — I am a big believer in team medicine, bringing thought leaders and practical people together.

Author: You are building this interdisciplinary team medicine model in the specialty care hospital that the City of Hope is, which I presume also exist in some form or other in some of the best teaching hospitals in the country. But how about the rest of the hospitals? How about the rest of the world? Are they also benefiting from these incredible advances in medical sciences?

Dr. Salgia: Oh absolutely! We strongly believe in having an international presence in education. We have partnered with the Einstein Hospital in Sao Paolo in Brazil, and we have a lot of these meetings together. We do a lot of teachings in other countries as well. We are part of national and international organizations where you want to empower your colleagues no matter where they are. Our national and international venue is the American Society of Clinical Oncology, (ASCO). Then there is the European Society of Medical Oncology, (ESMO), there's also this organization called IASLC or the International Association for the Study of Lung Cancer. Let's say we all come together and ask, what do the thought leaders think about EGFR.

[**Author's note:** A gene that encodes the epidermal growth factor receptor.]

If there are EGFR gene mutations in a patient, how would City of Hope or Dana Farber respond, versus France, Germany, India . . . so we all get together. I know that education is an important component of what we do. Publishing papers for describing the nuances of the various EGFR mutations is important. How should one think about mechanisms of resistance to the drug that targets mutated EGFR receptor proteins, those that result from activating mutations? You think about Erlotinib and the mechanism of resistance of T790M.

[**Author's note:** A mutation that has replaced the tyrosine residue at position 790 in the amino acid sequence of the normal EGFR protein by a methionine, which then makes the protein insensitive to the chemotherapy drug Erlotinib (see Suda et al. 2009).]

Then fast forward to the drug Osimertinib, which can overcome that resistance, but then you can have C797S mutation on top of that . . . you can then have MET amplification.

[**Author's note:** The gene *MET* encodes a tyrosine kinase enzyme; the gene is often amplified in certain non-small cell lung cancer tumors (see Schubart et al. 2021).]

Besides this world of genetic alterations, there is this whole world of so-called nongenetic alterations.

[**Author's note:** Nongenetic or epigenetic changes are due to modification of gene function without affecting the DNA base sequence. These changes include DNA base methylation, chromatin-bound proteins modified by methylation, acetylation, and so on.]

This is exactly where education becomes important. We also have an international office. If they in another country have a tough case, then the international office may get us these cases, and I can opine on them. But if you ask me, I will say the effort is in its infancy because it depends on politics too, and on insurance coverage, how do you pay for your time, right? We love to do charity work but at the end of the day we still must earn a living. So how do you balance these various roles? These are all quite important questions. I don't know if I have all the answers but I do know we'll have to do more.

Author: Your points raise the following issue. You and the team have decided on a course of therapy optimized for a patient. How often do you find that during the progress of the therapy you will have to change course? How dynamic is the system, so that if a change is needed it can be done and what are the factors?

Dr. Salgia: Days are gone when, like in some TV shows, you as a doctor will sit there and say, this is what we are going to do and this is your prognosis, and that's it. The end of statement. That's not what happens in the real world. It's an iterative function. We start off by saying, ok, here's where you are in your journey, and this is what we are going to do, and this is what our thinking is: you have an early-stage lung cancer, we can give you chemotherapy, immunotherapy, get you to surgery and hopefully to a cure. But we are also cautious to say, these are your chances that you'll do well but if you don't then we will attack that part of the journey when we get to it. If on the other hand somebody has let's say metastatic lung cancer and you are coming up for a plan for the first time, you discuss what it is going to be. But then you also say, yes there will likely be some resistance of the cancer to the therapy. Currently, most of our patients and their families are incredibly savvy. They have already done Dr Google searching. They will say, "but what about this mechanism of resistance?" I will always say, we will get to it when we get there. The more time we get, the more options we will have. We will then reevaluate and see if certain things still make sense. Be it radiation, be it surgery, be it more chemotherapy, or targeted therapy or immune therapy. It is a continuum for us, this journey, and we tell the patient that we are in this journey together. However, the predictive powers that you see on television don't exist. We go on generalities to say, we expect great response. But we are going to monitor you. We hope you will do well and you should; however, if we need to then we will change the course.

Author: How about 20 years ago? What has changed in the past 20 years?

Dr. Salgia: I have been in practice for about 30 years now. In the old days, we will give the statistics, this is what the data says . . . I have given up on all that because every patient is so different, every mutation's effect is so different. Even all the EGFR mutations don't have the same effects; L858R has different effects than exon 19 deletion. Exon 19 deletion versus 747 is different. Then you have exon 20 alterations; exon 18 alterations. A whole different ball game now. In the old days we'll say, oh you have an EGFR mutation. The more we learn the better we get, and more precise we get in our answers because our therapies are getting better and better. That's what I am excited about.

Author: Let me now transition to prevention. To what extent do you recommend testing the relatives who might be susceptible. In your case it might be environmental exposure or perhaps also genetic predisposition.

Dr. Salgia: Clearly, smoking is a big cause for lung cancer worldwide but we don't see that for southern California that much. The majority of our patients don't smoke. Or they gave it up long time ago, or had a very light smoking history. But worldwide, smoking is a huge issue. The prevention of smoking is an important issue for us in the lung cancer world. Then, vaping is another one. We are not quite sure but it sure causes a lot of lung damage. Can we prevent vaping? We talk to our patients and families regarding all of that. Then we also ask them about exposure, to things such as radon. In my practice in Boston, we used to see Chernobyl radiation-exposed patients. And Three-Mile Island, for example. Those are realities.

Author: As you obviously know, in New England and upstate New York there are high levels of radon in the basement of many houses.

Dr. Salgia: Yes, we talk about that too. But at the end of the day, family history matters. We are just realizing this now. In our institution, we like to do germline DNA sequencing as well, looking at the family history, whether we can trace the inheritance pattern of germline mutations and link them to lung cancer.

Author: How often would you do that . . . at what rate?

Dr. Salgia: In my clinic, we have achieved 100%. Believe it or not, we are finding a huge amount of information that I would have never thought possible even three years ago, based on germline DNA sequences, even for our lung cancer patients. What do you do with that information? To be determined, because lots of it we don't understand. Let's say somebody has Lynch syndrome or has EGFR T790M hereditary mutation. We have seen the presence of germline BRCA1 and BRCA2 mutations among some of our lung cancer patients in our clinic at City of Hope. We are very sensitive to these investigations. However, in the nation it is not being done regularly. So, we are trying to teach our outside colleagues to say, it [germline sequencing] should be done.

And prevention. You have the family history; you have got the history of smoking or vaping . . . are there other prevention strategies? Can we develop drugs to help prevent smoking? That's one of the holy grails for us — prevention. More work needs to be done on the prevention side, for lung cancers especially but also for other cancers. Surveillance, screening. Screening is important. Do you subject someone to [CT Computerized Tomography] screening? Do you do a [PET Positron Emission Tomography] scan? Do you do a blood test, like these liquid biopsies that are being touted? For sure in the high-risk populations, we do CT screening — that's our standard. I do think there's a role for liquid biopsies, looking for cell-free DNA, RNA, epigenetic markers in body fluids and urine. But that technology is not perfect yet.

Author: FDA [Federal Drug Administration] obviously has regulations in place as to how physicians will treat patients. I suppose medical associations also have some say about these things. But how much freedom do you have as a physician to do some of these things according to your own judgment?

Dr. Salgia: We are pretty good about working within the framework that has been defined. Be it the FDA in terms of the therapeutics we can give. We respect all of that; we don't deviate from that. There are guidelines. Guidelines are not meant to be absolute laws. You can take them, but you can also navigate them a little differently if you believe that certain patients will respond better. Sometimes the guideline is a little delayed . . . by about six months. What you know today is not necessarily reflected in the guideline now, so you use that information in your decision-making, because you want to give your patients the best outcome based on the best information available to you. If we have more data then we can go beyond just the guidelines. But you still must abide by the FDA guidelines, rules, and regulations. You'll also have to be very cognizant of the payers — the insurance companies. Let's say, coming back to your EGFR example, we want to give both Erlotinib and Ramucirumab for EGFR L858R mutant tumors in lung cancer, which is an approved regimen, the insurance company may say, no, you can only go with Osimertinib, then we will go with Osimertinib. In this way, we work with the regulatory agencies and the insurance companies.

[**Author's note:** The first drug, Erlotinib, inhibits the EGFR receptor mutant proteins that in cancer cells turn on intracellular signaling for tumor cell division even in the absence of growth stimulatory factors from the outside, and the second immunotherapy drug prevents the migration of blood vessels into the growing tumor and thus starves the tumor cells by inhibiting a different receptor, namely VEGFR2. The two drugs, Erlotinib and Ramucirumab, in combination can be highly effective for certain lung cancers (Nakagawa et al. 2019), though not yet demonstrated for other cancers. Osimertinib is yet another inhibitor of EGFR that works well with the EGFR L858R mutant (Ji et al. 2023).]

Author: One last question. One of the reasons I am writing this book is I that noticed an almost lamentable lack of appreciation among engineers, biotechnologists, and even some physicians of the role of chance, stochasticity and of evolution in modern medicine and related areas. Perhaps the current political philosophies are to some extent responsible for going back to deterministic ideals even though determinism had been demolished nearly a hundred years ago. What is your parting thought on this question? About how evolution plays a role in our understanding, upon which progress in medicine is based?

Dr. Salgia: Oh, I am a big believer in that. As you know, I also do chaos theory, I am a big believer in deterministic chaos and fractal analysis and the importance of breaking of symmetry in complex systems.

You know, whoever designed this universe has a huge role here. Could this be, as Einstein said, "God does not play dice," versus is it deterministic chaos? Is this evolution? Is this a component of all of those? Is this also manmade and man-driven? If you look at toluene, benzene, asbestos (as environmental carcinogens), are these part of determinism, part of evolution, man being a part of evolution . . . I think there are components of all the above . . . I must agree that all those components lead to diseases. That's what I work on. Diseases. My specialty is cancer. It is very clear that evolution plays a huge role. It is also very clear that manmade stuff is also very important. It's also clear we don't understand enough to be able to say, beyond just the stochasticity what else is there . . . in my view, we always must be hopeful. No matter how we got here, we'll have to say, can we make it better?

Author: Thank you for indulging me with such an illuminating conversation despite your very busy life.

Chapter 3

Cancer as an Evolutionary Disease

In this chapter, we examine how oncogenes and tumor suppressor gene mutations become drivers of cancer in a multihit mutation-selection process. If it is true that cancer is due to gene mutation, we should try to understand how many mutations are needed minimally to form cancer and which genes drive cancer. Moreover, we provide evidence and support for the hypothesis that cancer is a disease that has most of the properties of Darwinian evolution itself. More interestingly, we discuss the role of mutator genes in controlling the rate at which mutations (i.e., the source of genetically variant cells) occur in cancerous cells and their implication in the ability of cancer cells to outcompete normal body cells. We also grapple with the possibility that mechanisms other than genetic variation might be important for certain cancer types and whether the existence of such mechanisms might indicate that fundamentally non-Darwinian processes might be at work in these cancers. These observations together raise big questions on how we can better understand cancer using the Darwinian paradigm. Armed with a deeper understanding, we can more accurately predict who might be at a higher risk of cancer and how we might develop even better treatment. We also discuss some aspects of cancer that we don't yet understand and others that appear to contradict some of our standard views.

Important terms:

A *clone of cells* is a cluster of genetically identical cells descended from a single cell through a well-defined lineage without mixing with cells of other lineages. A *lineage* is a relationship of direct inheritance pattern — a diagram of who comes from whom, much like a *genealogy*. A *subclone* is a part of a clone of cells, some of which might have additional genetic variations. A *mutator* gene is one that is needed to keep the mutation rate down (a gene is often named for its "mutated" phenotype); if a mutator gene is itself mutated to a loss-of-function status, the mutation rate of the cell generally increases.

Ploidy is the number of chromosomal sets: diploidy is two sets of homologous chromosomes, one from the mother and one from the father. *Aneuploidy* is a change in the total number of chromosomes. Aneuploidy might be the loss or gain of a whole or a chunk of a chromosome.

The *cell cycle* is the observed clock-like program through which a cell "cycles" through its initial resting *G1* phase, then enters the phase of its chromosomal DNA replication or synthesis (*S phase*), progressing to a second "resting" phase *G2*, in which the newly replicated DNA is examined for errors in base pairing (which are repaired if any), followed by the mitotic *M* phase during which the cell divides, culminating in a reentry into the G1 phase. G0 is a quiescent resting phase from which a cell must reenter G1 to begin the S phase. A cell usually does not skip its ordered cell cycle phases; so-called *checkpoints* ensure the fidelity of each phase and the order.

Metastasis is the process by which cancer cells migrate from their original tumor location to invade other tissues and organs. This process involves gaining novel properties of "amoeboid" movement and homing into specific organs by processes of chemotaxis and avoiding immune surveillance.

Epigenetic processes are those that change the behavior of the genes without changing the linear array of DNA sequence of the genes.

In our imaginary parallel universe, we are in an afternoon party. Outside there is bright sunshine. Inside there are green sofas surrounded by wood-paneled walls and large windows. On one corner lay a bongo and drums. There was a blackboard on one wall with intersecting squiggles with arrows on the ends. The colorful characters are Richard Feynman, the genius American physicist, J. B. S. Haldane, the British-Indian polymath who made contributions to evolution, genetics, and biochemistry, and Charles Darwin again.

Feynman: What do you guys think about cancer? I got one. It was in my abdomen. I believe John here got it too.

Haldane: As always, you are correct Dick. I also had cancer in my gut, which did me in.

"So do not wait for aches and pains
To have a surgeon mend your drains;
If he says "cancer" you're a dunce
Unless you have it out at once,
For if you wait it's sure to swell,
And may have progeny as well.[1]"

Darwin: Progeny, did you say?

Feynman: Yes, he did. The secondary tumors spreading from the primary one are the progeny — that's what he means.

Darwin: Anything that leaves progeny will evolve. I had said that while on a sailboat a long time ago.

Haldane: Not so fast Darwin. You talked about species, not tumor cells.

Feynman: Haldane, you of all people must know that it doesn't matter whether it is species or cells or even computer programs. Anything that leaves progeny will evolve. Johnny von Neumann proved that. Now you have computer viruses that evolve. Genes, that is DNA or RNA, evolve, because they replicate and leave progeny DNA or RNA molecules. So do of course cancer cells. We all have genes that can mutate to make a cell cancerous. But how many cancerous mutations are needed? What are the

---

[1] Source: JBS Haldane. Cancer's a funny thing. *New Statesman*, 21 February 1964: 298.

numbers? You know the saying, "all science is numbers; the rest is stamp collecting".

Haldane: Three. Maybe four. The exact quote is "All science is either physics or stamp collecting". Obviously, I don't buy that. And Dick, don't call me Johnny. I'm always called Haldane.

Darwin: Three or four what?

Haldane: Three is the minimum number of mutated cancer genes you need to get cancer.

Darwin: How does one know that?

Haldane: That is a very interesting story, and that leads directly to how cancer can be treated.

Feynman: That's no longer of importance to us. I will play a Bossa Nova on the bongo. Have you guys heard "The Girl from Ipanema"?

The two gentlemen dance to the tune of bongo's beat.

***

In 1976, Peter Nowell (1928–2016), often credited as a discoverer of the chromosomal basis of cancer, proposed "a model for the evolution of tumor cell populations in terms of stepwise genetic modification" (Nowell 1976). In the abstract of that paper, Nowell wrote presciently:

> "*. . . each patient's cancer may require individual specific therapy, and even this may be thwarted by emergence of a genetically variant subline resistant to the treatment. More research should be directed toward understanding and controlling the evolutionary process in tumors before it reaches the late stage usually seen in clinical cancer.*"

Tools for determining genetic heterogeneity in tumor cells became available only when large-scale and inexpensive DNA sequencing from tumor fragments became possible in the early 2000s. How to tailor the therapy against the mutational spectrum of tumor cells in individual patients, however, had eluded any direct clinical impact until single-cell DNA sequencing of clinical samples became possible on a mass scale another 15 years later.

Over the past few years, the evolution of tumor cells has become an important paradigm for understanding and approaching personalized clinical treatment of cancer, especially of drug-resistant cancer

(Enriquez-Navas, Wojtkowiak, and Gatenby 2015; Fittall and Van Loo 2019; Gatenby and Brown 2020; Tsimberidou et al. 2020).

3.1 Estimates of the Minimum Number of Mutations Needed to Make Cancer

The idea that somatic mutations might trigger cancer has had a long history, mainly stemming from observations that the agents that cause mutation, such as x-rays and mutagenic chemicals, are the same agents that cause cancer in experimental animals. It is worth our while to delve into early work in this direction, if only to emphasize the value of thinking with nothing but a pencil and paper.

The first clear statement that a series of somatic mutations must occur successively to drive cancer and a method to estimate the minimum number of such mutations needed were published in the mid-1950s (Armitage and Doll 1954; Nordling 1953), and were later elaborated to refine the estimate (Fisher 1958). This was at a time when the proposal for a double-helical structure of DNA was barely a year old.

Peter Armitage (1924–) is an English mathematician who came of age during the Second World War, when he had contributed to the war effort, and later became a professor of statistics at Oxford University. Armitage used a simple method to estimate the minimum number of mutations needed to form cancer. He imagined that cancer is induced as a result of a series of mutations that occur in a clonally propagating cell (Armitage and Doll 1954). In other words, a mutation occurs first in a cell that continues to divide and gives rise to descendant cells, all carrying that same mutation — a clone. Later, a second chance mutation occurs, which is also propagated among the descendants of that cell, producing a subclone, and then a third one, and so on. If these mutations continue to accumulate among the descendant population of cells, the number of mutations that cause cancer should be related to the average age of people who first report cancer. If one could write down a quantitative relationship between the frequency of cancer incidence by age and the probability of 1, 2, 3, 4, and so on, mutations that might have occurred in each decade of human age, then one may compare demographic data on age and cancer incidence to see which probability of mutation fits the data.

In 1954, there was not enough data on cancer incidence by age. Armitage and Doll, therefore had to make the simplifying assumption that cancer incidence was equal to cancer death (with a few months of time lag between the two). This was not a very unrealistic assumption in those days, when the diagnosis of cancer was synonymous with a death sentence for most cancer patients. Armitage and Doll also made a second simplifying assumption: if a cancer cell arises due to the minimum number of cancer-causing mutations, then such a cancer cell always survives and develops a full-blown disease. Using the then available data on cancer deaths, not cancer incidence, Armitage and Doll estimated this minimum number of mutations to be 6, or on average six mutations per cancer death (see Chapter 15).

Today we have better information on cancer incidence, so we can revise this estimate. As it turns out, the number of new cases of cancer in the United States in the year 2021 was approximately 1.9 million (Siegel et al. 2022). Given the population of the United States in 2021, the frequency of new cancer incidence at any age since birth in the United States is approximately 1.9 million divided by 300 million, which provides a better estimate of the minimum number of mutations as 3 to 4 when used in the formula that Armitage and Doll provided.

It is, however, never a good idea to rely on only one indirect measurement. Can we check this out from a different perspective?

Shortly after Armitage and Doll's paper came out, John C. Fisher III (1920–2018), who had obtained a doctorate in mechanical engineering from MIT and at that time was a scientist at the General Electric Corporation in New York, came to a rather fundamental insight. He realized that while the final stage of cancerous growth of the cells is expected to be exponential, the effects of each of the clonal series of mutations prior to the final driver mutation should not cause exponential growth of the cancerous foci at these intermediate stages. In his own words (Fisher 1958), for a nonmalignant clone of cells:

> *"Since the environment at the perimeter of the clone is in an essentially steady state — a sheet of mutated cells in edgewise contact with a sheet of normal cells — the rate of spreading of the clone should remain approximately constant"*

According to John Fisher's model, if r is the radius of the clone, which increases at a steady rate, t_0 is the time the clone had started growing, and t is the time at which we are measuring the clone radius, then, $r = k(t_0 - t) = k\Delta t$, where k is a proportionality constant. Notice that the area of the clone will be, $a = \pi r^2 = \pi k^2 \Delta t^2$. Therefore, the number of cells, N, in the clone will be proportional to the area occupied by the clone, that is, $N \propto \Delta t^2$, π and k being constant.

John Fisher knew that pathologists had already described premalignant spread of tumor cells as occurring generally on two-dimensions . . . only the late malignant clones begin to spread deeper into the inner tissues from the surface of the epithelium, and that is when exponential growth of malignant tumors sets in. Therefore, Fisher did not consider the volume of the clone but only the area of the clone at this premalignant stage.

Obviously, the number of cells in the clone must somehow be related to the incidence of cancer.

The consequence of this power law (because the unknown variable is proportional to the known variable raised to a power, here, of 2) of the form $N \propto \Delta t^2$ is interesting. If one assumes that the mutation rate is constant per unit time and that mutations occur at random (according to the Darwinian hypothesis), at least in the premalignant phase, then in the population as a whole, cancer incidence (I), which is proportional to N within every person, should be a function of age raised to some power greater than or equal to 2. That is,

$$I \propto (age)^n$$

where $n = 2, 3, 4, \ldots$ and so on, are the number of mutations.

Fisher pointed out that if only one mutation could cause cancer, then most people would have cancer at a very early age. This relationship is not supported by epidemiological data. Fisher wrote:

> "*Cancers of this type, appearing at an early age in many members of the population, would be strongly selected against genetically, and should largely have been eliminated through the evolutionary process.*" (Fisher 1958)

However, if $n = 2$, then the number of clonally descendant cells produced because of the first mutation is proportional to (*age*), as well as the average number of cells in each of the clones is proportional to $(age)^2$. Epidemiological data of the time did not fit this model. John Fisher found that a value of n between 3 and 6 fitted the data better (Fisher 1958). A more accurate value of n can be estimated from recent demographic data (Siegel et al. 2022), which also appear to fit a value of $n \sim 3$.

The above considerations strongly imply that approximately three mutations are minimally necessary for the induction of cancer. Given the uncertainty inherent in demographic data and the need to extrapolate from a model involving tumor growth in single humans to whole populations, perhaps the real minimal number of mutations to trigger cancer ranges between roughly 3 and 6. In fact, the latest estimates based on direct sequencing of the whole genomes of tens of thousands of cancer cells of patients have shown that these estimates were remarkably accurate (Gerstung et al. 2020).

3.2 Massively Parallel DNA Sequencing Has Caused a Paradigm Shift in Medicine and Biotechnology

Before the mid-1970s, sequencing DNA was a big challenge, and only a few nucleotides could be read at a time. Fred Sanger (1918–2013) in Cambridge, UK, and Walter Gilbert (1932–) in Cambridge, United States, each developed simple methods to read long stretches of DNA.

Sanger, who had already won a Nobel Prize in Chemistry for his work on protein sequencing, came up with an elegant method. He realized that if you grow DNA chains in a test tube, you can stop them from growing by adding special "poison" (or base analog) molecules that mimic one of the four DNA bases. Sanger synthesized four such base analogs, one for each base: adenine (A), guanine (G), cytosine (C), and thymine (T).

He added one of these analogs to separate test tubes. In the tube with the A analog, the growing DNA chains stopped wherever there was an A in the sequence. In the tube with the G analog, chains stopped at Gs; similarly in tubes with C or T analogs.

By separating the DNA fragments by size, he could see a ladder of terminated chains. For example, in the tube with the A analog, all chains ended at As. By comparing the fragment sizes from each tube, the DNA sequence could be directly read off. This method made it possible to read long DNA sequences for the first time.

Gilbert used a different strategy. He devised four different chemical reagents that cleave at one of the four DNA bases (or approximately so). If you conducted these DNA cleavage reactions in four different tubes (ensuring partial cleavages) and size-separated these DNA samples side by side, again from the sizes of the spectrum of DNA fragment lengths, you could read off the DNA sequences.

Although both Sanger and Gilbert each shared one half of the Nobel Prize in Chemistry quickly following their discoveries (Sanger for the second time, one of only two persons so far to have received the Nobel Prize in Chemistry twice), their methods were not simple to perform by any standard. I had spent three painful months in the early 1980s to sequence only a few hundred base pairs of DNA. When everything worked as expected, it used to take about a week's effort to sequence about 200 nucleotides of DNA at that time, and everyone who did it successfully would take pride in carrying out an impressive feat.

A major technological breakthrough occurred in the late 1980s. Lloyd Smith (1954–), working in the laboratory of Leroy Hood (1938–) at the California Institute of Technology, managed to automate the task of DNA sequencing in 1986. This technology was immediately commercialized, which led to cheap and rapid DNA sequencing instruments affordable to most laboratories — all we had to do was to provide a piece of DNA to a service laboratory, which would then churn out the DNA sequences, as much as a few thousand DNA nucleotide pairs in a few hours, costing about one US dollar per nucleotide pair in the mid-1990s. This took the pride out of DNA sequencing but made it democratic. In two decades, advancements in DNA sequencing technology reduced the sequencing time to a few minutes per 100 bases, and millions of DNA bases could then be sequenced in parallel at the cost of a fraction of a cent (in US dollars). Today, the entire genome of a person ($\sim 6 \times 10^9$ nucleotide pairs) can be sequenced for a few hundred US dollars in a day.

The importance of DNA sequencing can be understood by noting that the genetic code is universal — by knowing the DNA sequence, it is possible to decipher the protein sequence encoded by the DNA through a direct reading of the encoded amino acid sequence off the table of codons. Once the protein sequence is known, one could potentially figure out how the protein folded in three dimensions, and its shape dictates its function. The universality of the genetic code is the direct result of Darwinian evolution, because alternate genetic codes can be devised in the laboratory (Ostrov et al. 2020), and yet these alternate codes are never found in nature. In other words, the universality of the genetic code on earth appears to be an accident of evolution — one that worked early on and happened to have survived.

In retrospect, this consequence of evolution, one that binds all living creatures to a single web of life that shares some important evolutionary properties such as the universality of the genetic code, is easy to see, and the impact of DNA sequencing appears reasonable. In the late 1980s, however, this potential was not readily appreciated, including by this author. By 1986, the longest whole genome DNA sequence known was that of the bacteriophage lambda (48,502 base pairs).

A larger challenge, however, still impeded progress. How does one know which DNA or protein sequence encodes what functional trait for the individual, say, for the lambda virus? It had taken 5 years of intense effort to sequence the lambda genome, which was completed in 1982. And yet in 1986 not all the lambda genes' functions were known. Most of us thought, if it took five years to sequence about 50,000 nucleotides of lambda, to sequence 3×10^9 nucleotides of the human genome would take 60 centuries. Even if we could accelerate DNA sequencing speed by 10 or even 100 times (though at the time we had no idea how), it would take at least 500 years to finish the full sequencing of the human genome.

A few exceptional people thought otherwise. Prominent among them were James Watson, Robert Sinsheimer (1920–2017), Charles DeLisi (1941–), Walter Gilbert, Sydney Brenner (1927–2019), and David Galas (1944–2023). Intuitively, they realized that advances in both DNA sequencing and computer power would rise to the challenge if they started the endeavor. The power of deciphering gene function from DNA sequences would also be accelerated, they thought, because computers would be able to find DNA and protein sequence relationships among many laboratory model organisms (whose genomes would also be sequenced), and the conservation of function in a Darwinian world would connect the dots — that which is true of a fly or a worm or a mouse protein might also be true for humans.

It was a bold step at the end of the 1980s to begin the mapping of DNA sequences of the human genome under the leadership of David Galas, who assumed the directorship of this effort through the US Department of Energy (Collins and Galas 1993). This effort soon galvanized the scientific community worldwide, and global efforts began on whole genome sequencing of not only the human genome, begun in 1990 by James Watson and subsequently led by Francis Collins (1950–), but also of two eukaryotic fungi, the bread mold and the fission yeast, of the insect *Drosophila melanogaster*, of a nematode *Caenorhabditis elegans*, and of a plant *Arabidopsis thaliana*. Soon a fierce competition emerged for completing the human genome sequence by an independent team of scientists led by Craig Venter (1946–2026). The dream was realized at an unprecedented speed, and the full sequencing of the genomes of these organisms within the next decade revolutionized how biology is done. The interested reader may like to read a beautifully written book on the history of this topic (Watson, Andrew, and Davies 2017).

In 1991, Walter Gilbert wrote in an influential editorial article in the journal *Nature* (Gilbert 1991):

> *"There is a malaise in biology. The growing excitement about the genome project is marred by a worry that something is wrong — a tension in the minds of many biologists reflected in the frequent declaration that sequencing is boring. And yet everyone is sequencing. What can be happening? Our paradigm is changing."*

Wally (Fig. 3.1), as Gilbert is affectionately known among his students and colleagues, explained that while scientists have been using laboratory model organisms such as mouse, Drosophila, *E. coli*, lambda bacteriophage, and so on, to painstakingly conduct experiments involving gene discovery, sequencing the DNA of the discovered genes, and doing

Figure 3.1 Walter Gilbert (left) and Dr. Katalin Kariko (right, see Chapter 8) on September 28, 2023. Gilbert did his PhD in theoretical physics under the Nobel laureate particle physicist Abdus Salam at the University of Cambridge. Gilbert became interested in molecular biology while a young assistant professor at Harvard. Besides contributing numerous seminal works in molecular biology, in 1977, he invented one of two direct DNA sequencing techniques for which Gilbert shared the 1980 Nobel Prize in Chemistry with Fred Sanger and Paul Berg. Dr. Kariko, a Hungarian émigré scientist in the United States, was instrumental in discovering how RNA molecules could be modified such that they do not induce an "allergic" response in humans, which became the cornerstone of RNA vaccines against COVID-19. She received the Nobel Prize in Chemistry in 2023, announced 5 days after this photo. [Image from the author's personal collection]

biochemistry to analyze functions of the encoded proteins, in the future it will be different:

> *"The new paradigm, now emerging, is that all the 'genes' will be known (in the sense of being resident in databases available electronically), and that the starting point of a biological investigation will be theoretical.* (Gilbert 1991)."

This paradigm shift in biology was possible not merely because of the invention of technical tools such as DNA sequencing and computer-aided search and alignment of protein sequences to discover relatedness (see Chapter 16). It was made possible fundamentally due to the underlying deeper reality that all organisms are related by a common descent, and therefore, predictably, genes in various organisms related by sequence are in fact related by function as well.

If we knew the function of the *BMP4* gene, for example, as determining the pattern of bone growth in a mouse or in a finch from the Galapagos Islands, we might guess that its function in humans is likely related to bone growth. Thirty years later, Wally's conjecture has become the dominant paradigm in medicine and biotechnology. In the rest of this book, we will continue to use this paradigm over and over without explicitly mentioning how we discover a particular gene's function. Most often the underlying basis is the same: DNA and protein sequence conservation defines the function of a gene based on the gene-relative's experimentally determined function in a laboratory organism. We then seek evidence of confirmation or refutation of the inferred function in humans.

3.3 Some Genes Are Frequently Mutated in Tumors

So far, we have assumed that the probability of mutation is constant per unit time (per act of cell division, to be precise). This assumption might not always be true. Are there genes that control the biological mechanisms by which mutations arise? If so, when some of these controlling genes are themselves mutated by chance, shouldn't the overall mutation rate (in all genes) be affected?

Indeed, the presence of certain gene mutations in precancerous cells in the process of cancer development is associated with increases in the rate of mutation itself. One famous case involves mutations in the tumor suppressor *TP53* gene. We have already encountered this gene in previous chapters. Its product is important for repairing broken DNA.

DNA, as the precious "tape of life," the cell's informational repository, needs to be maintained in a relatively intact state even though it is under constant chemical and mechanical abuses that cause frequent damage to its integrity. These ravages need to be repaired, and genes have evolved to repair the damages. The product of the *TP53* gene, the p53 protein, has a dual role, as a sort of molecular sentinel. If the chromosomal DNA is broken, p53 signals the cell to stop dividing and induce the repair of the broken DNA. If the broken DNA ends are not healed within a reasonable period, the p53 protein triggers cell death by unleashing the activation of a series of enzymes known as caspases, which literally begin to eat up cellular proteins from the inside of the cell. The cell thus commits suicide, as it were. Evidently it is better for a multicellular organism to let die one of its constituent cells harboring a broken chromosomal DNA than to keep it alive. This phenomenon is termed programmed cell death or apoptosis. We shall later see why this might be important and why cells have evolved this capacity.

Other genes needed for DNA repair include those required for correcting mispaired DNA duplexes (e.g., *MSH2-6, MLH1, MLH3*), or for repairing double-chained DNA breaks (*e.g., BLM, ATM, BRCA1, BRCA2, etc.*) (Fig. 3.2).

Such genes as above, in which inactivating mutations could increase the spontaneous rate of mutation itself, are often termed "mutator" genes because of the abnormally high mutation frequency in cells carrying mutations in these former genes.

To recapitulate, we saw in Chapter 1 that several of these genes (e.g., *TP53*, *BRCA1*) are classified as tumor suppressor genes. The reason for such a classification is because of their demonstrated role in suppressing tumorigenesis. If any such gene is mutated, then the probability of subsequent mutations in any other gene in these cells also rises.

With the observation that the mutation rate itself may change due to the clonal inheritance of "mutator gene" mutations, one may anticipate that by sequencing the whole genomes of many different cancer types, one might observe repeated occurrences of these mutations (Hanahan and Weinberg 2011). In other words, some of the earliest mutations to occur before the clonal expansion of cancer cells could be in these "mutator" genes.

As expected, people who inherit one or two already mutated alleles of a mutator gene from their parents have higher cancer rates than those without such mutations. This is because their higher general mutation rate increases the chance of cancer-inducing mutations (Fig. 3.3).

An interesting discovery was made by studying glioblastoma tumors, an aggressive type of brain cancer. Patients who had already undergone surgery and chemotherapy showed a high number of mutations in DNA

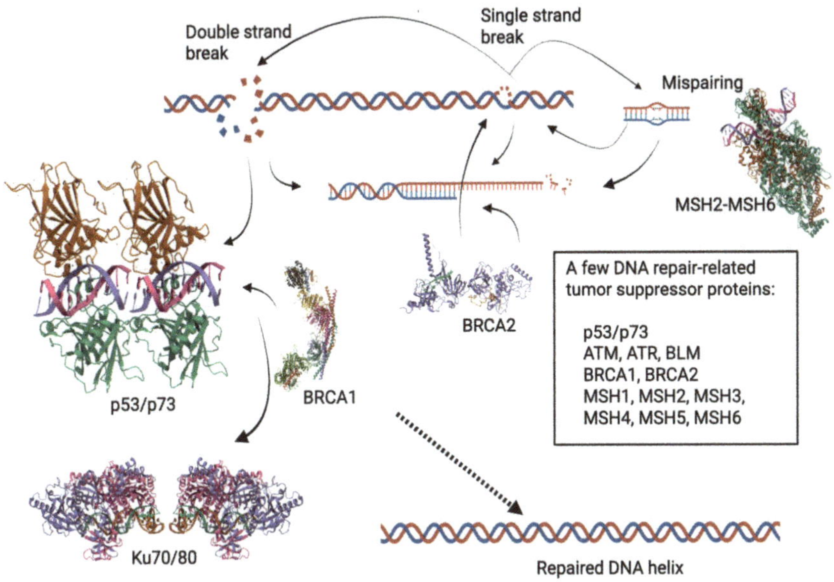

Figure 3.2 Many DNA repair genes (and their encoded proteins) are tumor suppressors. Among them are p53 and its close sequence relative p73, breast cancer susceptibility proteins BRCA1, BRCA2, DNA mismatch repair proteins MSH1-6, DNA double chain repair proteins Ku70 and Ku80, ATM, ATR, BLM, and so on. Various forms of DNA damage are repaired by engaging the functions of many of these proteins. Only a subset of these proteins/genes are mentioned. Note that whereas p53 is a confirmed tumor suppressor, which is mutated in over 70% of solid tumor cancer samples, the case for p73 as a tumor suppressor is still controversial. While the latter's gene is also frequently mutated in cancer cells, a deletion of this gene does not increase the frequency of tumors in mouse, whereas a deletion of the gene encoding p53 (*TP53*) does. [Drawn in BioRender under educational license: www.biorender.com]

repair genes in parts of the tumor that grew back subsequently. These mutations seemed to have appeared following treatments with radiation and a common chemotherapy drug used for glioblastoma, temozolomide (Wang et al. 2016). Remarkably, nearly all recurrent tumors that had become resistant to the chemotherapy had at least one mutation in a DNA damage repair gene (i.e., in a mutator gene). In other words, mutator genes are more frequently mutated in glioblastoma tumors that return later than in the initial glioblastoma tumors.

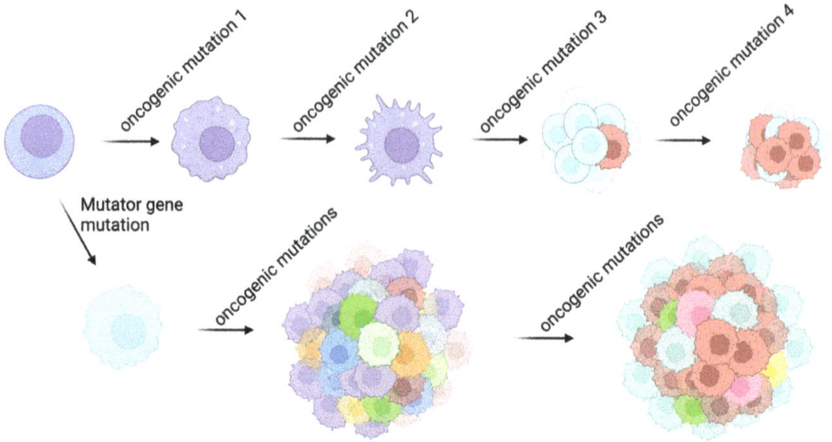

Figure 3.3 An early mutator gene mutation greatly accelerates the chance of additional oncogenic mutations. Colors represent different gene mutants. Cancer cells are in red. Precancerous cells with sequential mutations could generate a cancer cell when one or more oncogenic mutations accumulate. If, however, an early event is a mutator gene mutation, then mutations arise frequently, and therefore the chance of an oncogenic mutation is high. [This figure was drawn using Biorender.com under an academic license]

Temozolomide is known to be a mutagen by itself, causing DNA base changes in cells exposed to this chemical. Thus, if the recurrent tumor cells had by chance suffered a mutation in one or more mutator genes, the rate of further mutation will accelerate in the presence of sublethal concentrations of temozolomide. Some of these later mutations might then cause the emergence of cancer cells becoming resistant to the drug.

An important question arises from the above results: does the chemotherapy itself induce mutations, which could drive further evolution of cancer cells? There was insufficient support for this scenario (Wang et al. 2016); the results were consistent with the possibility of unrelated chance events that had caused a mutation in a DNA damage repair gene, which subsequently caused additional mutations and the eventual production of temozolomide resistance independently of the treatment with temozolomide.

By contrast, some evidence exists that treatment with thiopurine, another mutagenic chemotherapy drug, of acute lymphoblastic leukemia (or ALL) patients might itself induce the development of resistance to the drug (Li et al. 2020). Among the genes found to be mutated in drug-resistant ALL cells were *TP53, MSH2, MSH6,* and *PMS2* — all are known mutator genes. None of the cancer genomes sequenced from non-relapsed ALL patients had mutations in these genes. Therefore, these results suggest that chemotherapy treatment itself might have caused the cancer to return, producing drug-resistant ALL in these patients.

The above results are suggestive but not conclusive for establishing a cause-and-effect relationship. Proper evidence will need to enumerate the rate of mutagenesis induced by chemotherapy relative to that occurring spontaneously in the cancer cells and the strength of selection (see Chapter 14) on the sensitive versus the resistant cells.

We noted in Chapter 1 that so far about 40 confirmed proto-oncogenes have been described, any one of which could be a driver mutation for oncogenesis. In addition, there are at least 30 confirmed tumor suppressor genes (Chai 2008). These latter numbers, however, might be as high as ~500. Tumor suppressor genes such as *TP53*, *RB1*, and several of the DNA repair/recombination genes mentioned above serve as "brakes" on the evolution of tumors to the metastatic[2] stage and further to drug-resistant cancer. Some tumor suppressors are also parts of the cell–cell as well as intracellular signaling machineries. Some examples of the latter are *NF1*, *NF2*, and *VHL* gene products.

A major process of tumorigenesis is an initial step in which aneuploidy (the loss or gain of a chromosome or a segment of a chromosome) occurs, possibly through accidental defects in chromosome segregation during cell division (Fig. 3.4). If any one of the homologous pairs of chromosomes (of which there are 23 pairs in human females, including 2 X chromosomes, and 22 pairs plus X and Y in human males) or its segmental loss overlaps one of the tumor suppressor genes, which could have been mutated, then such a cell will have lost some of its DNA damage repair abilities. If this cell continues to replicate its DNA and divide, then there would be a chance that some of its clonally descendant cells would incur further mutations.

[2]Metastasis means migration of cancer from its site of origin to a location not physically adjacent to the original tumor. Thus, a metastatic tumor is one that develops at a site distant from the original tumor.

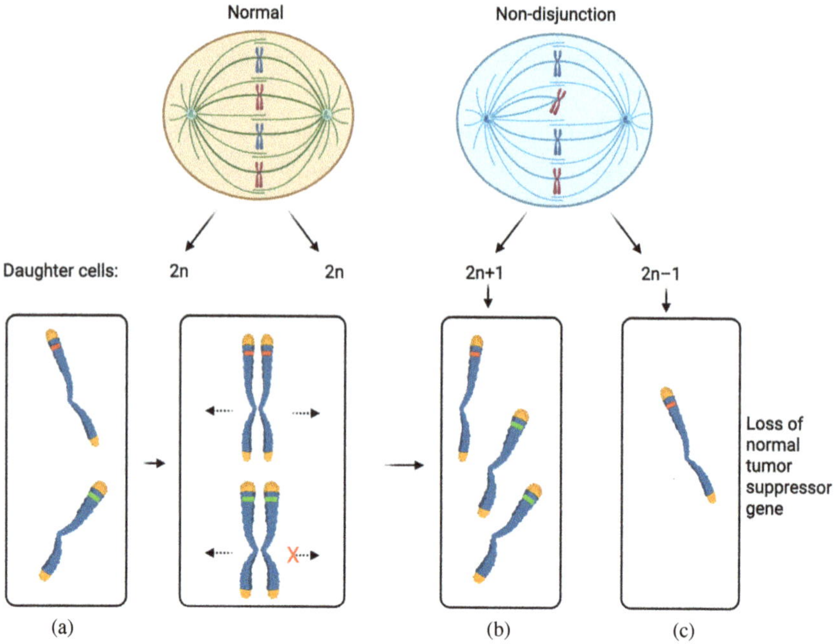

Figure 3.4 Abnormal segregation of chromosomes. An individual's cell is heterozygous for a tumor suppressor gene mutation because one of the two homologous chromosomes, one inherited from the father and the other from the mother, carries a mutant allele (red bar) of the gene. After an abnormal mitotic chromosome segregation during a division of cell (a), both sister chromatids (the duplicated copies of each original chromosome) containing the normal copy (green bar) of the tumor suppressor gene may infrequently segregate into one of two daughter cells (cell (b)) whereas the other daughter cell (c) may get only the mutant (red bar) copy. In such a case, the daughter with $2n-1$ number of chromosomes is left without the benefit of a normal tumor suppressor gene and may become more prone to suffering further insults, leading it to becoming cancerous eventually. Note that such a process might not always involve the whole chromosome but may more frequently involve a chromosome arm or a short segment of a chromosome. Note: the homologous set of only one chromosome is shown here. [This figure was drawn using Biorender.com under an academic license]

If, on the other hand, this cell does not divide, possibly because it has progressed into a differentiated state, then there is little chance for generating further mutations. That is because mutations are supposed to occur only during DNA replication.

Let us now summarize a few important aspects of cancer development:

First, cancer is rarely due to a single mutation, generally requiring at least three and possibly about six mutations in cancer-critical genes, such as oncogenes and tumor suppressor genes.

Second, the precise order of mutations in these genes is important, because depending on which mutations occur early may often set the trajectory of evolution through further mutations and selection.

Third, even if a cancer-critical mutation might occur in a cell, there are several mechanisms by which the fate of this cell can veer away, for example, by cell death or by cell differentiation, from ultimately forming a cancerous lesion.

In the next section, we will examine the landscape of cancer-causing mutations from actual patients and thus understand whether all our expectations are borne out by observation on real patients.

3.4 Cancer Mutational Landscape — "Pan-Cancer Genomes"

Questions of great significance for understanding the mechanisms by which cancer develops are the type and frequency of mutations that are associated with the genomes of cancer cells and the rates of occurrence of such mutations. To get a comprehensive view, a global consortium of researchers analyzed whole genome sequences of over 2,600 tumor samples from 2,834 donors with 38 tumor types. Most samples were from primary cancerous lesions, with a nearly 1:1 male-to-female ratio, and donors aged 1 to 90 years ("Pan-Cancer Analysis of Whole Genomes" 2020). This pioneering work is now known as the "Pan-Cancer Genomes" study.

What kind of mutations were observed? Broadly, a mutation could be a change in a DNA nucleotide or base — replacement, deletion, or insertion of a single nucleotide. Such variations are called single nucleotide variations (or SNVs). There could also be an insertion or deletion of multiple base pairs of DNA together, which is termed an "indel." Finally, there could be a structural variation of the chromosome itself, where a large chunk of the chromosome spanning thousands or millions of base pairs of DNA, or even a whole chromosome, is deleted, inserted, or translocated to another chromosome. These latter mutations involving whole or large chunks of chromosomes are called structural variants (or SVs). An examination of numerous whole cancer genome sequences spanning all cancer types, considered together with the age of the DNA donors, suggested that SNVs tend to occur at a frequency of about 190 mutations per tumor per year, indels at an average frequency of about 22 per tumor per year, and SVs at about 1.5 per tumor per year ("Pan-Cancer Analysis of Whole Genomes" 2020).

3.4.1 *It is difficult to determine which mutation drives a specific tumor*

In the study described in the previous section, which of these mutations drove the respective cancer lesion in the corresponding donor patient? Answering this question poses considerable challenge. In principle, one could answer this if one had access to time series samples from tumors. For understandable practical reasons, this global study did not have access to tumors collected early as well as at later times from the same respective donors. Thus, judgment calls as to which mutation might have acted as the driver mutation in a particular tumor were at least partly subjective, based on our current knowledge of how various gene mutations can drive cancer in experimental organisms, such as mice or in somatic cell mutagenesis experiments using cultured cells in laboratory conditions.

The reader will guess that if a mutation is identified within a known tumor suppressor or a mutator gene, or an activating mutation in an oncogene, then such a mutation with high likelihood could be a driver of cancer. This is usually the operational definition of a driver mutation. It is important to recognize, however, that such a pronouncement is ultimately subjective. If a mutation is observed in a larger number of cells in a particular tumor sample than another mutation, then the former is thought to have arisen earlier only if one assumes that there was no significant selective advantage imparted by that mutation. If there was a significant selective advantage, then even a late-originating mutation in the context of an earlier mutation might be over-represented among the tumor cells.

It was observed that a vast majority of patients had at least one identifiable "early" mutation in the *TP53* gene, followed by mutational lesions in the *CDKN2A* gene (475 donors). More remarkably, about 17% of patients had inherited at least one copy of a "presumed" cancer driver mutation from one of his or her parents, suggesting that these patients' cancer cells might have lost the unmutated copy of the gene (inherited from the other parent). The unmutated copy would normally have suppressed or buffered the effects of the mutated copy. However, some cells having lost the unmutated copy of the gene had progressed along the cancerous pathway due to the loss of protection provided by that normal gene.

3.4.2 *There may be different drivers of cancer in different organs*

There were several curious observations associated with the cancer genome sequences described above. For example, SV mutations that overlapped the presumed driver gene mutations were significantly more frequently observed in tumors from breast and ovarian cancer patients than in those from other cancer patients. By contrast, presumed driver single-nucleotide variant (SNV) mutations were more frequently associated with colorectal carcinoma and B-cell lymphoma patients. Even more importantly, signs of clustered multiple mutational events were associated with a vast majority of cancer patients: 77% of patients with *TP53* mutation had at least two mutations in known driver genes. In fact, 96% of the two presumed driver mutations were both in the *TP53* gene itself! Why so?

A significant frequency of clustered SNV driver mutations in aggressive skin cancer, melanoma, cells appeared to have been frequently followed by an SV mutation (in these cases, amplification of the chromosomal region) spanning thousands to millions of base pairs of a chromosome within which the driver mutations had occurred, suggesting that the clustering is due to the amplification of the mutated sequences in a subsequent event. Thus, the order of events in such cases was probably the following: an initial driver gene mutation occurred. This was followed by amplification of the chromosomal region that overlapped the mutated genes. One can only surmise that perhaps a selection mechanism had favored such co-occurrences.

The above course of events is consistent with a general mechanism of new gene evolution (Ohno, Wolf, and Atkin 1968): amplify an essential[3] gene's copy, then random mutations will occur in the redundant copies with little impact on viability. Some of these redundant copies may eventually gain a novel function — here, the cancerous state provides a selective growth advantage relative to other surrounding cells.

About 16% of the tumors in the cancer tissues donated by patients in this study exhibited mutations in genes related to cellular aging mechanisms. Curiously, many of these cells also had defective chromosomal DNA ends, which are known as telomeres. Telomeres protect chromosome ends from erosion during cell division and are also related to cellular aging processes (see Chapter 4). If a telomere is eroded, a cell dies. Abnormal repeated DNA sequences were detected in these cells at their telomeres, suggesting a possible connection of cancer with avoiding the normal cell death processes. Telomeres in most eukaryotes contain repeated DNA sequences, but the exact DNA sequence of the repeats in a species of organism is generally specific to that species and does not change in normal circumstances. Therefore, the above-described change in the telomere repeated DNA sequence motif in cancer cells suggests that perhaps associated with cancerous development there is a change in the molecular structure of the telomere formation apparatus in these cells. This change might underly a profound shift in the cell's biological property, including perhaps aging-related properties. Cancer cells often do not age as normal cells do — they do not cease their divisions in response to a shortening of their telomeres as normal cells generally do. Perhaps these cancerous cells have overcome the natural aging processes by having their telomere maintenance pathways altered or decoupled from other cellular control systems.

A significantly large fraction of these tumors also had a mutation in the tumor-suppressor gene termed *Retinoblastoma* (*RB1*) (Fig. 3.5). Purely based on its high prevalence, one might presume that *RB1* mutations could have played a part in driving tumorigenicity in these cases. These results imply, but do not prove, that an early mutation in the

[3]The essentiality of a gene simply means that its activity is essential for life.

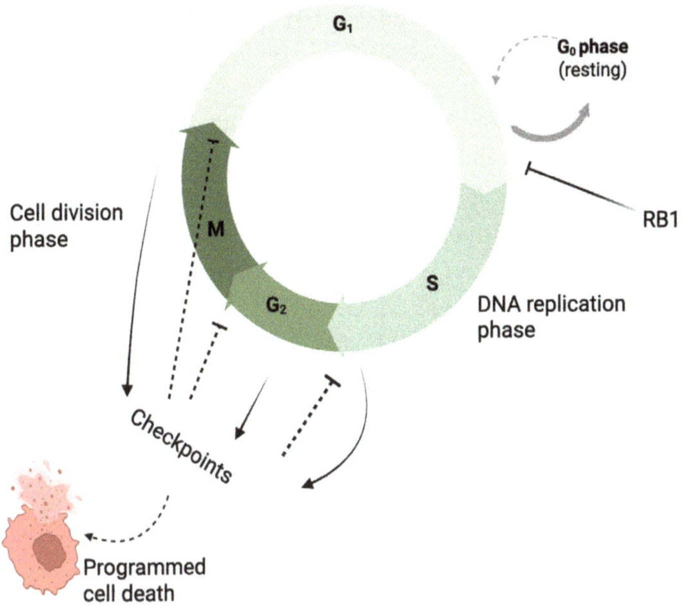

Figure 3.5 The cell cycle is under exquisite control. It starts from the resting G1 phase by entering the DNA replication, or S phase. This transition (the G1-S checkpoint or the restriction point) is inhibited by the action of the tumor suppressor protein RB1. Cellular events at each stage of the cell cycle are monitored and are under control of multiple genes, which are collectively known as checkpoint genes (separate gene sets for various phases). If the checkpoints are not properly executed, the cell is triggered to die by a specialized pathway termed programmed cell death (PCD). The PCD pathway is also controlled by several tumor suppressor genes, including *TP53*. [This figure was drawn using Biorender.com under an academic license]

presumed driver gene *RB1*, which would enable the cells to enter mitotic division without being inhibited by the product of a normal *RB1* gene, became immortalized, that is, become impervious to the normal cellular aging process.

An interesting observation was that about 10% of the cancer samples mentioned above had no identifiable mutation in any recognizable driver gene. This observation alone guarantees that the future holds much promise for understanding previously unknown aspects of cancer biology and thus further transform how we treat cancer patients.

3.5 Cancer Cells with Mutator Mutations Evolve Rapidly

Just how fast does a cancer cell evolve? To attempt to answer this question, one may assume that some property or properties of a cancer cell undergo natural selection, such that a fraction of the mutant cells with an emergent phenotypic property under selection grow faster than other cells (see Chapters 14 and 15 for a quantitative description of this process).

Favorable selection might occur due to a rare mutation activating an oncogene like *SRC*, *RAS*, or *BRAF*, leading to cancer-related changes such as independence from growth factors, insensitivity to cell death signals, immune evasion, or metabolic shifts. This oncogenic mutation is assumed to be the primary driver mutation.

Additionally, a person might inherit a cancer-predisposing mutation in a mutator gene. Most mutations are recessive, so the presence of a normal gene copy masks the effect. However, during cell division, abnormal chromosome segregation (aneuploidy) might cause loss of the normal allele. Although the chance of specific aneuploidy is low (0.1%–0.8%), it can occur over numerous cell divisions.

If this occurs in a precancerous cell, the mutation rate increases because the mutator gene's normal function is lost. The order of events is unimportant; what matters is an early genetic event that is propagated to the descendant cells. If another oncogene is activated by mutation, the resulting cell clone may gain yet more growth advantage. The activated oncogene might enable these cells to divide more frequently or make them ignore the growth inhibitory signals that affect the normal cells. The reader can see that any one of these molecular events might provide a selective advantage to the clone of cells descending from this genetically altered cell relative to the surrounding normal cells.

Some individuals inherit mutant alleles of a cancer-susceptibility gene in both copies, one from each parent. Such cases are observed more frequently in certain ethnic communities where the prevalence of deleterious mutant alleles in mutator genes such as *BRCA1* or *BRCA2* is relatively high (Table 3.1) (Hall et al. 2009; Lynce et al. 2015; Roa et al. 1996; Szabo and King 1995).

Table 3.1 Deleterious mutant allele frequencies of *BRCA1* and *BRCA2* genes in women who have breast or ovarian cancer or are at high risk for cancer as determined by clinical evaluation.

Breast and/or ovarian cancer at-risk women ancestral origin	*BRCA1* deleterious mutant allele frequency	*BRCA2* deleterious mutant allele frequency
Western European	6.9%	5.2%
African	10.2%	5.7%
Latin American	9.6%	5.4%
African American	7.4%	6.0%
Asian	6.3%	6.3%
Ashkenazi Jewish	50%–90%	50%–90%

3.6 Fewer the Stem Cell Divisions, the Higher the Cancer Occurrence

Interestingly, the number of carcinogenic mutational hits appeared to vary among different organ-specific cancer types. For example, certain ovarian adenocarcinoma and glioblastoma have the smallest number of recognizable carcinogenic mutational lesions (or hits) associated with them (approximately two hits per cell), colon and lung adenocarcinomas had approximately three carcinogenic hits per cell, whereas hepatocellular carcinoma had on average about eight such hits (Fig. 3.6).

What do the number of carcinogenic hits associated with different cancer sites imply? These numbers correlate well with the number of stem cell divisions associated with the respective organs where the cancer formed (Fig. 3.6) (Anandakrishnan et al. 2019). This suggests that organs exhibiting more stem cell divisions over the lifetime are somewhat more resistant to cancer than organs that exhibit fewer stem cell divisions over the lifetime. This is because on average more hits are required to produce cancer in the former organs. This interpretation is consistent with an early conjecture by John Cairns (1922–2018), who had proposed that cancer might be kept in check in multicellular organisms through the restricting of most cell divisions in an individual to stem cells (Cairns 1975; Cairns 2002).

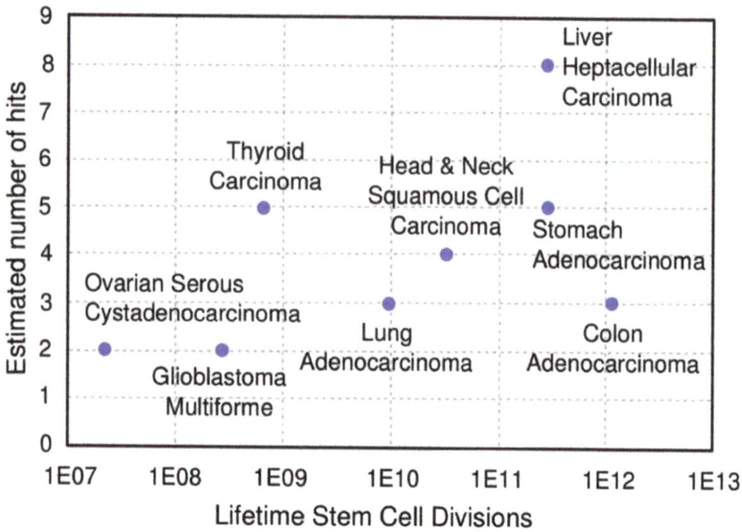

Figure 3.6 Number of somatic stem cell divisions in various organs over an average lifetime plotted against the estimated number of mutational hits that produce cancer in that organ. The letter E on the x-axis labels represents the exponent or the power of 10. [This diagram is reproduced here from Anandakrishnan et al. (2019) under the Creative Commons Attribution License]

Pan-cancer genome studies showed that organs that have intrinsically fewer stem cell divisions over a lifetime tend to develop cancer types that frequently exhibit abnormal telomere length and telomere sequence than those organs with more stem cell divisions over a lifetime. Thus, when they do become cancerous, even if rarely, they tend to acquire a property that bypasses the normal telomerase-dependent senescence mechanisms, which is heralded as abnormal telomere lengths and/or telomere sequences in these cells.

The pan-cancer genome sequencing effort (Section 3.4) found that the frequency of carcinogenic mutations weakly correlated with patient age, suggesting that the probability of mutation does not change significantly over time (see Chapter 15). Older patients accumulate more mutations, but the mutation rate itself may not increase with age. However, as discussed before, cells with mutator gene mutations accumulate mutations faster than those without.

Therefore, while the total number of mutations increases with age, the likelihood of cancer-causing mutations per unit time remains relatively constant unless a patient has inherited cancer-predisposing mutations.

These insights allow for the classification of tumor cell genomes into subgroups with varying malignancy potential, aiding in clinical decisions regarding follow-up and treatment options. Clinical decisions on follow-up frequencies and periods as well as therapeutic options can thus be better informed using these approaches.

3.7 Asymmetric Division of Stem Cells

3.7.1 *The immortal chain hypothesis*

Stem cells (though not the induced pluripotent stem cells[4]) have been observed by some to possess an interesting property — asymmetric cell division. One of two daughter stem cells appears to retain its stemness through multiple rounds of division, whereas the other daughter may differentiate to specific cell fates (Clevers 2005) (Fig. 3.7). However, that property is likely not expressed in all stem cell divisions. A stem cell might be able to switch between these two states of symmetric and asymmetric divisions, and it has been suggested that the symmetric mode might be related to cancer progression (Morrison and Kimble 2006).

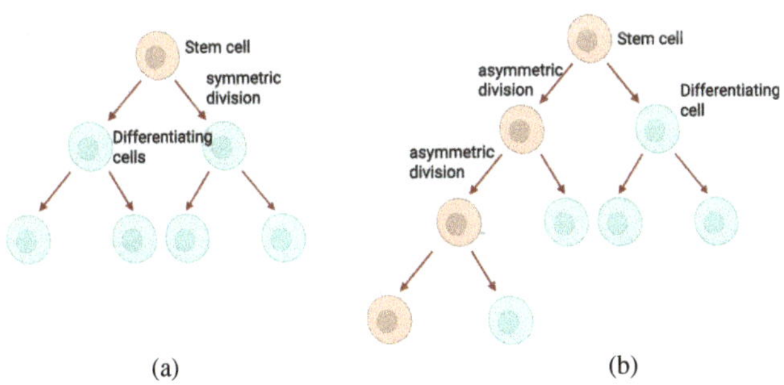

Figure 3.7 Modes of stem cell division. (a) Symmetric division. (b) Asymmetric division. [This figure was drawn using Biorender.com under an academic license]

[4]Induced pluripotent stem cells (iPSCs) are somatic cells that are converted to a stem cell-like property in which these cells regain the ability to differentiate into many different cell types.

How does asymmetric cell division occur? A related but perhaps more cogent question becomes, what is different between the two daughter cells immediately upon division? One possibility is an asymmetric division of the cytoplasm (which includes either a polarity-specifying molecule or an asymmetric arrangement of the mitotic apparatus) that somehow determines the subsequent fate (Morrison and Kimble 2006): strong evidence exists in its favor from studies on *Drosophila* and the nematode *Caenorhabditis elegans*. These organisms, however, do not get cancer!

Another possibility is that cells carrying one of the original four DNA single chains (or strands) from each of the two chromosomal sets of DNA helices inherited at birth retain their full stem cell characteristics (Fig. 3.8) (Cairns 1975).

For organisms with multiple chromosomes, this raises a question: how does the cell division apparatus recognize and asymmetrically distribute newly synthesized duplexes? There must be a mechanism to distinguish original DNA strands from newly replicated ones, possibly through epigenetic marks like DNA base methylation. Could cells that retain at least one original DNA strand be marked as special? Starting from the fertilized egg with 23 chromosome pairs, consider what happens after the second semiconservative replication. At this point, there are 46 DNA duplexes, each containing one original parental strand and one newly synthesized strand. If these original strands segregate randomly in subsequent divisions, there are 2^{23} possible ways to distribute them among daughter cells. Of these possibilities, only one results in a cell inheriting no original strands at all. This means there are, $2^{23} - 1 = 8,388,607$ distinct configurations where a cell retains at least one original parental strand. Through successive cell divisions, this could theoretically give rise to approximately 8 million different descendant cells, each inheriting at least one original DNA strand from the zygote in a unique chromosomal configuration. This theoretical calculation assumes no somatic recombination and uniform random segregation across all cell lineages, which may not reflect actual biological processes. However, it illustrates the potential diversity of DNA inheritance patterns. Given that the human body contains an estimated 300 million to 1 billion somatic stem cells across all tissues, could these cells that retain original parental DNA strands preferentially become the stem cell population?

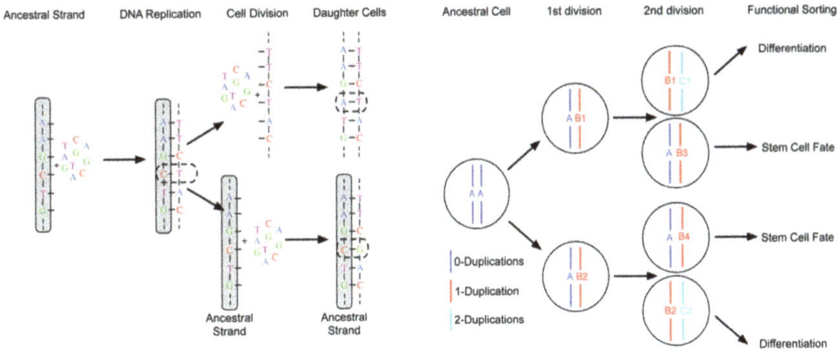

(a) (b)

Figure 3.8 The immortal chain (strand) hypothesis. Cairns (a) (1975) suggested that one mechanism by which multicellular organisms might protect themselves from the ravages of mutation is to retain one of the original four chains of DNA inherited from the parents in a small reserve of cells throughout their lives. Since the best guarantee against a deleterious mutation is to keep as a "blueprint" one of the four original DNA chains in some cells. Obviously these four chains were part of two DNA molecules that had successfully participated in the previous generation of reproduction and had survived through embryogenesis. (b) A small reserve of cells might be formed by an asymmetric division in which the duplex helix that contains the atoms of one of the two chains of either paternal or maternal DNA duplex never (or rarely) differentiates into tissues or organs. Instead, it keeps itself in an undifferentiated state. Thus, daughter cells that do not carry an ancestral chain are subject to mutation. Whereas the daughter cells that carry an ancestral chain always have a choice of repairing any mutation in favor of the original chain. If there is a problem with some descendant cells in an organ, it might be possible to overcome this by producing more descendant cells from further (still asymmetric) divisions of the cell that carries an ancestral chain. There have been evidences both against and in favor of this model (see text). The jury is still out. [This diagram is reproduced here from Werner and Sottoriva (2018) under the Creative Commons Attribution License]

Some studies with mammalian cells are in apparent support of this so-called immortal chain hypothesis (Conboy, Conboy, and Rebo 2015; Conboy, Karasov, and Rando 2007; Potten, Owen, and Booth 2002), whereas others have provided contrary evidence (Kuroki and Murakami 1989) and critique (Legraverend, Escobar, and Jay 2010) of these results.[5]

[5] It should be noted that most cancer biologists ignore this topic of research. I, however, think that there is merit to pursue these questions because much is unknown.

An unpleasant aspect of the immortal chain hypothesis, despite its obvious elegance, is the practical difficulty of testing it. Is the cell being observed truly a stem cell? How can we be certain of that? Does the method for monitoring asymmetric DNA chain segregation robustly sample all chromosomes in a cell, or does it monitor only a subset of chromosomes per cell? Does the detection method itself perturb the stem cell behavior? These concerns have been challenging to address. Thus, currently there is no clear experimental evidence as to which of these above models, namely, the polar cytoplasm model or the immortal chain model, is correct. It is possible that both might be true, at least for certain cell types, because the models are not mutually exclusive: a cell could utilize both the polar cytoplasm and the immortal DNA chain mechanism simultaneously for asymmetric division.

A recent study used deep sequencing of healthy single cells to analyze mutation frequencies in the two DNA chains. The results indirectly support the idea that human tissues show widespread asymmetric DNA chain segregation (Werner and Sottoriva 2018). While this result is intriguing in its consistency with the immortal chain hypothesis, no mechanism for asymmetric segregation has yet been provided. The consistency of observations with expectations of a model is not sufficient evidence for accepting the model because an alternate future model might also predict an identical result.

In summary, somatic stem cells, which come from symmetrically dividing embryonic cells, usually divide asymmetrically. This allows some cells to keep a "reference" DNA chain (perhaps some of the original atoms of the parental DNA chains), preserving a kind of evolutionary memory of a recent mating. During organ development, like in blood or neural tissue formation, these stem cells may switch to symmetric division to grow their numbers and differentiate all at once. Stem cells that divide symmetrically may be more likely to develop cancer. The tumor suppressor p53 might help control the switch between asymmetric and symmetric division (Merok et al. 2002; Rambhatla et al. 2005).

3.7.2 *Is DNA repair always good for a cell?*

It is widely accepted that high-fidelity DNA damage repair protects all cells against cancer. However, a provocative counterargument suggests the opposite for stem cells (Cairns 2002). If a DNA chain from either parent or early embryonic stem cell division must remain relatively unchanged, it is crucial to reduce DNA repair enzyme activity targeting that chain (Cairns 2006). If DNA repair occurs, the repaired chain could lose its epigenetic marks. This would be very difficult — or even impossible — if the damage involves a double-chain DNA break or a long single-chain exposure overlapping the original DNA chain. Based on this idea, Cairns suggested that stem cells might have intrinsically diminished DNA repair activity. Here, we modify this idea slightly. We propose that DNA repair in stem cells should avoid processes that involve double-chain or single-chain gap repair, especially in cells that divide asymmetrically. This is consistent with the observation that the two DNA chains should have different levels of mutations if the chains are segregated asymmetrically (Werner and Sottoriva 2018).

Cairns (2002) had argued that the role of p53 as a tumor suppressor is consistent with this prediction — one of the important functions of p53 is to trigger programmed cell death (Fig. 3.9). It might be a better strategy to kill an asymmetrically dividing stem cell than having to repair its broken chromosome using a non-immortal DNA chain as the template. Otherwise, it could produce homozygosity for a mutant gene that may potentially cause cancer.

Consistent with the above arguments, certain types of DNA repair, like nucleotide base excision repair that uses one DNA chain as a template to correct mispaired nucleotides, are less prevalent in some adult stem cells. These cells die if even low levels of DNA base mismatches are produced (Roth and Samson 2002). Now, the mutation rate in embryonic stem cells is low, suggesting that mismatched DNA might not be repaired and cells with these DNA lesions just die off. However, DNA double-chain break repair activity, which is thought to correct some of these mismatches, appears to be more efficient in embryonic stem cells than in adult cells, potentially contradicting this idea (Frosina 2010).

In science, controversy is the mother of insights and new discoveries. The idea that somatic stem cells might be special with respect to DNA segregation as well as DNA damage repair may or may not be true. This unusual proposal, even if ultimately shown to be wrong, is important for the process of understanding the complex phenomena of how cancer cells are formed. Such controversies have the potential to stimulate new insights for the future treatment of cancer.

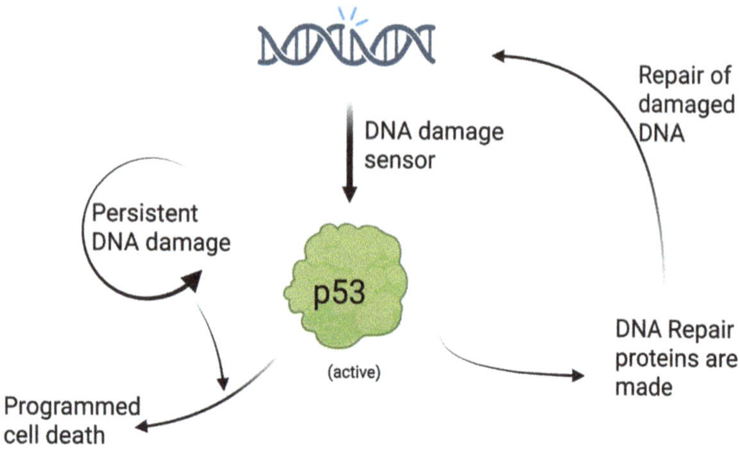

Figure 3.9 Tumor suppressor p53 protein has multiple functions. It is activated by DNA damage (both double-chain breaks and single-chain lesions) to induce the activation of DNA damage repair processes. If the lesions persist, the protein assumes a different state in which it triggers programmed cell death, thus eliminating the cell. Evolution has found it more expedient to remove a cell with persistent DNA lesions than to risk mutations, perhaps cancer. [This figure was drawn using Biorender.com under an academic license]

3.8 Mutation Selection or Epigenetic Adaptation?

We have so far attempted to describe how purely Darwinian thinking has helped shape the understanding of cancer. It is possible, however, that not all manifestations of cancer can be explained by a purely Darwinian evolutionary paradigm.

There is strong experimental evidence to suggest that gene expression in general and gene expression patterns in cancerous states in particular exhibit non-genetic components — "somewhat" heritable states stable over multiple "cellular" generations have been demonstrated (Baylin et al. 2001; Casado-Pelaez, Bueno-Costa, and Esteller 2022; Esteller 2002; Jones and Laird 1999; Virani et al. 2012). These so-called "epigenetic" states are associated with defined signatures of gene expression patterns that cannot be attributed to the consequence of a gene mutation.

A key idea in applying Darwinian evolution to cancer biology is that mutations that make cancer cells resistant to drugs usually happen on their own, not because of the treatment itself. This has important consequences for physicians, as they want to avoid treatments that might trigger drug-resistant cancer cells.

However, it is common for cancer patients who initially respond well to treatment to experience a relapse, with their cancer becoming resistant to the drug. The question is, does the drug cause resistant cancer cells to develop, or do these cells appear by chance and then survive and grow because of the drug, as Darwinian evolution suggests?

Most experiments show that while some drugs can cause mutations, the resistance mutations likely arise on their own. These mutations happen more often if a "mutator gene" had already mutated earlier.

But that might not be the whole story. Recent experiments suggest that in some cases, cancer cells' resistance to chemotherapy might not come from mutations. Instead, it could be due to epigenetic adaptation rather than genetic mutations that make the cells resistant to the drug (see Chapter 15, Section 15.2, Figure 15.2).

In a series of intriguing experiments were conducted with an aggressive type of skin cancer (melanoma) cells in test tube cultures. The melanoma cells were initially sensitive to an anti-cancer drug. Upon exposure to the drug, however, drug-resistant cells emerged rapidly (Emert et al. 2021; Goyal et al. 2023; Shaffer et al. 2017; 2020; Torre et al. 2021). It appeared that an exposure to the drug was associated with the emergence of resistant cells rather than the drug selecting for a few pre-existing drug-resistant cells in the culture.

The above unexpected observation did not fit into our classical idea of Darwinian evolution: the spontaneous generation of variants and the selection for a variant's fitness by the environment being independent of each other. Here the environment, namely, the drug, was apparently inducing the appearance of resistant clones of cells.

There is an important detail to consider, however, when interpreting these experimental results. What makes these results different from a purely Lamarckian process — where the drug itself might directly cause the appearance of resistant cells in a predictable way — is the following:

It was observed that among the skin cancer cells, some rare individual cells tended to switch to multiple new gene expression (epigenetic) states, but only in the continued presence of the drug. These altered epigenetic states involved coordinated changes in the expression levels (i.e., the amounts of their corresponding mRNA molecules) of many genes together and were temporary. Only a few cells in the population made this switch, and it seemed to happen randomly. Interestingly, some of these new states, but not all, were passed down through several generations of cell division. A few of these inherited states matched the states seen among the drug-resistant cells (Emert et al. 2021; Goyal et al. 2023). However, unless continuously exposed to the drug, They cells reverted to the drug-sensitive state after a few generations of cell division.

Once again Darwin wins over Lamarck: there was no evidence of continued stable inheritance of an acquired character in the absence of continuous selection by the drug. The molecular nature of these gene expression state transitions is still a mystery. Exciting developments are surely forthcoming in this direction.

In summary, cancer is rarely caused by a single gene mutation, generally requiring at least two to three mutations, each of which either activates an oncogene or abrogates the function of a tumor suppressor gene. The rate of mutation in cells that are on their way to the cancerous state might be higher than in normal cells, because cells might be selected for a higher mutation rate due to the inactivation of a DNA repair gene. These numerous genetically variant cells become the playing field of Darwinian evolution, in which natural selection for a higher rate of cell division, resistance to growth inhibitory signals, bypass of the needs for growth-stimulatory signals, abrogation of cell death mechanisms, and cell survival through additional processes (such as increased nutrient supply through tissue remodeling) all contribute to the emergence of fulminant cancerous conditions. The exact molecular events leading to cancerous properties, however, are still being debated, where competing theories are stimulating further insights, and hopefully these will lead to the innovation of more effective cancer diagnosis and treatment.

Chapter 4

On Aging and DNA Ends

The central problem we consider in this chapter is whether aging is a biological process, and if so, is it under natural selection? Cellular aging is the fundamental process of aging that afflicts all living cells in our body, which is invisible to us without special biological experimentations. We delve into the molecular mechanisms behind cellular aging and show that it is an innately genetic process and therefore must be evolving. We see how the mechanisms of cellular aging were deciphered by researchers using mental constructs, or models, provoked by the Darwinian evolutionary theory of similarity of genes due to a common descent of organisms. The fascinating account of how these discoveries were made by the protagonists emphasizes the central role of Darwinian evolutionary theory in modern biomedicine. Cellular aging, however, is not the same as organismal aging. Organismal aging is what we commonly observe as the symptoms of growing old. The latter, generally regarded as a gradual decline in health, is often accompanied by chronic diseases, such as type II diabetes or age-related dementia, that surface with the advancing age of individuals. While it might be impossible to eliminate cellular aging, understanding the biological basis of chronic diseases might in principle extend the productive life of individuals. Are gene variants that contribute to the susceptibility or resistance to chronic diseases selected in evolution? What can we do to recognize, understand, and perhaps one day counteract the harmful effects of human gene variants that contribute to the onset or progression of chronic diseases?

Important terms:

Eukaryotes are organisms that sport a nucleus (which contains the genome) in each of their cells. Sometimes a eukaryote is a single-celled organism (such as the baker's yeast, the malarial parasite, and *Tetrahymena pyriformis*) and sometimes is a multicellular organism, such as humans. Organisms that do not sport a nucleus in their cells are collectively called *prokaryotes*. Prokaryotes are of two main types: *Eubacteria* (such as the familiar *Escherichia coli*) and *archaebacteria* (such as some bacteria that live in geothermal springs or other inhospitable places).

Enzymes are catalysts that increase the forward rate of a chemical reaction; every chemical reaction is in principle reversible, and if it is allowed to proceed without interference, it will reach an equilibrium concentration of products and reactants. Enzymes only accelerate the rate of the forward reaction but do not affect the equilibrium ratio of the products to the reactants. Enzymes are not consumed during their catalytic process but are reused repeatedly. Most enzymes are proteins; some enzymes are RNA.

DNA or RNA replication generally involves successive and sequential polymerization of units of *nucleic acids* (DNA or RNA), here simplified as DNA or RNA bases, on the *template* chain, one *nucleotide* (which is composed of one unit of phosphate, a sugar, and a base) at a time in a directional manner. Polymerization that enables the extension of the chain or, equivalently, the chain of *polynucleotides*, can occur only along the so-called 5′ to 3′ direction. The newly synthesized chain is complementary in the sequence of bases to the old (template) chain.

A *primer* in the context of DNA or RNA synthesis is a short piece of DNA or RNA chain with a 3′ end free for the addition of further nucleotides once the primer sequence is positioned by complementary base pairing on the template chain.

Telomeres are special DNA ends of organisms that have linear chromosomes. Eubacteria and many archaebacteria have circular chromosomes, so they have no telomeres. Many other archaebacteria and all eukaryotes have linear chromosomes, and each linear chromosome has two telomeres, one on each end.

> *Stem cells* are a group of cells that are kept in reserve in the body of multicellular animals. These cells generally have a slower rate of division than their surrounding cells, and they retain the property of changing into other cell types. Stem cells are the generalists in a human body. Thus, wear and tear of an organ, composed of many cells of specialized function, can often be replenished by division and subsequent specialization (also called *differentiation*) of these cells. This ability is not limitless, and not all stem cells can be transformed into all specialized cells.

In our imaginary parallel universe, on a seaside bench are seated three people, one in a flowing white toga, one in a three-piece suit and a bow tie, and the third in a one-piece black swimsuit wrapped under a red, white, and navy blue beach towel. Our characters are Epicurus, the ancient Greek philosopher and physician; Rosalind Franklin, the British chemist who first took X-ray pictures of DNA based on which the double-helical structure of DNA was proposed; and Charles Darwin again.

Epicurus: It is not the young who should be considered fortunate but the old who lived well.

Franklin: I suppose you are right. They tell me had I lived a few more years they would have given me the Nobel.

Darwin: Rosalind, you did exceptionally well with the short time you had. Your untimely demise was unfortunate for the world of science. Your later work on RNA viruses was extraordinarily exciting. I heard Francis and Odile became your friends, and you went sailing together, right?

Franklin: Yes, Charlie, we did. It was marvelous. Oh youth! She is nimble and full of sport, like summer brave. You know, I never could imagine what secrets my X-ray photo of wet DNA fiber would yield. The very essence of youth located on the ends of my DNA molecules!

Epicurus: True. Now we know the secret of living well. If your DNA ends are kept well in your stem cells you are more likely to live well and long.

Franklin: Unless you have cancer, like I did. You already know that cancer is mere chance, like random variation in evolution, which Charles here had proposed.

Epicurus: But also false, because chromosome ends don't look like your standard DNA, and because aging is more complex than just DNA.

Darwin: Well said. I don't think we can ever avoid aging because after the end of our reproductive age, there is no selection on genes. You can't get better genes to live longer through natural selection unless you can also extend the reproductive age.

Epicurus: Unless we are also good grandparents, who rear well-adapted kids. These kids adapt better because their grandparents' genes, which they inherit, make their grandparents live long enough to teach them survival skills.

Franklin: We should swim when we can.

The two elderly gentlemen observe wistfully as the young woman dives into the water in a graceful arc.

***

At the end of the first two decades of the 21st century, we are told that we might be on the verge of extending human life span indefinitely. Drumming this claim, we also see companies peddling anti-aging creams and potions backed by allegedly impressive science and nutritional advice based on DNA analysis of gut bacteria, all promising to defy aging and age-related degenerative processes.

In the Roman Empire, the average life expectancy at birth was about 25 to 30 years (Scheidel 2001). Much of that was due to high child mortality rates, but that was not the whole story. In the Middle Ages in Europe, life expectancy at birth rose to about 33 years, and in the early 1900s it had reached 55 years (Finch 2010). Life expectancy at 20 years of age, however, had not changed much since the Roman Empire — it had remained at about 50 to 60 years range until about 1900. In 2024, the average life expectancy at birth in Western Europe was about 80 years for males and about 85 years for females.[1] Interestingly, the average life expectancy at 20 years of people in the hunter–gatherer societies, despite

---

[1] https://www.who.int/data/gho/data/indicators/indicator-details/GHO/life-expectancy-at-birth-(years)

their chronic exposure to the elements or perhaps because of it, is also about 60 to 70 years — slightly lower but not too far from that of the technological Western world (Blurton Jones, Hawkes, and O'Connell 2002; Gurven and Kaplan 2007; Robson and Kaplan 2003). How so?

It has been suggested that the carbohydrate-rich diet made possible by agriculture, the consequent ready availability of abundant calories to most people, and the conditions of sedentary urban life that resulted together had conspired to reduce the average human lifespan (Diamond 1987; Hariri 2015) from its nearly universal optimum in the 60 to 70 years range to about 50 years or less in most pre-18th-century civilizations. Once those factors are taken care of by healthy nutrition and exercise, human lifespan appears to increase to that range again on average but no more. Is it possible to live significantly longer than the 70 to 80 years range on average and to live well?

4.1 Why We Age

How much longer can humans live and still lead a productive life? While much of the increases in life expectancy at birth since the Roman Empire in the Western world has been due to reducing childhood mortality and greater access to science-based primary care in affluent nations, further extension of life expectancy appears to rest on preventing aging *per se*.

Is aging a consequence of the physicochemical processes of wear and tear at the level of molecules? Or is aging fundamentally biological? If aging is a biological process, has it been selected by evolution to restrict human life on average to the 70 to 80 years range?

As it turns out, the process of aging in all eukaryotic organisms is related by a fundamental biological process having to do with the ends of their chromosomes. This molecular mechanism has been highly conserved in evolution. While it is certainly not the only mechanism of cellular aging, it is one of the most important ones. We will now dig into the history of discoveries that were central to the understanding of how aging occurs, to illustrate how the principle of Darwinian conservation of inherited characteristics had motivated these seminal discoveries.

4.2 Chromosome Ends Are Peculiar

In the summer of 1990, a small group of graduate students and scientists met for their weekly seminar in the auditorium of the Whitehead Institute. That was a special day, as the visiting speaker was Elizabeth Blackburn (1948–), who, with her PhD student Carol Greider (1961–), had recently identified an enzyme that produced the terminal structure of eukaryotic chromosomes — the telomeres. The enzyme was named the telomerase.

The local expert on telomeres was Jack Szostak (1952–), at Massachusetts General Hospital just across the Charles River, who had collaborated with Blackburn 6 years ago in some of the first seminal experiments that probed the molecular structures of the ends of eukaryotic chromosomes. He, however, was not present at the talk that day.

If my memory serves right, it was also the week of the biologist Mark Ptashne's (1940–) birthday, who was then a professor at Harvard and famous not only for his three-decades-old experiments that had revealed how genetic switches work but also for his colorful personality. That evening, before the cutting of the cake, Ptashne dazzled the small audience of students and professors with a bravura performance on his Stradivarius violin. A little later, Blackburn spoke about Greider's intricate series of experiments that had shown how linear eukaryotic chromosomes survive a curious folly of evolution with the help of an ancient enzyme.

DNA replication has never managed to evolve an enzyme for starting its business of replicating at the very ends of a DNA molecule. It is unclear why, but because of this, if chromosomes were linear double-helical DNA molecules, with free 5′ and 3′ ends at both termini, then they are predicted to be progressively eroded from both ends at every replicative cycle (Fig. 4.1). This prediction was first articulated by none other than James Watson, the co-discoverer of DNA's double helical structure. In his 1972 paper, Watson pointed out (Watson 1972) that most prokaryotes had solved this so-called DNA-end replication problem by making their chromosomes circular, at least during the time of DNA replication — a circle has no end.

A connection between the predicted progressive shortening of eukaryotic chromosome ends and cellular time-keeping (i.e., aging) was made independently nearly simultaneously with Watson by a Soviet scientist,

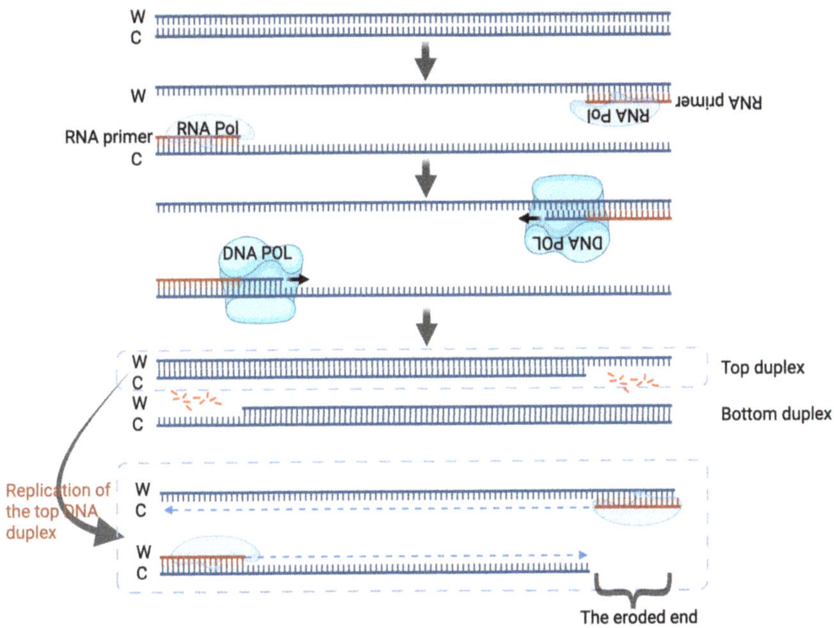

Figure 4.1 The chromosome end replication problem. DNA polymerases cannot start DNA synthesis at the ends of linear duplex DNA. It needs a primer, which is provided by a small piece of RNA made by an RNA polymerase. Using this RNA as the primer, DNA polymerase copies the complementary chain. Each of the RNA primers is then degraded, leaving a single chain at one of the two ends. At the next round of replication, a repeat of the process would shorten one of the two chains in each of the daughter cells. Chains are labeled as W (5’ terminus to the right, 3’ to the right) and C (opposite polarity and complementary to W) for their respective polarities. The dashed box represents one of two daughter cells. The other daughter is omitted for simplicity. [This figure was drawn using Biorender.com under an academic license]

Alexey Olovnikov (1936–2022) (Olovnikov 1973). Long before these specific insights, however, Barbara McClintock (1902–1992), working with maize plants, had shown that if a chromosome of maize is broken, the two broken ends internal to the chromosome become highly “reactive,” apparently inducing additional breakages in the chromosomes, suggesting that the normally unreactive DNA ends of a linear chromosome must be different from the DNA ends produced willy-nilly due to breakage within the chromosome. Putting these observations together, Olovnikov suggested that the two normal ends of linear eukaryotic chromosomes (i.e., the telomeres) must be protected somehow and therefore must have a peculiar structure. But what was so peculiar about them?

In 1980, Blackburn was looking at chromosomes of a tiny pond-dwelling single-celled animal, *Tetrahymena thermophila.* This organism has a strange life cycle, at one stage of which it amplifies a specific small chromosome many, many fold. This provided Blackburn with a tool to study the telomere, because every cell provided many copies of the terminal structures of chromosomes rather than the usual four per chromosome pair in most diploid organisms. She also took an intellectual leap: there was already an organism — a bacterial virus called lambda — which had a linear DNA chromosome that circularizes before it begins replication. This circularization was shown to be due to perfectly complementary or "sticky" ends at the two termini of lambda's linear chromosome (Shalitin and Stahl 1965; Streisinger, Emrich, and Stahl 1967). Blackburn wondered whether evolution might have preserved a similar strategy for replicating the chromosomes of this protozoan.

Protozoa and bacteria had diverged at least 2.7 billion years ago. If the two processes — one adopted by a prokaryotic virus and the other by a single-celled eukaryote — were similar, it would imply a remarkable generality of molecular mechanisms. Perhaps the solution is so simple it can be repeated many times in different contexts. Darwin's idea that evolution repeatedly uses a particular solution to a fitness problem in many different species that are related by descent is fundamentally at work behind this simple expectation.

The solution in this case, however, turned out to be quite different. The mechanism used by the Lambda virus was distinct and much simpler. *Tetrahymena* uses a very different strategy to stabilize its linear chromosomes that never circularize. That had to do with an unusual, repeated DNA sequence at its telomeres. Nonetheless, the motivation that drove Blackburn's initial experiments was guided solidly by Darwinian thinking, without which there was no simple expectation that would have prompted Blackburn to risk her early career by choosing a pond-dwelling protozoan that few scientists were working on.

A series of elegant experiments conducted in the respective laboratories of Blackburn and Szostak suggested (Blackburn and Szostak 1984) that the ends of eukaryotic DNA helices might have hairpin bends. In effect, the two chains of the DNA double helix might become joined by a covalent bond linking the two chains, forming a hairpin bend, as it were, held together by intra-chain base pairing (Fig. 4.2a). Curiously, however, there appeared to be a specific sequence requirement: 5′TTGGGG3′, repeated many times on one chain, and, of course, its complementary sequence, 5′CCCCAA3′, on the other, on *Tetrahymena* telomeres.

Remarkably, the telomeric sequences isolated from the protozoan *Tetrahymena* also worked perfectly to stabilize chromosome ends in a fungus, the baker's yeast, *Saccharomyces cerevisiae*! *Tetrahymena* and fungi had shared their last common ancestor about a billion years ago, and yet yeast, the fungus, could successfully use the other organism's telomere to stabilize its own chromosome (Szostak and Blackburn 1982) (Fig. 4.2b). This experiment — to test whether *Tetrahymena* telomeres would work in yeast — was audacious because it required a belief that nearly a billion years of evolution would conserve the same mechanism between *Tetrahymena* and yeast. In the words of the protagonists (Blackburn, Greider, and Szostak 2006):

> *The idea was so far-fetched that it seemed outlandish: to test whether the telomere replication mechanisms are conserved between such evolutionarily distant species.*

Yet, their experiments based on this idea worked! In brief, *Tetrahymena* telomeres with 5′TTGGGG3′ repeats stabilized the ends of a linearized yeast DNA, which didn't have its own telomeres (therefore would normally be unstable) (Fig. 4.2b). Moreover, yeast-specific telomeric repeat sequences, 5′$T(G)_{1\text{-}3}$ repeats, were added on to the *Tetrahymena* telomeric repeats in subsequent generations, suggesting that an enzyme encoded in the yeast genome could recognize the *Tetrahymena* telomeric repeat sequences as if it were its own scaffold to extend.

To Blackburn and Szostak, this remarkable evolutionary conservation in two organisms from across the phylogenetic kingdom, whose last common ancestors lived a billion years ago, suggested that there must be an ancient enzyme that is responsible for making telomeres to stabilize all eukaryotic chromosomes. As if one couldn't do without such a wise old bird as this enzyme appeared to be!

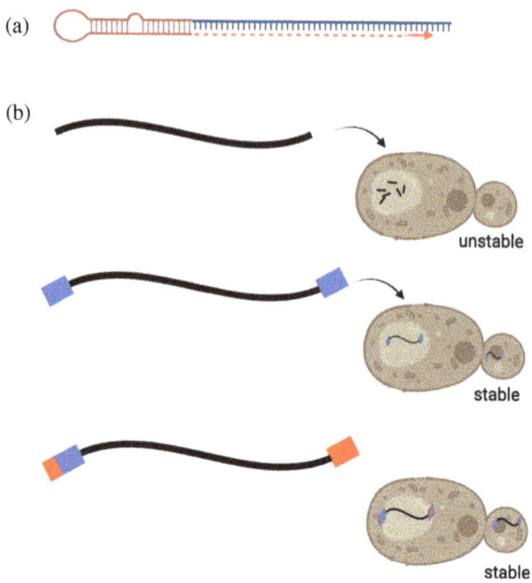

Figure 4.2 Telomeres at chromosome ends. (a) Blackburn and Szostak suggested that telomeres might have hairpin bends. The hairpin bend allows the foldback DNA to serve as a terminal primer, thus solving the end-replication problem. This proposal turned out to be incorrect in details. (b) Blackburn, Greider, and Szostak introduced linear DNA fragments (wavy line) into yeast. As expected, these were unstable and could not be maintained. When telomere repeat DNA isolated from *Tetrahymena* (blue boxes) was attached to the linear DNA ends, these stabilized the linear DNA: they could replicate and were stably maintained in yeast cells during multiple rounds of cell division. *Tetrahymena* had diverged from yeast over a billion years ago; it is remarkable that *Tetrahymena*'s telomeres worked in yeast. When they replaced the *Tetrahymena* telomeres with random genomic DNA fragments from yeast, some of these fragments stabilized the linear DNA (red boxes). These turned out to have DNA sequence repeats similar but not identical to *Tetrahymena* telomeric repeats (see text). [This figure was drawn using Biorender.com under an academic license]

On that summer evening at the Whitehead Institute, Blackburn described one elegant experiment after another, conducted by Carol Greider, which enabled the understanding of how telomere structures are formed.

4.3 An Enzyme Like No Other

Greider did indeed find an enzyme in *Tetrahymena* extracts that extended the existing *Tetrahymena* telomeres at DNA ends in the test tube (Greider and Blackburn 1985). Remarkably, the same enzyme from *Tetrahymena* added telomeres to the ends of yeast chromosomes too, but what the enzyme added was not the yeast-specific telomere sequence 5′T(G)$_{1-3}$ but the related yet different *Tetrahymena*-specific telomere sequence, 5′TTGGGG3′ repeats (Greider and Blackburn 1987). A billion-year-old molecular memory — somewhat diversified in its finished product, yet it remembered the class of substrates — the telomeres — unerringly!

This suggests that at least in the case of telomeres, it is not contingent upon the organism to solve a general biological problem, namely, stabilizing of chromosomal ends, by one chemically unique solution — it could choose either a yeast type or a *Tetrahymena* type of telomere sequence to stabilize its chromosome — but is guided to one solution or the other by the nature of the gene that encoded the enzyme. The gene is the stuff that is fashioned in evolution.

Even more exciting was the finding that Greider's enzyme had two integral components — a protein part and an RNA part (Greider and Blackburn 1987). The RNA component provided a template, but in addition, the peculiar geometrical folding of the RNA is essential for its activity. This enzyme provides an example of rare enzymes made of both RNA and proteins. The enzyme utilizes the RNA component as a template to synthesize DNA complementary to the RNA template at the 5′ terminus of eukaryotic chromosomes and thus solves the end replication problem. Since the enzyme synthesizes the telomeric DNA, the enzyme was named telomerase, of which the catalytic protein component is named telomerase reverse transcriptase (or TERT) and the RNA part is the telomerase RNA component (or TERC) (Fig. 4.3).

4.4 What Is True of a Protozoan or Yeast Is True of a Human

The TERT protein sequence is highly conserved among all eukaryotic lineages and even in some bacteria. The gene encoding this protein is thought to have evolved from some ancestral jumping genes termed retrotransposable elements, which are together often thought of as parasitic

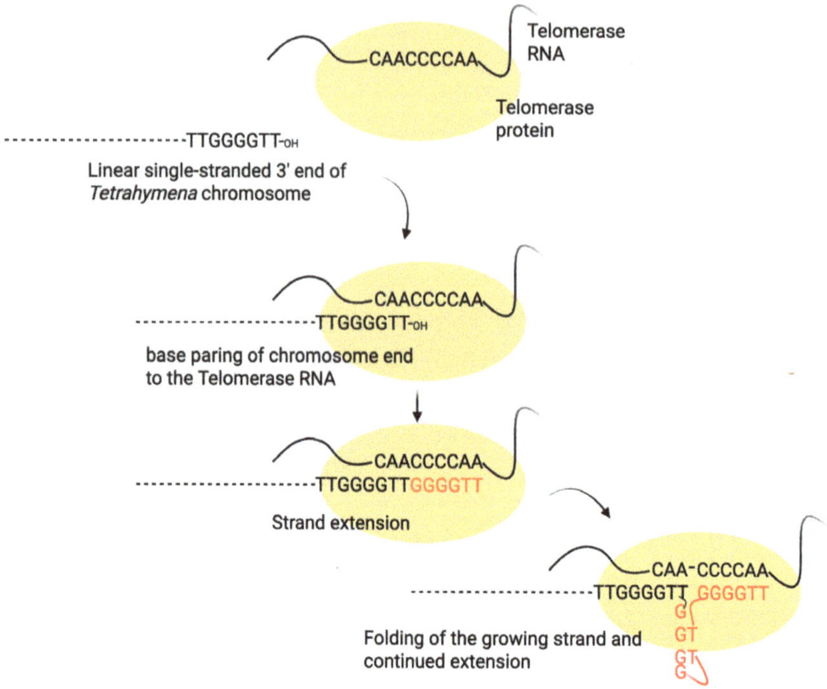

Figure 4.3 Telomerase extends the free 3′ single-chained end of a chromosome. The addition of nucleotides at the terminus is highly processive, meaning that the TERT enzyme tenaciously grasps the chromosome end and keeps adding nucleotides while the newly added nucleotides (red) fold into a 3-D structure that facilitates further chain displacement off the RNA template, such that further DNA nucleotides can be added. Note that the enzymatic activity of TERT is in principle a reverse-transcriptase activity (polymerizes DNA on an RNA template). Note also that the enzyme is not very choosey about the DNA sequence at the end, so long as a few nucleotides (in this case -GTT) are present. It will add nucleotide sequences complementary to the TERC RNA template. [This figure was drawn using Biorender.com under an academic license]

genes (Kidwell and Lisch 2001). The human TERT protein's gene, *hTERT*, has similar activities in the presence of its template RNA component TERC in extending human telomeres. Its activity is important for maintaining chromosome stability in healthy growing cells. However, its TERC component has a slightly different sequence, such that the human telomere sequence is a repeat of 5′TTAGGG3′.

The first clue that telomerase might be important for cellular aging came from experiments when the gene for TERT in yeast was mutated.

The chromosomes of the mutant yeast cells began to be shortened from their ends, and when a critically shortened length of the chromosome was reached, the yeast cells stopped dividing. Adding back the catalytically active yeast telomerase gene rescued the defect.

Interestingly, when non-tumorous human cells are cultured in the laboratory, they stop growing after a variable number of DNA replicative cycles. When researchers measured the amount of TERT protein or its mRNA in these cultured human cells, they found that *hTERT* gene expression was reduced or undetectable in cells that had stopped growing, while its expression level was relatively high in "younger" cultured cells. Adding back the *hTERT* gene (the human version) made the human cells begin to divide again. The length of telomeres in the human cells correlated with hTERT mRNA activity — long in cells with high hTERT mRNA and short in cells with low or undetectable hTERT mRNA. Thus, telomerase activity correlated inversely with replicative aging in cultured human cells, as it did in yeast cells. Even more so, drastically reduced telomere length in human and non-human mammalian cells triggered cell death (so-called, programmed cell death, or apoptosis).

Cultured human cells might be artificial; does TERT behave in a similar manner in whole animals or in humans?

First, mice with half the capacity to produce the TERT protein exhibited widespread telomere shortening, a reduced capacity to regenerate tissues, and a drastically reduced lifespan. Second, a mutation in the RNA component, TERC, of the human telomerase causes an inherited disease of premature aging, dyskeratosis congenita, in which young patients develop skin abnormalities generally associated with advanced aging and often have bone marrow failure. Third, it was found that most types of differentiated cells in humans have low or undetectable levels of hTERT mRNA. By contrast, somatic stem cells express high levels of hTERT mRNA and protein and contain robust length distribution of telomeres (stem cells are supposed to be ageless). Fourth, monitoring of over a hundred people over the age of 60 years indicated that the chance of death due to cardiovascular and infectious diseases correlated strongly with the extent of shortening of telomeres in their blood cells, suggesting that age-related disorders could be related to telomere shortening in blood cells. Shortened telomere lengths in pancreatic beta cells or their precursors have also been reportedly associated with the onset of type II diabetes, though the jury is still out on whether telomere shortening is a cause or a consequence of the disease.

In science, when trying to show a cause-and-effect relationship — like between telomere length maintenance and aging-related changes in humans — it's not enough to rely on correlation alone. For a conclusion to be convincing, there needs to be a clear mechanism that explains how telomere shortening in blood cells contributes to biological aging. Currently, such a mechanism has not been fully established. Additionally, it must be shown that fixing telomere dysfunction can reverse biological aging. Without this evidence, the connection remains uncertain (De Meyer et al. 2018).

In summary, telomere shortening is positively correlated with cellular aging and age-related disorders, and cellular aging is in turn positively correlated with reduced activity of the telomerase enzyme. It is unclear whether these correlations imply a cause-and-effect relationship. If indeed the two are mechanistically related, then is it possible to reverse cellular aging by reinvigorating the lost activity of telomerase?

4.5 Telomerase Expression Immortalizes Cells

We mentioned before that mature or differentiated cells turn off the expression of the hTERT gene and that only somatic stem cells or progenitors of differentiated cells retain the capacity of telomerase expression. Aging is generally thought to be associated with the exhaustion of the somatic stem cell repertoire in the body. Does this imply that if we can turn on telomerase expression artificially in certain cells, we can renew somatic stem cells indefinitely toward reversing the effects of aging?

Indeed, turning on hTERT expression in human fibroblasts, the prevalent cell type in the connective tissue, immortalizes these cells under *in vitro* culture conditions.

Curiously, unlike in humans, fully grown mice do not turn off their TERT genes in most of their cells. The average telomere length in mice is longer than in humans. Mice, however, do not live quite as long as humans. Aging, therefore, is more complex than merely the loss of telomerase activity.

As in human fibroblasts, turning hTERT expression on in human epithelial cells isolated from gastric mucosal lining also immortalized these cells. However, these epithelial cells did not retain their full characteristics of epithelial cells — they were converted into cells with ambiguous biological properties, some of which resemble the properties of cancerous cells. Telomerase expression is often high in many types of cancer cells; turning on hTERT expression is a frequent switch that propels the process of carcinogenesis forward.

Could it be that in humans, telomerase expression is turned down in adults, especially in differentiated cells, because given the long lifespan of our species, too much telomerase could increase the chance of cancer? Could evolution have selected for this transition to reduced telomerase levels only in relatively long-lived mammals to reduce the risk of cancer? Indeed, a study of human telomere length distribution in more than 400,000 people suggested that longer than average telomere length was associated with an increased risk of cancer, but shorter than average telomere length was associated with an increased risk of cardiovascular and metabolic disorders (The Telomeres Mendelian Randomization Collaboration 2017).

While these studies demonstrate only correlation, the results are consistent with the idea that telomere length balance might be important — too long the telomeres, one gets cancer; too short, and one gets metabolic or heart diseases!

4.6 No Magic Bullet against Aging

The discovery of telomeres and their connection to aging has been a remarkable journey, and the reader by now has seen how a reliance on the simple Darwinian idea of conservation of molecular characteristics opened new areas of medical research. The next step, the translation of these findings to human aging, is yet uncertain and is under intense investigation. If further research allows the making of a "magic bullet" that could delay intrinsic human limits to longevity as defined by telomerase and telomere shortening, how long would a human be expected to live?

Human death rate has two components: (1) an age-independent component that includes accidental deaths and deaths due to exposures to infections, war, or other similar factors; (2) an age-dependent component. In countries with efficient healthcare, public policies that reduce accidental deaths (such as more stringent regulatory laws), and political processes that discourage war and violence, the first component can potentially be reduced to a minimum, even close to zero on average. The age-dependent component is what is relevant for our discussion here. How much can we reduce the age-dependent component?

The age-dependent component can be described as an exponential function of the form (Kirkwood 2015):

$$p_x = \alpha e^{\beta x}, \tag{4.1}$$

where, p_x is the probability of death at age x, and α and β are parameters that can be estimated from population studies.[2] This is also called Gompertz' law, after the name of Benjamin Gompertz (1779–1865), who discovered this relationship in 1825 during his work as an "analyst" in an insurance agency.

A plot of the function in Fig. 4.4 shows that as age increases, the probability of death accelerates (the tangent to the curve gets steeper). The parameter α is related to what is known as the "initial vulnerability to death" at birth. Given perfect prenatal and neonatal care in an imaginary country and given perfect delivery and perfect protection against infectious diseases, the initial risk of death at birth might be related to genes that affect neonatal viability. The parameter β is related to the rate of rise of the age-dependent risk of death as age progresses. For our discussion in this chapter, we are mainly concerned with this parameter. Ignoring age-dependent rates of exposure to environmental hazards, such as wars, guns, driving, boxing, skydiving, and so on, the value of β encapsulates the probability of occurrence of various bad diseases, such as cancer, metabolic disorders, and progressive degeneration of bodily function, which must be influenced by genes as well as those genes' interactions

---

[2] We will derive this equation, also called a growth equation, in Chapter 13.

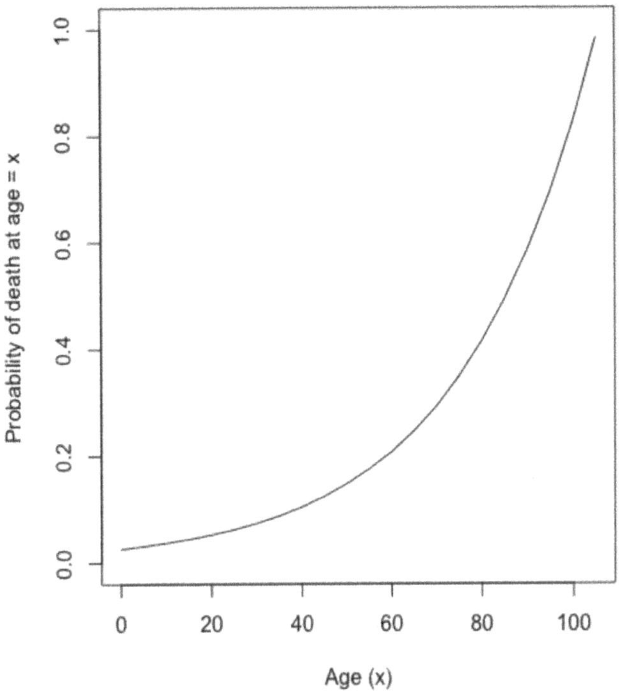

Figure 4.4 A plot of the expected probability of death at a certain age according to Gompertz law (Equation 4.1) for an initial vulnerability to death at birth, α, of 2.6% and a 3.45% constant rate of increase of the risk of death, β, with age. According to published data from the United Nations, 2021 global average for α was approximately 2.6% per live birth. The shape of the curve shows that the rate of rise of the risk of death (*i.e.*, the slope of the tangent to the curve at any age) rises slowly at first, then accelerates fast above age 60.

with environmental conditions. Are there disease-causing variants of genes segregating in the population? Has every one of us by chance inherited a subset of these numerous gene variants, some of which considered in aggregate determine the average value of β for a given human subpopulation?

Could functions of some of these genes be modulated, perhaps by new drugs, to "flatten the curve" (making the tangent to the curve flatter by retuning the value of β) of death?

4.7 Chronic Diseases of Aging

Some age gracefully, remaining relatively healthy late into their lives; others are not so lucky. Aging is generally associated with a slow erosion of muscle mass, bone density, immunity, and certain cognitive skills such as the ability to acquire a new language or to retain a recent memory. Conversely, aging may increase the percentage of fat in the body through increased production of fat cells. It is also accompanied by an increased propensity to succumb to chronic diseases such as type 2 diabetes, hypertension, osteoporosis, or arthritis.

The degree to or the rate at which different people become afflicted with one or more of the numerous chronic conditions associated with aging might be partly genetically determined, and not all of these might have direct relations to telomeres.

Two main ways to think about aging besides telomeric aging are the cumulative burden of biomolecular wear and tear and the exhaustion of the somatic stem cell pools in various tissues. Moreover, aging generally sets in when the age of fecundity ends. There is possibly an evolutionary reason for this: reproductive capacity is one of the principal traits upon which natural selection acts. Therefore, most organisms past their reproductive age, dependent on the limited pool of healthy germ cells — eggs, their numbers fixed at birth in humans, and sperms, whose telomeric aging sets a limit to their number — have little selective advantage at the individual level to prolong their lives.

Some social organisms, however, including humans, might have gained an evolutionary advantage by serving as the caretaker of subsequent generations: grandparents likely played an important role in human societies by passing down lived experiences. Cycles of alternating drought and flood, whose periods exceeded an average human generation; the "songlines" of ancient migratory routes, of waterholes, "myths" of pestilences past, or of marauders across the mountains — all these factors and many more likely played a role in human survival through the collective memory that shapes the next generation's adaptive strategies. Thus, rare gene alleles that had helped extend lives of younger tribe members past the bearers' reproductive age might have been selected in early human societies. In the case of most genes, however, there would be little natural selection beyond the time of reproductive senescence because the selected traits cannot be passed on to the next generation from the survivors.[3] Evolution might therefore have had little opportunity to fix within the human populations alleles that might have given us lives vastly longer beyond the decline of our reproductive capacity. By little I mean only about 300,000 or so years, or about 10,000 human generations, since the discovery of fire, a relatively short generation time for evolution.

Are we therefore destined to gradually erode and die after the end of our reproductive competence? Which genes control the period of human reproductive capacity? What are the genes that guard against the onset of chronic diseases? These questions are intimately linked to the question of aging in humans, and perhaps both are related by the same genes! If true, can we find a chemical agent — an elixir of life — that can both extend our lives and ward off chronic diseases or other age-related decline in health and vigor?

4.7.1 *Progressive accumulation of molecular errors*

It has been speculated that at least one factor of aging may reflect a progressive accumulation of molecular errors, because all biochemical processes are error prone. For example, DNA has a relatively constant error rate during replication, which might indeed increase (i.e., have more mutations per base pair) if genes responsible for correcting these errors become themselves mutated first (see Chapter 3).

Accumulation of relatively long-lived proteins harboring oxidative damage (in its component amino acids, for example, relatively commonly, oxidation of proline residues) might be another important molecular cause of aging. Accumulation of defective proteins could in principle enhance the rate of further errors through several different mechanisms. For example, defects in protein folding or degradation machinery could cause further accumulation of defects. This is because the normal function of the protein folding machines (chaperones and chaperonins) is to prevent the accumulation of unfolded or incorrectly folded proteins and to target the misfolded proteins for removal by the protein degradation machines

[3] See Chapter 14 for a quantitative description of this process.

(proteasomes). This latter mechanism appears to be involved in progressive degenerative diseases, such as Alzheimer's disease and Parkinson's disease, among others.

A large fraction of the genomic DNA is transcribed into RNA, but most of these RNA molecules are rapidly degraded, resulting in a rapid turnover of RNA nucleotide precursor pools. Accumulation of protein synthesis/degradation errors could potentially clog the RNA turnover mechanisms and thus might choke the cellular biosynthetic apparatus.

Can the rate of accumulation of errors in RNA and protein synthesis be slowed and thus perhaps increase lifespan? Studies with animals have shown that food components rich in antioxidants are generally beneficial and do extend lifetime in these organisms under laboratory conditions. But how do these work, and can we understand the process better?

More than a century ago it was found that the metabolic rate, which is the amount of oxygen consumed and converted to carbon dioxide per unit time by respiration, correlates inversely with the lifespan of most organisms. This led to the speculation that perhaps oxidative respiration causes the production of some factor that reduces lifespan. One of the suspects is the so-called reactive oxygen species ($O_2^{\bullet -}$ or ROS), which is an oxygen molecule with an unpaired electron generated by the leakage of electrons from the cytochrome oxidase electron transport chain of mitochondria:

$$O_2 + e^- \rightarrow O_2^{\bullet -}$$

ROS is highly reactive and can readily oxidize DNA, RNA, proteins, lipids, and other biological molecules. The inverse correlation of organismic lifespan with the rate of generation of ROS is even better.

This is a paradox of life: the same oxygen that is the essence of our aerobic life is the agent that reduces our lifespan. But if so, there must be evolutionary mechanisms to cope with ROS too.

Indeed, an enzyme called superoxide dismutase, SOD in short, seems to be the main enzyme that converts ROS to hydrogen peroxide, which is subsequently converted to water by another enzyme named catalase:

$$2O_2^{\bullet -} + 2H^+ \rightarrow H_2O_2 + O_2$$
$$2H_2O_2 \rightarrow 2H_2O + O_2$$

Since ROS induces stress response pathways in most organisms, is it possible that mutant organisms that are more resistant to stress might remove ROS (e.g., by the actions of SOD and catalases) more efficiently or perhaps even produce less ROS and therefore also have a higher lifespan? If evolution has fashioned enzymes to reduce ROS, then some of the genes involved in the production and control of these enzymes must be under natural selection. Any gene under such a selection might then be identified by mutating the gene and selecting for shorter or longer lifespan.

This was exactly the case (Finkel and Holbrook 2000) with the small nematode worm *Caenorhabditis elegans*, a favorite organism for geneticists studying the nervous system because it has only 302 neurons (compared to billions in humans). When these worms were treated with mutagens and selected for stress resistance or longer lifespan, three important genes were identified: *age-1*, *daf-2*, and *daf-16*. Worms with a mutation in the *age-1* gene lived about 65% longer than normal worms. Worms with a mutation in *daf-2* lived twice as long as normal worms. Both mutants also showed greater resistance to stress from ultraviolet light, an oxidative agent called paraquat (which produces ROS), and heat. In contrast, worms with a mutation in *daf-16* showed the opposite effects. This mutation removed the stress resistance and lifespan extension seen in worms with *age-1* and *daf-2* mutations (Finkel and Holbrook 2000).

Using our evolutionary paradigm, we next ask whether the corresponding protein's sequence relatives are present in humans. You bet they are. The human homolog of the Daf-2 protein encodes the human insulin receptor protein — the cell surface protein to which insulin binds and thus increases glucose (the fuel of oxidative respiration) uptake by cells.

The insulin signaling pathway in humans is functionally analogous to the *DAF-2/DAF-16* (TGF-β signaling) pathway in *C. elegans*. Mutations in these pathways lead to increased lifespan in both humans and worms. Similarly, TGF-β signaling is conserved and plays roles in development, stress response, and aging across species.

Reducing sugars such as glucose react with proteins, which is the basis of the clinical hemoglobin HbA1c test widely conducted to assess whether someone might have a chronically high blood sugar level. It is anticipated that a chronically high blood glucose level would damage proteins, and the effects of the damage might be cumulative.

When the pancreatic hormone insulin binds to the insulin receptor on the cell surface, the cell begins to transport more glucose into the interior of the cells, especially in liver, muscle, and fat cells, and convert the sugar into stored forms or break it down to produce energy and other materials for upkeep. Could the *daf-2/daf-16* pathway help maintain sugar balance in worms, and is that why mutating this gene increases lifespan?

There are likely other mechanisms behind longevity besides sugar metabolism. The human homolog of the worm *age-1* gene encodes yet another component of a signaling pathway present on the cell membrane, and the *daf-16* human homolog (named FOXO) encodes a gene regulatory protein that appears to regulate at least a hundred genes. These include genes involved in apoptosis[4] (e.g., *BCL2L11*, *BBC3*), cell cycle arrest (e.g., *CDKN1A*, *CDKN1B*), DNA repair (e.g., *GADD45A*), and metabolic processes (e.g., *PDK4*, *G6PC1*).

The connection between genes involved in insulin signaling and metabolic processes that are regulated by TGF-β and other signaling cascades is exciting, because we now might have a handle on understanding how reactive oxygen species — a result of metabolism — might in principle be related to lifespan control. If a single gene mutation can double the nematode worm's lifespan, can we manipulate human genes to accomplish a similar feat? Should we?

4.7.2 *Progressive accumulation of inflammatory immune cells*

Surely men are somewhat different from worms, although they are evolutionary relatives. One important difference is the presence of a sophisticated immune system in humans and other mammals that are not found in the nematode worm. There is increasing evidence that immune cells that cause inflammation as a byproduct of fighting infection might also conspire to reduce the human lifespan. There is also overwhelming evidence

---

[4]Apoptosis or programmed cell death, is an internal program in all mammalian cells, which, if actuated, will kill the cell from within. See Section 4.7.3.

of a change in immune response with age — a progressive loss of immune response to vaccination and a progressive increase in markers of inflammation (Fulop et al. 2018).

It appears that with age certain white blood cells that are normally attracted to chemicals from infectious organisms lose their mobility, their ability to migrate to sites of infection, and their ability to engulf infectious organisms. Immune cells that make antibodies are depleted. Conversely, immune cells that induce inflammation by secreting certain chemicals to which other cells react by secreting chemicals responsible for cellular damage response display increased sensitivity. Aging tissues that participate in producing specific types of immune cells are affected in subtly different ways, and there is increasing evidence of DNA damage accumulation with age that seemingly accelerates this process of immune senescence (Li et al. 2023).

Intriguingly, immune-related aging appears to be dependent on the composition and prevalence of intestinal as well as skin-borne microorganisms. The composition of skin microbiome (the community of microbes) correlates rather accurately with a person's chronological age (Huang et al. 2020). Intestinal microbiome composition appears to correlate with a number of chronic disease states as well as with the longevity of people who have already lived beyond the median age of their respective ethnic communities (Li et al. 2023). These recent findings have prompted a hunt for the role of microbial flora associated with human aging. We humans appear to be more intimate partners of a complex ecosystem than we had previously appreciated, and evidently the evolution of human longevity has influenced and is being influenced by the larger community of organisms that live with, upon, or within us. As always with science, however, these are still at the level of correlation — we need further investigations to determine causation.

4.7.3 *Progressive depletion of stem cells*

Mitochondrial damage due to the liberation of oxidative free radicals during aerobic respiration (which occurs within the mitochondria) is a frequent source of cellular dysfunction, which ultimately leads to the breakdown of the mitochondrial membrane followed by the liberation of cytochrome oxidase into the cytoplasm. The latter has the consequence of triggering the activation of a series of protein-degrading enzymes known collectively as caspases, which kill the cell from within — a phenomenon known as programmed cell death or apoptosis. Progressive removal of cells from organs due to apoptosis depletes organ volume as well as the pool of somatic stem cells in these organs, both of which contribute to organismic aging, leading to the exponential rise in the probability of death.

If a chemical agent can reduce the progression of age-related decline, it must target a genetic basis that might have been selected in evolution for determining the average lifespan. How conserved such genetic processes might have been over evolution?

Again, recent studies with the nematode worm *C. elegans* have indicated that metformin, a widely used first-line antidiabetic medication, has a demonstrable effect on delaying aging and age-related decline in muscle performance of nematodes (Onken et al. 2022); however, the effect is dependent on the genetic background of the worms tested. The genetic diversity in the panel of worms that were tested is thought to be higher than the genetic diversity that exists in human populations (Teterina et al. 2022). Therefore, it is quite possible that if metformin has a similar lifespan extension activity in humans, the genes that account for its action should be identifiable. Metformin acts through complex, still unclear, mechanisms by modulating energy metabolism in cells (Foretz et al. 2014; Kukidome et al. 2006; Pernicova and Korbonits 2014), but it also has some adverse side effects. As is clear from the above, its effects on lifespan, if any, will likely be dependent on genetic factors that might or might not be inherited by an individual. Thus, it is unlikely that every individual will benefit from such a drug.

4.7.4 *Parabiosis, platelets, and the aging brain*

Recent studies have explored the potential of young mice blood to counteract aging in older mice. In one approach, the circulatory systems of young and old mice are connected surgically — a technique known as heterochronic parabiosis — allowing them to share blood over several months (Zhang et al. 2023). This procedure led to notable rejuvenation in older mice, including extended lifespans and improved biomarkers associated with aging, while the younger mice continued to behave their real age. What was in the young blood that reversed aging?

Another study focused on small extracellular vesicles (sEVs), tiny vesicles surrounded by plasma membranes enclosing some molecular components, including signaling proteins and RNA, present in the plasma of young mice (Chen et al. 2024). When these sEVs were introduced into older mice, they enhanced mitochondrial energy metabolism, effectively reversing age-related functional decline. How exactly they might be exerting their age-reversing effects is currently unknown.

Inflammation of the brain and associated neural processes accompanies aging in several different ethnic populations (Butts et al. 2024). Blood platelets, the cells that are responsible for the clotting of blood when an injury occurs are emerging as an important factor in age-related inflammation of the brain (Rodriguez Moore et al. 2024).

When a protein called CXCL4, released by platelet cells in young mice, was introduced into older mice, it slowed down aging-related changes in the hippocampus. This brain region, which is often one of the first to be affected by age-related degeneration, a showed improvements, and cognitive decline in older mice was reversed (Schroer et al. 2023). The signaling molecule CXCL4 is part of the chemokine family of cell signaling molecules that promote the migration of immune cells (such as the white blood cells) to sites of infection, inflammation, or injury. The term "chemokine" derives from their function as "chemoattractant cytokines," meaning they guide cells by generating a chemical gradient for movement, a process called "chemotaxis." Whereas genes encoding chemokines are widely conserved in vertebrates, this gene family appears to be rapidly evolving in the mammalian lineages.

These results imply that human longevity might be understandable and, per chance, extendable beyond that which has been fashioned over at least 200,000 years of human prehistory.

In summary, from a discussion of factors that influence human mortality, we noted that there appears to be a biological limit to human lifespan. At the level of cells and genes, aging is intimately related to the progressive erosion of the ends of chromosomes; however, cellular aging is not the same as organismal aging. Organismal aging is influenced by genes as well as environments, with genes playing significant roles through modulating age-related diseases, including the propensity for inflammation. Current research activities are centered on understanding the molecular basis of aging.

Chapter 5

Evolution Within Us

A remarkable property of the human immune system is that it can somehow generate immunity, if given a fair chance, to any invading organism. While it remains in the realm of science fiction, I bet it can develop immunity against a Martian pathogen, were one to exist and infect a human, if the hapless astronaut were given sufficient time and opportunity. What kind of chance? A vaccine, for example, is a small dose of a portion of the pathogen's constituent body part that is almost certain to be some kind of polymeric compound with a specific shape required for performing a certain vital function for the Martian. The question then is the following: how does the human body produce immunity against these foreign molecules — how does it recognize something as foreign, and how does it mount a defense against it? As it appears, the human immune system is an evolving system that works by generating random variant genes, whose protein products, antibodies, may or may not bind to the foreign bodies. Binding increases the chance of occluding and thus preventing the foreign molecules from functioning. If they don't bind, their corresponding variant genes are eliminated — if they do bind, then the corresponding antibody genes are amplified and further variants are produced. Those that bind more tightly, their genes are even further amplified. This, in principle, is Darwinian evolution by the independent processes of variation and natural selection occurring within our bodies. This chapter describes this story.

***

Important terms:
Human or mammalian immunity to foreign invaders has two components: *innate immunity* and *adaptive immunity*. *Innate immunity* is the primary line of defense in the human body against pathogens, which is due to a group of cells circulating in the blood which can recognize the surface features of invading organisms, such as viruses and bacteria, and engulf them, followed by digesting these invaders. Some of these innate immune cells, such as the *natural killer (NK) cells*, do not directly engulf the pathogenic entities, but instead, they recognize host cells that are compromised by invading pathogens and destroy these compromised cells. They also produce localized inflammation, which recruits additional immune cells to the site of infection. *Adaptive immunity*, by contrast, requires a prior exposure to the pathogens. In response to the first exposure, the adaptive immune cells learn to recognize the pathogen and retain a memory of that pathogen. When a second exposure to a similar pathogen occurs, some adaptive immune cells release a neutralizing reagent called an *antibody*. Antibodies are proteins tailor-made to specifically bind to the *antigen*, i.e., the foreign molecule (usually a protein, but can be any large molecule, including DNA, RNA, carbohydrate, or even a nonbiological molecule with a sufficiently complex surface structure). By binding specifically and tightly to the foreign molecule, antibodies effectively neutralize the latter. *Lymphocytes* are a group of white blood cells that can be broadly divided into two groups, *B cells* and *T cells*. B cells make antibodies through maturation to *plasma cells*, primarily in the bone marrow. T cells mature in the thymus and do not differentiate into antibody-secreting cells. Instead, they are part of the cellular immunity: by recognizing through their receptors foreign or abnormal cells, they kill the latter. T cells are of three types. *Helper T cells* communicate with B cells through secreting chemical signals (*cytokines*) and activate the latter to begin to make antibodies; *cytotoxic T cells* kill infected cells and/or cancer cells; *regulatory T cells* prevent autoimmunity by suppressing overactive immune reactions. A *receptor* is a protein that sticks out on the

surface of a cell to which a specific molecule (a *ligand*) binds and causes a biological effect on the cell that displays such a receptor. A *BCR* is one to which an antigen might bind and cause the B cell to proliferate. A *clone* is a collection of genetically identical descendants of a defined cell. *Monoclonal* refers to molecules (e.g., antibodies) that are derived from a single lineage of clones (of B cells); *polyclonal* defines multiple clonal origins.

In our imaginary parallel universe, two gentlemen were seated by a low table on rattan chairs. A woman in a dark gown entered presently and took the empty seat. This time, our characters are the Danish physicist Niels Bohr, who was considered the father of the atomic structure; Madame Curie, the Polish physical chemist who discovered radioactivity; and Macfarlane Burnet, the Australian virologist/immunologist who first proposed that the human immune system evolves in response to infections by viruses and bacteria.

Bohr: Madame, you are all wet!

Curie: Never mind Niels. I am used to it. My hair will dry out momentarily. It's only a drizzle. Young Burnett, tell me, what is exciting in the world of immunology . . . I find it fascinating that no matter what the intruder molecule is, the body always finds a reagent to neutralize it! It doesn't make sense to a chemist.

Burnett: Indeed. Chemistry can only explain part of it — the part of the exact physical chemistry of how the neutralization works. The chemical principle, as you well know, is simply the fact that two molecules bind to each other because of complementarity, which Professor Bohr had speculated a long time ago. The surfaces of the two molecules — that of the invader molecule and the defensive molecule — are complementary in the sense of a key is to its hole in the lock, or a plaster-cast is to a statue.

Bohr: Precisely. There are interatomic forces between the molecular surfaces that fit together precisely. Immunologists call the invaders, the antigens and the defensive molecules antibodies. In the specific interaction between an antigen and its cognate antibody, geometrical shapes, electrical and thermodynamic factors all play important roles. But what chemistry doesn't explain is how the immune system can raise an antibody against any, like millions or billions, of different invading molecules. How could the genome hold so much information?

Burnett: You are quite right Professor Bohr; chemistry alone can't explain it. The genes that we get from our parents can't possibly provide us with the imprint of billions of different antibodies that are potentially needed to neutralize all the possible foreign invading molecules. What happens is that the immune system is like a Darwinian world of its own. Cells evolve new varieties of gene sequences, every such immune cell making a unique type of gene sequence at random, and one of these by sheer chance makes an antibody that binds tightly to an antigen. When that happens, the cell that makes that specific antibody becomes more evolutionarily fit — it produces more progeny cells, all of which make the same antibody. So, in a little while you get a lot the same evolved cells that together make a whopping amount of the neutralizing antibodies against the invading antigen.

Curie: Fascinating. As fascinating as radioactivity was in my time. I hear that you can now make antibodies against nearly anything! What can you do with them?

Burnett: You can make vaccines against any infectious disease — so far it has been successful against most infectious diseases except HIV-AIDS. But you can also make antibodies against cancer cells.

Bohr: How does that work?

Burnett: Some diseased cells have surface proteins that are not found in normal cells. So, you can make antibodies that can target these cancer antigens. For example, nowadays an antibody is available which targets the abnormal immune cells of aplastic anemia. This medicine is called anti-thymocyte globulin or ATG for short.

Curie: Very interesting. Only if ATG was available in my time, perhaps I could have been with my granddaughter Hélène a little while longer. Oh well . . .

Outside, it rained hard and a canvas awning flapped in the wind.

***

I was a second-year grad student when I met Sir McFarlane Burnet (1899–1985). He had won a Nobel Prize several decades ago, and he showed up to give a talk when he unveiled our shiny new department building at Monash University, near Melbourne, Australia. He decided to swing by our lab for a visit. He was this tiny, energetic, friendly man, already in his 80s, who had retired by then after an amazing career as a virologist and an immunologist.

We gave him a quick rundown of what we were working on, and he threw in a few questions and some chitchat. Once he departed, my advisor Ron Skurray, with a cheeky grin, said something like, "*I'd be lucky if I had even a tiny bit of his brainpower left if I hit his age!*"

For the first time in the late 1950s and 1960s, Burnett and independently, Niels Jerne (1911–1994) had provided a hypothesis, the clonal selection theory, unifying Darwinian evolution with how we humans fight infectious diseases. Coupled with these insights, pioneering efforts in the 1960s and 1970s by Cesar Milstein (1927–2002) demonstrated the genetic basis of practically limitless kinds of variant antibody molecules that can be produced within the immune cells. Strikingly, these researchers provided a mechanistic explanation for evolution by natural selection occurring within our individual bodies — as if an individual human (or any multicellular organism) is but cells that have come together in an interacting community to make us who we are. These cells play by the same rules that nature engages to fashion the individuals of different species of organisms.

This idea — the clonal selection theory of antibody production — has been demonstrated to be true beyond doubt and has become a linchpin of modern medicine with impacts ranging from vaccine development to personalized treatment of cancer. Multibillion-dollar pharmaceutical companies, including the likes of Amgen, Bristol Myers Squibb, Johnson & Johnson, Hoffman-LaRoche, Regeneron, and Bayer, have drugs for previously untreatable diseases made of monoclonal antibodies, which were first conceived and constructed by Milstein. The idea has also had a far-reaching impact on other aspects of biotechnology, including the making of artificial enzymes for medicinal and medical research use.

In a nutshell, Burnett's and Jerne's clonal selection theory is Darwinian evolution of antibody-producing cells within the human body. Starting with a limited diversity of antibody genes inherited from the parents, the immune cells generate further genetic diversity by processes of hypermutation and DNA recombination confined to a few sections of this limited set of inherited antibody genes, based on the previously mildly successful antibody sequences.

Each of these cells expresses on its surface the corresponding antibody molecule, but with a twist — this form of the antibody proteins is membrane-bound. These proteins project out of the cell surface like antennae, ready to bind to a foreign protein if the latter has a "complementary" protein surface (we still don't understand the exact rules of this apparent complementarity, though we do know a lot). Some of these immune cells just happen to bind to a foreign antigen, first only weakly. This binding activates these immune cells to generate further mutations within the antibody coding genes (termed somatic hypermutation) and to clonally proliferate (akin to natural selection) and generate further antibody protein variants by producing further sequence variations. Those cells with new variant genes, producing the corresponding antibody protein (cell surface bound) that happen to bind with higher affinity to the antigen survive and proliferate more and those with lesser affinity are killed — survival of the fittest in cellular terms (Fig. 5.1). In this manner, the antibody sequences migrate over the fitness landscape (Plate 2, Part 1) to a higher fitness peak: fitness here is equivalent to the extent of cell proliferation proportional to the binding strength of the circulating antigen to the cell surface antibody. Thus, a "generalist" antibody sequence with mild affinity eventually improves to sequences with a higher and higher fitness values (higher affinities) by the occurrence of further variations built upon the former.

5.1 B-Lymphocytes Make Antibodies

Circulating within the blood is a type of white blood cells called lymphocytes, which fight foreign invaders and constitute the main components of our adaptive immune system. The invaders — viruses, bacteria, or any other foreign organisms, or inert materials such as proteins and other complex macromolecules such as complex carbohydrates and nucleic acids — induce lymphocytes to either engulf and digest these or to produce antibody molecules that bind to these organisms or macromolecules to neutralize

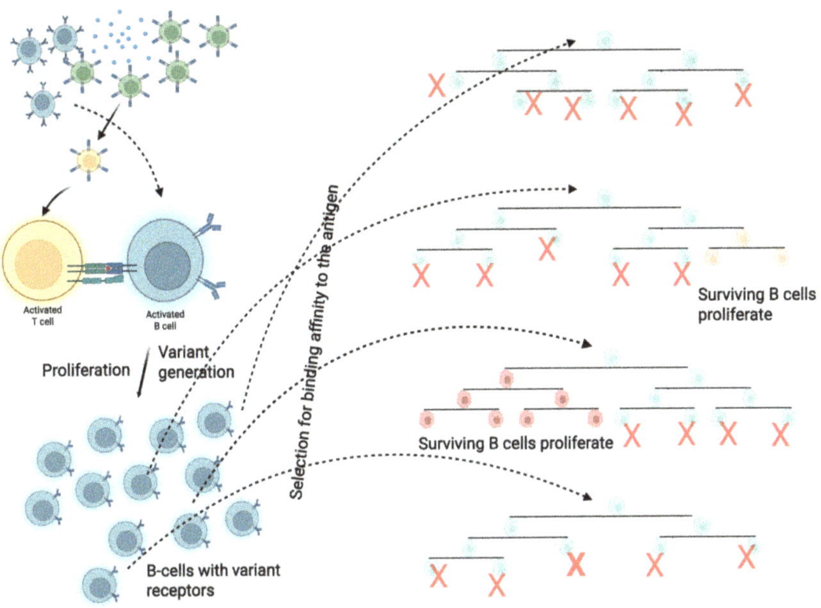

Figure 5.1 A schematic summary of main events in adaptive immune response. The immune cells expressing variant cell surface receptors that bind weakly to a foreign antigen (T and B cells) become stimulated by the binding of the antigen molecules to the surface receptors. These cells cooperate to intensify the stimulation. The stimulated cells proliferate and generate further diversity in receptor genes. Those cells with receptor genes that make weak-binding receptors to the antigen are killed, whereas those cells expressing high-affinity receptors survive and further proliferate. These then differentiate to make antibody-producing cells. [This figure was drawn using Biorender.com under an academic license]

these. These mechanisms are distinct from the innate immunity conferred by the ability of certain other white blood cells with invariant (and inherited) surface receptors to recognize and engulf invading organisms or to stimulate the adaptive immune cells.

Let us remember that the ability of one macromolecule, such as a protein, to "recognize" another implies that these molecules selectively bind to each other. This binding ability requires complementary surfaces, which include topological shape complementarity, complementarity in surface charge distribution, and complementarity of weak interactions (hydrogen bonding, van der Waal's interaction, hydrophobic–hydrophobic

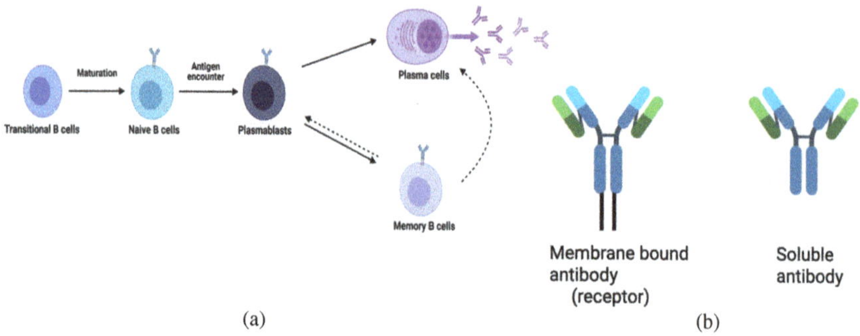

Figure 5.2 Summary of antibody production by B cells. (a) Naïve B cells display BCRs, each with a variant structure (hardcoded in the genome). The receptor binds to an antigen, which signals the cell to produce further genetic diversity due to mutation in the receptor gene, and the resulting clone of cells is selected further to proliferate in response to tighter binding to the antigen. This "maturation" process ultimately leads to differentiation of the B cell to plasma cells that produce the soluble version of that antibody. Memory B cells are the long-lived B cells with mature receptors (B cell receptors or BCRs), which can differentiate into plasma cells if the same antigen is encountered later. (b) membrane-bound and soluble forms of the antibody. [Panel A is adapted from Upasani, Rodenhuis-Zybert, and Cantaert (2021) under the Creative Commons Attributional License agreement.]

interactions, etc.). Since the diversity of invading macromolecules could be vast, practically infinite, how is it that our adaptive immune system, namely the lymphocytes, can recognize any of these macromolecules as nonself or foreign entities?

B cells, a major class of lymphocytes, produce antibodies, which are protein molecules fashioned to specifically bind and neutralize a foreign macromolecule, the antigen (Fig. 5.2; Upasani, Rodenhuis-Zybert, and Cantaert 2021).

Unless the individual had previously encountered a foreign antigen, there is usually no sign that a specific antibody for neutralizing that antigen exists in that individual. Upon prior exposure to the foreign antigen, specific and selective antibody molecules appear in the bloodstream. Upon a future exposure to the same antigen, more antibodies of the same type are produced, and these help to neutralize the antigen.

Since the possible diversity of potential foreign antigens vastly outnumbers the genetic diversity inherited from parents, antibody molecules with the required specificity must be generated *de novo* upon exposure to the foreign macromolecule. The immune system learns to fashion the exact antibody molecules in response to an antigen.

A group of blood cells cooperates to enable this process of immune learning. As the first step, the antigen protein (let's assume it is a protein, although nucleic acids or complex carbohydrate molecules can also serve as antigens) is engulfed by specialized antigen-presenting cells or APCs, which include macrophages and dendritic cells circulating in the blood. APCs cleave the foreign antigens into small fragments and ferry these fragments to their cell surface — a process known as antigen presentation (Box 5.1; Fig. 5.3).

Box 5.1

Antigen presentation involves the following steps:

1. Antigen Capture: The APCs internalize exogenous antigens through various mechanisms, such as phagocytosis, pinocytosis, or receptor-mediated endocytosis.
2. Antigen Processing: Within the endosomes and lysosomes of the APC, the internalized antigens are degraded into smaller peptide fragments through enzymatic cleavage.
3. Peptide Loading: Specialized molecules, known by the fancy name MHC class II, are in the endosomal and lysosomal compartments of the APC. The processed peptide fragments then bind to the MHC class II molecules in these compartments, forming stable MHC class II-peptide complexes.
4. Migration to the Cell Surface: The MHC class II-peptide complexes are transported to the cell surface of the APC, where they are displayed for recognition by another class of lymphocytes called helper T cells.
5. The TCR on helper T cells interacts with both the peptide fragment and the MHC class II molecule on the surface of the APC. The binding of T cell receptors or (TCRs) to both components of the MHC class II-peptide complex is essential for the activation of helper T cells, which, in turn, plays a central role in coordinating and regulating the immune response to the antigen.

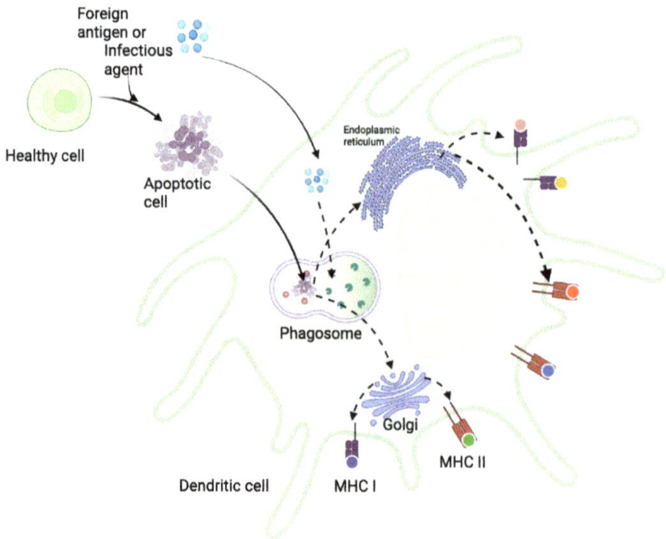

(a)

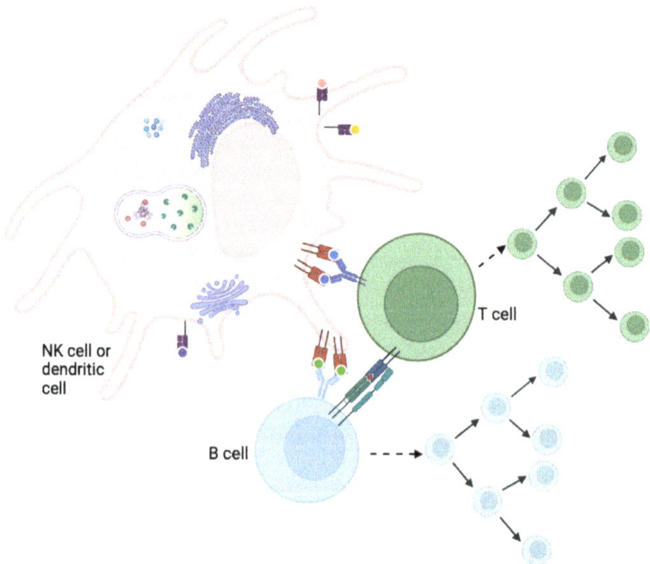

(b)

Figure 5.3 (*Continued*)

The APCs thus serve as the chefs, as it were, by processing the foreign antigens into little chunks and displaying these chunks on the surface for other cells, such as helper T cells (see definition in the Important Terms Box), to interact with these chunks of proteins. The interaction is mediated primarily by the displayed peptides binding to a special group of receptor proteins on the T cell surface, aided by a series of additional membrane-bound proteins and secretory signals emitted by these cells (Fig. 5.3). As a result of this interaction, the T cell becomes "activated," such that the helper T cell begins to secrete signaling molecules to stimulate B cells, which also are attracted to this "conference of cells," to proliferate.

The question then arises as to how a helper T cell "knows" that the fragment displayed on the APC surface is a fragment of a foreign antigen and how it leads to a specific type of B cell for subsequent activation. Unless first exposed to an antigen as an initially naïve B cell, how would a T cell receptor (TCR) have a complementary surface to which the foreign antigen might bind? The answer appears to be the ingredient on which Darwinian natural selection and evolution occur.

5.2 Antibody Diversity Is Immense

Genes that encode antibody molecules are the same as those that make the T and B cell receptor (BCR) proteins. An individual inherits one set of BCR genes consisting of one heavy chain gene and two light chain genes, one set of TCR genes consisting of one set of TCR-alpha genes, and one set of TCR-beta genes, from each parent. Every heavy, light, alpha, and

Figure 5.3 Summary of T and B cell activation. See text for a description. (a) Activation of dendritic cells for antigen presentation. Viruses or other foreign molecules, or even host cells that have been compromised by the infection of pathogens, are engulfed by dendritic cells. The englufed engulfed contents are digested by intracellular organelles called phagosomes, and short segments of foreign protein fragments are then transported to the cell membrane in complex with receptor proteins termed MHC I and MHC II adapters. The foreign protein fragments are thus displayed by the dendritic cells. (b) The dendritic cells (or the NK cells which perform similar functions) interact with B and T cells through their MHC I and II adaptors displaying the foreign protein fragments. In response, the B and T cells are triggered to divide and diversify their antibody coding genes. See Box 5.1 for elaboration. [This figure was drawn using Biorender.com under an academic license]

beta gene segment contains four regions of interest for us: a constant region that has a particular sequence that does not change during somatic cell division; 40 to 45 segments of the variable (*V*) chain gene region associated with BCR-heavy and TCR-beta; 30 to 40 segments of variable (*V*) chain gene region associated with BCR-light and TCR-alpha; a smaller number of Diversity (*D*) region associated with each of BCR-heavy and TCR-beta; and a small number of Joining (*J*) region associated with BCR-light and TCR-alpha (Fig. 5.4; Market and Papavasiliou 2003). Thus, from our parents, we inherit a foundation of approximately 45 × 2 × 40 × 2 = 7,200 different variations in the antibody repertoire. How do we get further gene variants based on this foundation?

Somatic intrachromosomal recombination occurs between the repeated copies of *V* and *J* gene segments, producing expansion and contraction of the number of *V(D)J* sequences in clones of descendant cells within an individual. For any individual, the number of these gene copies in somatic cells is, therefore, variable. Finally, the *D* region undergoes somatic hypermutation in response to T and B cell activation, leading to further variation in antibody gene sequences. In every clone of B and T cells, therefore, there is a unique pattern of *V(D)J* heavy/beta chain joined to a unique pattern of *V/J* light/alpha chain. It has been estimated that the total diversity thus generated could potentially form at least 10^{15}, likely as many as 10^{18}, unique antibody molecules. Let us delve a little deeper into how that feat is achieved.

5.3 Darwinian Natural Selection on B-Lymphocytes

Due to the immense diversity of TCRs, each of which is made up of a *V(D)J* region (Fig. 5.4) of a beta chain gene and a *VJ* region of an alpha chain gene, with at least a few million to a hundred million different variable sequences, there are always some receptor molecules encoded by a clone of cells descending from a single T cell, which are able to bind at least with a low affinity to a foreign antigen fragment displayed on an APC surface. This low-affinity binding of APC with a T cell is augmented by other cell surface molecules, which also trigger the secretion of cell signaling molecules by the APC as well as the T cell that are clustered together by surface attachment. Thus activated, the T cell progeny,

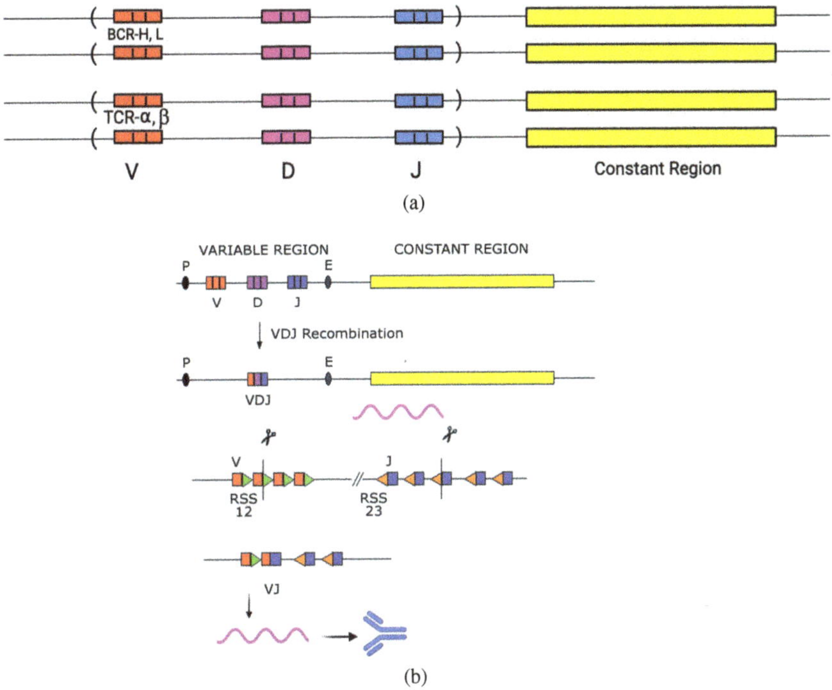

Figure 5.4 Schematic diagram of immunoglobulin coding regions and *V(D)J* recombination during maturation of B and T cells. (a) Each person inherits two sets of BCR loci and two sets of TCR loci. (b) DNA sequences at *V*, *D*, and *J* regions are partially homologous, enabling recombination to occur in two steps. For the heavy chain, *VDJ* regions are all present; for the light chain, only a *VJ* recombinant junction is present. P is the promoter, and E is the enhancer of transcription. [Panel (b) is adapted from Market and Papavasilou (2003) under the Creative Commons Attributional License agreement.]

secreting signaling molecules while proliferating, migrate into lymphatic follicles where B cells reside.

Naïve B cells, meanwhile, have also encountered the foreign antigen. These cells also have APC-like activity, by which they internalize and chop up the antigen and display these fragments to the cell surface. The activated T cells that have migrated into the follicles now recruit follicular dendritic cells and the B cells with surface display of the same antigen fragments that had activated those T cells. This leads the follicular dendritic cells and the T cells to activate the B cell to undergo further mutagenesis targeted at the *D*-region of its antibody gene.

Very interestingly, mutations do not occur willy-nilly in the B cell genome. There are highly targeted mutagenesis events at a chromosomal locus within the antibody gene that produces the initially low-affinity antibody. The mechanism by which this occurs is fascinating, and there are some hints that this mammalian system of immune-diversity process might have evolved from lowly bacteria! First, a DNA cleavage enzyme cuts the chromosome over the *D* region to produce a broken chromosome with two ends. This is followed by error-prone rejoining of the two DNA ends. The error, which randomly introduces and/or deletes nucleotides at the *V(D)J* junction, generates further diversity.

There is an additional step when mutations occur in the variable chain coding region. An enzyme is activated during the B cell maturation process that deaminates cytosine residues targeted to the variable chain coding sequences, producing uracil and thus producing U:G mismatches. If the mismatches persist during DNA replication, adenine is incorporated against U, causing in the next replication cycle a T:A base pairing (thus, a C to T mutation). These mutations produce further variations upon the major variations initiated by the DNA cleavage at the junction, producing further variant subclones of B cells.

Meanwhile, cells that do not make a productive or higher-affinity antibody undergo death because of a signal of death emitted by the activated T cells and the follicular dendritic cells in the complex. By contrast, those B cells that happen to generate by the random mutagenesis targeted at the D locus an antibody variable chain protein that binds more tightly to the displayed antigen can overcome the signal of cell death through activating death-suppressive pathways. These simultaneously turn on a program of rapid cell proliferation.

This cycle of "affinity maturation" of the *V(D)J* junction sequence for its encoded antibody sequences continues for several rounds of replication, leading to the survival of the fittest — the death defying, rapidly proliferative — B cells, and to a final genetic rearrangement in which the antibody gene "matures" to encode a secreted form of the protein.

Sometimes, these B cells differentiate into another cellular state called plasma cells or into the memory B cells for longer-term preservation of the immune memory.

The above is a simplified summary of what we know today about how antibodies are produced. Central to this process is Darwinian natural selection — the clonal selection process that was first proposed by Burnet and, independently, by Jerne, in simplified and speculative form.

During her lifetime, an individual generates a vast array of antibodies against many different foreign antigens. We speak of the antibody repertoire of an individual, which refers to the diverse collection of antibodies produced by the immune system in response to exposure to various foreign antigens. The antibody repertoire is a critical aspect of the adaptive immune response of an individual and plays a central role in defending the body against pathogens such as bacteria, viruses, and other harmful agents. Over time, this continuous exposure to antigens and the subsequent production of specific antibodies result in an individual's antibody repertoire becoming more extensive and specialized. Every person develops a unique antibody repertoire, which reflects the individual's history of exposure to various pathogens and environmental factors. The diversity and specificity of an individual's antibody repertoire allow the immune system to mount a targeted and effective response to protect against a wide range of potential threats to health. Conversely, if the antibody repertoire is not sufficiently diverse, then the person remains susceptible to infectious diseases of many kinds.

5.4 Immune System in Biotechnology

The natural diversity of antibodies and the ease with which they can be selected by using the natural diversity generation apparatus of mammalian spleen cells was immediately translated to human use in the form of two main technologies. The first was the monoclonal antibody technology pioneered by Cesar Milstein and his then postdoctoral researcher Georges J. F. Köhler (1946–1995) (Köhler and Milstein 1975; Leavy 2016). The two shared the Nobel with Jerne in 1984.

A monoclonal antibody is a unique antibody that binds with high selectivity and affinity to a precise region of a target molecule. It can be raised readily in the laboratory by injecting a target molecule into the mammalian bloodstream (e.g., mice), followed by isolating lymph cells that secrete these antibodies, separating the single cells, and propagating the descendants of each of these "clones" of cells in culture to generate a single antibody type. This contrasts with polyclonal antibodies that constitute a collection of multiple "clonotypes," each of which is a unique antibody, but this mixture of antibodies constitutes molecules with many different variable chain sequences. Monoclonal antibodies are useful both as reagents for highly sensitive and accurate detection of proteins (e.g., cancer-related proteins in cancer cells) and as therapeutic agents to kill unwanted cells, including cancer cells. Monoclonal antibodies also serve as the first line of defense against emerging viral diseases such as Ebola or COVID before vaccines can be developed (Kelley 2020). In this latter application, a preparation of a unique monoclonal antibody acts like a guided torpedo that precisely aims at a target protein of the virus and neutralizes it. Some of these monoclonal antibodies are also useful as highly sensitive and specific reagents for the detection of viruses, bacteria, or other agents. Some are used to bind and thus occlude or, in some cases, inactivate disease-specific cell surface proteins, such as abnormal antigens that mark the surfaces of cancer cells. Sometimes, they are used to fight off the body's own immune system — the first successful antibody drug was Muromonab, which was introduced in 1986 to prevent immune rejection of tissue grafts in humans. Autoimmune diseases have since become a target of monoclonal antibody therapeutics (Chan and Carter 2010).

To manufacture monoclonal antibodies, one nowadays clones the encoding gene sequences of the respective antibodies in a test tube. These genes are then introduced into cells, which are cultured in large volumes and are made to produce industrial quantities of these monoclonal antibodies.

Intriguingly, despite their surface electrical charges, some antibodies possess the capacity for cellular internalization, thereby extending their reach to intracellular antigens emblematic of pathological cellular states. This useful feature equips them with the ability to neutralize and eliminate diseased cells by targeting intracellular components, effectively disrupting their pathological progression (Leavy 2010).

Perhaps the most remarkable achievement lies in the realm of cancer therapy, wherein monoclonal antibody treatments of certain cancers have truly caused a paradigm shift (Beck et al. 2010). By improving the prognosis of certain forms of cancer from a once fatal verdict into a manageable chronic condition and occasionally evoking complete remission, these therapeutic agents have redefined the clinical landscape. Yet another use of monoclonal antibodies was prompted by the idea that a subset of the protein sequence diversity could indeed have a wider biological function than merely binding to an antigen — could antibody variable chain regions be evolved to embody enzyme activities?

Indeed, catalytic antibody enzymes, or Abzymes, occur in nature, especially in disease conditions such as lupus. Antibodies have now been successfully engineered by selecting for a desired enzyme activity through successive rounds of selection and amplification resembling the Darwinian natural selection process, but the entire process has been carried out in the test tube (Guo et al. 2006; Hanson, Nishiyama, and Paul 2005; Karlstrom et al. 2000; Rader et al. 2003). As we already mentioned, the possible variant space of antibody molecules is about 10^{15} to 10^{18}. However, direct measurement of natural antibody diversity by sequencing has shown that the practically achieved diversity is significantly smaller — approaching only 10^{9} (Briney et al. 2019). An open question is whether the unexplored diversity space, if achieved synthetically, could provide us with novel functional molecules not yet explored by natural selection.

A fascinating recent therapeutic use of immune cells is the so-called chimeric antigen receptor (CAR) T cells for unleashing killer (cytotoxic) T cells against malignant hematological cancer cells (Fig. 5.5). These cancer cells (e.g., acute lymphoblastic leukemia, multiple myeloma, B cell non-Hodgkin's lymphoma, etc.) express certain antigens on their surfaces. Antibody genes that encode parts of highly specific antibody proteins against one or more of these cancer-cell antigens are artificially synthesized in the test tube and fused to a cell surface signaling protein, making the gene encoding the CAR, by genetic engineering. The cancer patient's own normal (i.e., noncancerous) cytotoxic T cells are first isolated, and the engineered gene is then inserted into these T cells. These engineered T cells (the CAR T-cells) are allowed to divide many times in cell cultures, and then the patient receives an infusion of these cells. The infused CAR T-cells find the patient's own cancer cells and kill the latter. The killing is initiated by specific cancer-antigen binding to the CAR antibody, which activates intracellular signaling in the engineered

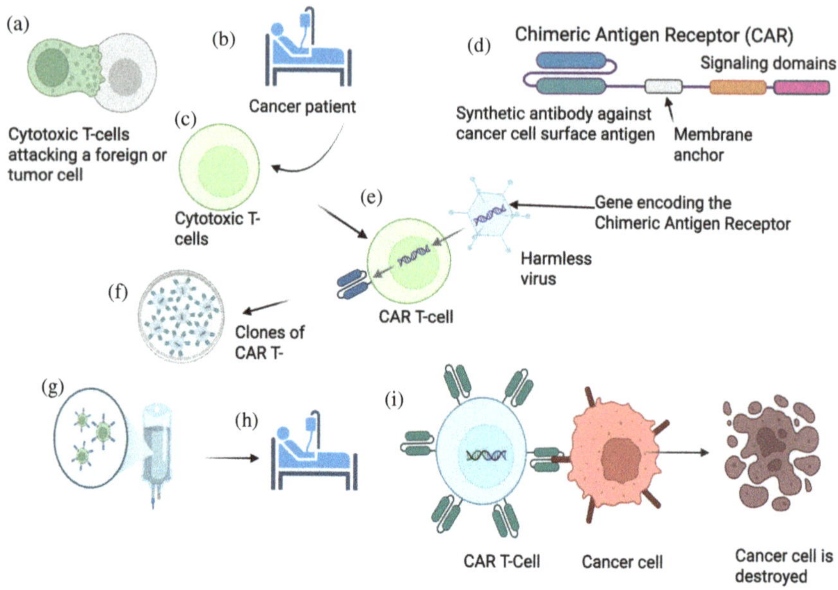

Figure 5.5 CAR T-cell therapy for blood cancer. (a) A specialized T cell lymphocyte displays receptor proteins on its surface, which allow it to bind to antigens on foreign or tumor cells. This binding triggers a signal within the T cell that allows the T cell to kill the other cell. (b, c) A cancer patient's own T cells are collected. (d) A gene encoding the CAR, which has an antibody made against the cancer cell surface antigen (blue), a membrane anchor domain (gray), and intracellular signaling domains (brown, purple) is constructed, and (e) inserted into the patient's T cells by delivering the gene through a harmless virus. (f) The T cells that now express the CAR are proliferated in cultures, activated for cytotoxic properties, and then infused into the same cancer patient. (i) The CAR T-cells now attack the cancer cells displaying the antigens that bind to the CAR and kill the cancer cells. [This figure was drawn using Biorender.com under an academic license]

cytotoxic T cells. These immunotherapies are "living drugs" personalized to the individual patients. Over the past decade in developed countries, this form of immunotherapy has become relatively common for treating aggressive hematological cancer.

Within a few decades of their discovery, the combined industrial use of antibody reagents and therapeutics has morphed into hundreds of billion-dollar industries. Darwin thus has been influential not only on our philosophy but also on our health and our pocketbook. It is important to note here that the public had invested in the conduct of research by a few bold and imaginative thinkers in a few countries, who were primarily motivated by curiosity — how to explain the puzzle of the enormous diversity of the human immune system — by funding them through a small fraction of our tax money. It is hard to determine exactly how much the total funding was in the 1950s through the 1980s to these "discovery-research" activities that were conducted in a few universities and institutions of basic science by a handful of individuals. Some of these early events were also catalyzed by social upheavals during the cold war era — Milstein had to leave Argentina where his colleagues were being politically persecuted for their liberal views, and found a receptive environment in Cambridge, England. In any event, it was likely no more than 100 million US dollars (in current value), probably a fifth as much, spread over about six different countries. By contrast, in 2023 alone, monoclonal antibody technology was valued at about 220 billion US dollars ("Monoclonal Antibodies Market Size & Share Report, 2030," n.d.). That is over 2,000-fold direct return on investment without counting the indirect effects on the economy that the saving of productive lives contributed, which otherwise would have been lost.

Such is the return of public investment in basic research.

Chapter 6

Measuring Evolutionary Conservation

In previous chapters we have seen how certain instinctive notions of evolutionary biology had impacted the discovery of telomeres and telomerase and how experiments guided by the genealogy of vertebrate phyla made possible the advance of a revolutionary idea in cancer biology. We have also seen how even within the human body, defensive genes evolve to tackle invading organisms. These understandings, and the consequent impacts on medicine and biotechnology, are built upon biological instincts honed on evolutionary theory. Progress in evolutionary biology, however, has contributed much more than merely training the intuitions of biologists. Science is largely about measurement. Many things can appear to be possible intuitively, but upon precise mathematical formulation, science can be very powerful in rejecting some implausible ideas and extending the range of the plausible ones. In this chapter, we will dive into a few important concepts that allow us to estimate crucial parameters such as evolutionary distance between a pair of proteins or a pair of genes. We will show with quantitative precision that our intuitions about evolution are consistent with many unrelated observations.

***

In our parallel universe, a Victorian-era drawing room is adorned with curtains and oak furniture. JBS Haldane, a tall man built like a boxer, who was one of the smartest biologists in the past century, reclines in an armchair; a man in a Hawaiian shirt and a pair of shorts, Stephen Jay Gould, an evolutionary biologist and a prolific writer on evolution, gesticulates animatedly; and a slim man in dark tailcoats, T. H. Huxley, a biologist from the 19th century who was once called "Darwin's bulldog," paces the room restlessly.

Haldane: Ah, the wonders of modern science. DNA and protein sequences have given us a ruler for evolutionary distance. In my day, we measured such distances by fossils and morphology. Now, it's all about sequence similarity.

Gould: Indeed, J.B.S., but I'd argue the tree of life is more than just homology. Evolution is the grand history of interrelated life. Homology tells us *how* things are related, but not always *why*.

Huxley: Both of you are too cautious. Homology gives us a quantifiable backbone for understanding life's unity. Darwin himself would have been thrilled to see such a precise measure of descent.

Haldane: True, protein and DNA sequence homology can reveal evolutionary distances, but these are only proxies. Evolution is messy — horizontal gene transfer, convergent evolution — they muddy the waters.

Gould: Precisely.

Huxley: Homology is a cornerstone of modern medicine! Look at how it enables us to model human diseases in mice. Shared genetic sequences let us study cancer, diabetes, even Alzheimer's.

Haldane: And yet, Huxley, not without limitations. A mouse is not a man. The evolutionary distance matters. A gene conserved in mice may not function identically in humans.

Gould: Or consider viruses. RNA sequencing has revolutionized how we track pandemics and develop treatments. The homology between a virus's spike protein and antibodies guides vaccine design. Yet, unexpected variants emerge, tweaking the sequence to evade immunity.

Huxley: Surely, that's where evolution becomes predictable. Mutations happen, natural selection prunes. Our understanding of sequence divergence allows us to anticipate and adapt.

Haldane: Predictable, but only within bounds. Neutral mutations accumulate, genetic drift induces biases, and selection is but one player in the game.

Gould: And let's not forget the anthropocentric lens. When we compare genomes, we decide what's important — proteins, non-coding DNA, regulatory sequences.

Huxley: So, homology is both a tool and a mirror. It reflects our evolutionary connections but also our biases in interpreting them.

Haldane: Exactly. From phylogenetics to CRISPR to personalized medicine, these techniques use evolutionary homology as the core concept. We must understand its limitations.

Haldane: Cheers to the proteins and genes that brought us here. Without them, we'd still be bickering about fossils.

6.1 Homology and the Conservation of Inherited Characters

An important consequence of Darwinian evolution, as viewed in its modern synthesis, is that all living organisms are related by descent. While life might have had multiple origins on earth from billions of interactions among complex organic moieties — originally called "coacervate droplets" when they were first proposed independently by J. B. S. Haldane (1892–1964) and A. I. Oparin (1894–1980) — what matters for this discussion is that only one lineage of life happens to have survived today, and the rest of the lineages are extinct.

This assertion is no handwaving argument; it is one that is provable using simple algebra. We provide the algebra in Part III (Chapter 15), but here we simply state the result, namely, the probability (P) that two distinct individuals in any population (containing N individuals) had two separate ancestors decreases with the number of generations, k, separating them (going backward in time):

$$P = \left(1 - \frac{1}{N}\right)^k \tag{6.1}$$

For large values of k this probability converges to zero (Fig. 6.1). In other words, after a sufficiently large number of generations, the probability of two individuals in an extant generation having two separate (distinct) and independent ancestors becomes practically zero. Since terrestrial life had originated at least 3.6 to 4 billion years ago and many microbes, which represent descendants of the simplest forms of life, had

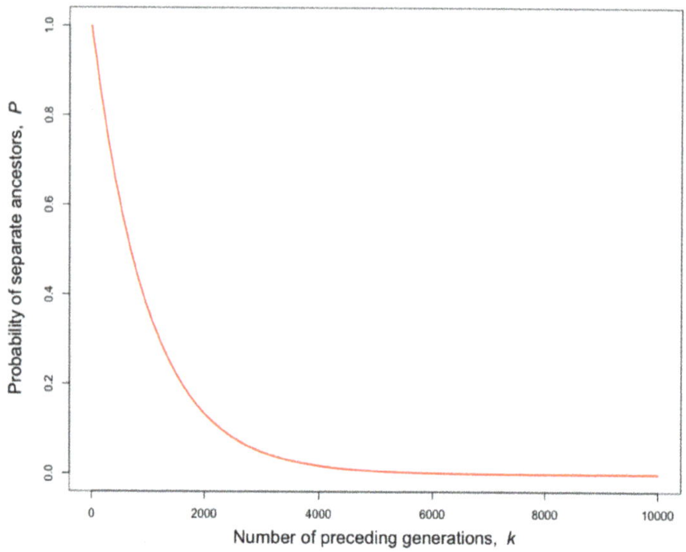

Figure 6.1 Probability that any two separate individuals have distinctly different ancestors plotted as a function of the number of generations, for $N = 1{,}000$ (Equation 6.1).

undergone numerous generations (e.g., the generation time of *Escherichia coli* is only 20 minutes in favorable growing conditions, providing over 25,000 generations in a year), it is nearly certain that even if there were multiple independent origins of life, all extant life appears to descend from one last common ancestral population. All others were either competed out (i.e., gone extinct) or had coalesced with other lineages, leaving no trace of their independent origins.

Therefore, there must be a genealogy of inheritance that can be traced for all extant organisms from a previously thriving but now possibly extinct ancestral organism. Whereas we showed above this mathematical consequence for self-fertilizing organisms in which only one parent is sufficient to produce progeny, it can be shown, with the help of more sophisticated algebra, that a similar expression holds for more familiar organisms, including humans, which require two parents to generate an offspring.

The existence of a genealogical relationship among extant organisms implies that certain characteristics are "conserved" among extant organisms. Importantly, the extent of conservation provides a metric by which to scale the evolutionary time that separates two different organisms' respective phylogenetic positions so long as the biological characteristic that is being measured is quantifiable. The less the evolutionary conservation of a characteristic, the further away are the two organisms along the evolutionary scale.

Note that the apparent conservation of a characteristic in two different organisms may not always imply that the characteristic property has been co-inherited.

Evolution might have adopted the same or a similar solution by chance, perhaps repeatedly, in organisms living in geographically isolated spaces. Or perhaps because that characteristic is physically an optimal solution to a particular functional problem. Either of these two possibilities might be true, and there is no easy way to distinguish between them.

Therefore, the presence of a particular characteristic in two otherwise distantly related organisms might be due to convergent evolution, where the same or a similar adaptation has occurred. It therefore is important to provide a quantitative metric for evolutionary distance, such that it is possible to estimate the probability of co-inheritance of a characteristic.

An important insight, originating with the legendary American chemist Linus Pauling (1901–1994), is that proteins can be considered as genetic traits or characteristics on which evolution can act, because proteins perform important functions on which life depends.

A particularly compelling example is hemoglobin. This protein carries oxygen to tissues and organs. Patients suffering from sickle cell disorder have a slightly different (a variant) hemoglobin protein. A single amino acid change at the sixth position of the beta chain (hemoglobin of an adult person is a tetramer of two alpha and two beta chains) of the hemoglobin molecule — from the normal glutamic acid residue to the abnormal valine residue — is sufficient to cause a major change in the structure of the hemoglobin protein. Sickle cell hemoglobin aggregates under certain conditions. Due to this abnormal property, red blood cells of a person containing two copies of the mutant gene (containing the codon GTA instead of the normal GAA at the sixth codon of the encoding gene) exhibit altered shape, adhesiveness, and rigidity. These can occlude narrow blood capillaries and restrict blood flow to tissues, causing tissue damage. Patients suffer intense pain associated with frequent infections, all contributing to significant morbidity and early death due to organ failure.

The instance of this major inherited disease was one of the first to illustrate how a change in a protein's sequence leads to a major disease. Subsequently, through numerous studies, it became an established dogma that the sequence of amino acids in a protein guides the three-dimensional structure of the latter and thus controls the characteristics of organisms, which contributes to the organism's survival and flourishing, that is, adaptation.

Obviously, if protein function contributes to adaptation, then its structure should be conserved in evolution. In the next section, we show how this idea led to a sensitive metric for measuring evolutionary distance as well as evolutionary time.

6.2 Sequence Conservation of Proteins Provides a Metric for Measuring Evolutionary Distance

Russell Doolittle (1931–2019), while working in the early 1960s at the Karolinska Institute of Sweden, realized that the rate at which amino acid sequences of a protein in various organisms change over time is related to the evolutionary distance separating the organisms.

Doolittle, a marathon runner, didn't leave this idea idly by. He followed this up as he started an academic position at the University of California in San Diego and helped develop one of the most useful techniques in modern medicine and biotechnology: determination of the phylogenetic distance between organisms or species, based on the conservation of amino acid sequences among their proteins.

We will provide here a general illustration of this principle, leaving the reader to explore the technique in more detail in Chapter 16.

6.2.1 *Protein structure determines function*

One of Linus Pauling's key ideas was that a protein's three-dimensional structure determines its function. He suggested that replacing an amino acid at a specific spot in the protein chain with any of the other 19 natural amino acids might change the protein's properties. In some cases, its function would be disrupted completely, and in others, it might affect its function only subtly or not at all in any measurable way.

The reason for this is that the 20 naturally occurring amino acids can be grouped by their physicochemical properties into separate similarity classes: nonpolar aliphatic [glycine (Gly or in the one letter code G), alanine (Ala or A), valine (Val or V), leucine (Leu or L), isoleucine (Ile or I), methionine (Met or M), and proline (Pro or P)]; aromatic hydrophobic [(phenylalanine (Phe or F), tyrosine (Tyr or Y) and tryptophan (Trp or W)]; polar but uncharged [serine (Ser or S), threonine (Thr or T), cysteine (Cys or C), asparagine (Asn or N), glutamine (Gln or Q)]; polar acidic [(glutamic acid (Glu or E) and aspartic acid (Asp or D)]; and polar basic [(arginine (Arg or R), lysine (Lys or K), histidine (His or H)] (Fig. 6.2).

In a protein, the replacement of an amino acid of one physicochemical group with another member of the same group (say, replacing an arginine with a lysine) might not always produce a drastic structural change and therefore might not lead to a profound change in the protein's function. By contrast, replacement with an amino acid belonging to a different group

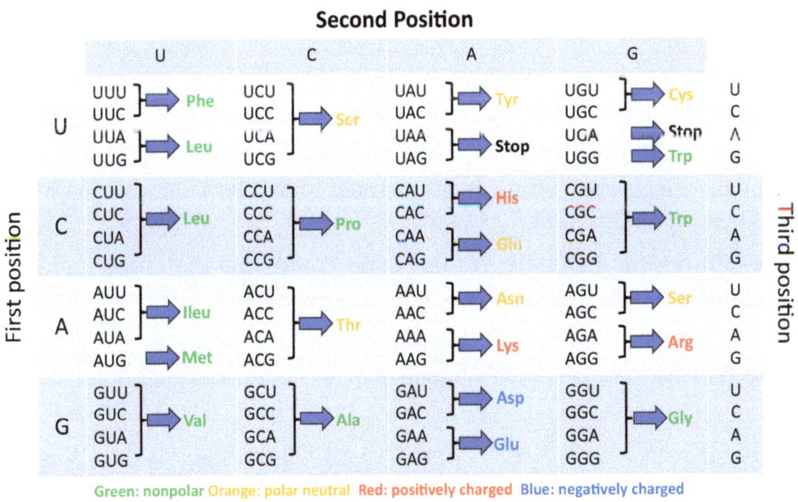

Figure 6.2 The table of codons. Amino acids of similar physicochemical properties (color-coded) tend to be encoded by related codons. Each amino acid is labeled by the first three letters of its name (except isoleucine, Ile), respectively.

(e.g., replacement of a leucine by arginine) might often produce a drastic alteration of protein structure and therefore may cause a large change in the function of the protein.

6.2.2 *The extent of sequence divergence between two proteins related by sequence conservation is measurable*

If two protein sequences (e.g., that of the hemoglobin protein's alpha chain from a human and that from a chimpanzee or mouse) are aligned with the assumption that the two molecules are similar such that the alignment achieves the maximum overlap or congruency of their respective sequences, then the mismatches between the two molecules can be given a score. This score essentially reflects the expected penalty to the structural alteration that might result from the respective changes (the amino acids are written in the generally used one-letter code). A short stretch of this alignment is shown below:

Human:	ADKTNVKAAWGKVGAHAGEY
Chimpanzee:	ADKTNVKAAWGKVGGHAAEY

Obviously, not all changes would be scored the same: replacement with amino acids of the same group should be penalized less than with amino acids from different groups; an amino acid residue with a small side chain when replaced by an amino acid with a bulkier side chain may disrupt the protein structure more than when it is replaced by a smaller side chain residue belonging to the same group, and therefore should be penalized less. A similarity score *S* of a pair of aligned sequences of amino acids (or a pair of whole proteins) can be calculated as the sum of all the penalty scores of the aligned pair of sequences (see Chapter 16, Section 16.2).

It is possible to define the extent of similarity, that is, the similarity score, *S*, between a pair of aligned proteins using a few simple postulates: (a) the replacement of one amino acid by another is a chance event that occurs with a constant probability per unit time, (b) every replacement is an independent event that does not influence the probability of another or subsequent replacement, and (c) the replacement themselves occur independently of adaptive need. Each postulate follows the fundamental axioms of Darwinian theory of the occurrence of spontaneous independent variation. By defining the similarity metric according to these principles, one can then estimate the time (i.e., the evolutionary time) it has taken to evolve from one protein sequence to another by random chance (see Chapter 16 for further details).

6.3 A Relation between Protein Similarity Score and Evolutionary Distance

The evolutionary distance between two aligned protein sequences is defined as (see Chapter 16, Section 16.3 for more details):

$$D = -\ln S \times 100 \tag{6.2}$$

The negative sign ensures that the higher the similarity score for an alignment between a pair of sequences, the closer is the evolutionary distance between the two (Fig. 6.3). In other words, the more the probability of obtaining an alignment given a pair of protein sequences relative to the probability of obtaining that alignment by random chance, the numerically higher (more positive) is the value of S, and therefore the smaller is the evolutionary distance. This obviously makes evolutionary sense, because the sequences of a protein inherited by two species of organisms from their respective direct common ancestral organism are more likely to be like each other than any two arbitrarily chosen protein sequences.

Remarkably, it is now possible to correlate the evolutionary distance D of pairs of conserved proteins with the estimated time of divergence of those pairs of organisms based on fossil records and thus provide a mapping of D on the chronological time of evolution, in units of millions of years before the present. This is the reason for calling the distance metric D also a measure of evolutionary divergence.

Fossil records can also provide a measure of evolutionary divergence time in at least three different ways. First, fossils are often found in layers

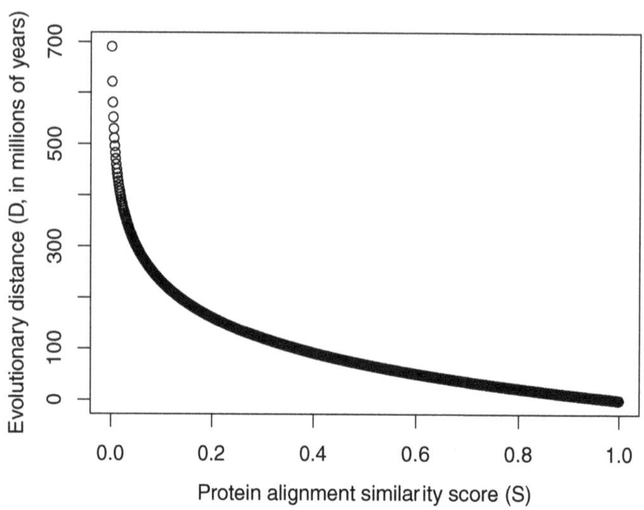

Figure 6.3 Relation between similarity score, S, and evolutionary distance, D. This relationship provides a kind of "molecular clock," relating the protein sequence similarity score to the evolutionary distance in terms of millions of years.

of sedimentary rock, which form distinct strata. By studying the relative position of fossils in different strata, paleontologists can determine their relative ages. This provides a chronological sequence of species, allowing them to infer the order of appearance and disappearance of organisms over time. Second, certain fossils are known (through prior studies) to have existed only during specific time intervals. These fossils, called index fossils, provide useful markers for dating the rocks in which they are found. By identifying index fossils in different sedimentary layers, scientists can estimate the age of the surrounding fossils based on the known age of the index fossils. Finally, radiometric dating relies on the decay of radioactive isotopes present in rocks and fossils. By measuring the ratio of parent isotopes to their decay products, scientists can determine the age of the material. For instance, carbon-14 dating is used for relatively young fossils (up to about 60,000 years old), while other isotopes with longer half-lives, such as potassium–argon dating or uranium–lead dating, are employed for older fossils. A plot of the evolutionary divergence time estimated from fossil records of New World primates against the evolutionary distance

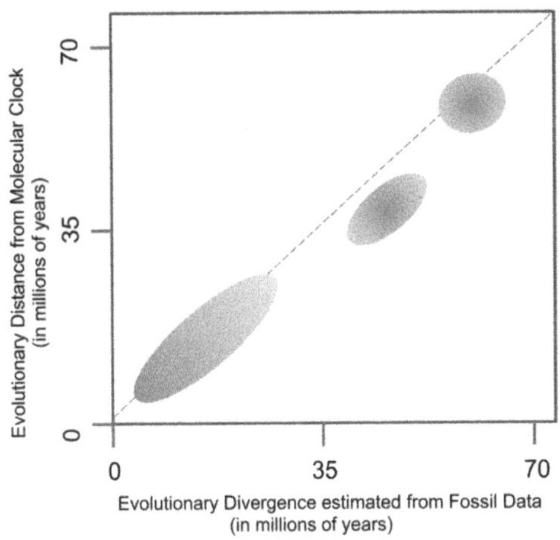

Figure 6.4 Approximate linear relationship between evolutionary distance estimated from the molecular clock (see Fig. 6.3) based on protein sequence divergence and evolutionary divergence deduced from morphological data in fossils of New World Primates. This figure has been adapted and redrawn from Schrago et al. (2013). Similar relations in other groups of organisms have been demonstrated.

estimated from the molecular clock is shown in Fig. 6.4 (Schrago, Mello, and Soares 2013). The diagram suggests strong support for the accuracy of the molecular clock.

More importantly, let us ponder the significance of what these results imply. The similarity metric *S* was defined according to the strict Darwinian postulates of random and independent occurrence of heritable variants (here, amino acid sequence variants of proteins). We are simply stating this assertion here, but you will need to read the mathematical derivation described in Chapter 16, especially Section 16.3, to convince yourself of the veracity of this assertion. Based on that definition, an evolutionary distance metric (i.e., the molecular clock) was derived. The distance metric provided the evolutionary divergence times between pairs of proteins made by the corresponding pairs of organisms. These molecular divergence times correlate closely with the evolutionary divergence times between these pairs of organisms estimated by paleontologists from fossils! One would be hard-pressed to find another explanation that produces such a remarkable numerical correlation between two such disparate observations as the location of primate fossils on geologic strata and the diversity of proteins synthesized by primates.

6.4 Implications of the Evolutionary Distance Metric for Medicine and Biotechnology

The congruence of evolutionary distances estimated from protein sequence alignment and from fossil data is remarkable from two angles. First, protein sequence divergence can serve as a surrogate for estimating the time for evolutionary divergence of organisms where fossils are not available. Second, principles by which the molecular clock was derived (Chapter 16) imply that proteins exhibiting relatively low sequence divergence between two organisms are likely to encode similar functions.

This latter consequence is the basis for discovering the function of previously unknown, say, human, proteins using the information gleaned from studying animal models in the laboratory.

If a protein in a laboratory organism has a highly conserved sequence homolog in humans, then it is probable that at the molecular level the two proteins perform a similar function. For example, if a protein is discovered by laboratory experiments to promote cancer in mouse, and the human genome encodes a protein of high sequence similarity (though of unknown function), then it is possible that this human version of the protein also promotes cancer in humans.

If, however, one chooses pairs of protein sequences willy-nilly from two different organisms, the evolutionary distances estimated from such pairs of proteins might be quite far because some proteins evolve significantly slower than other proteins because of constraints put on their structures by their functions. For example, a protein essential for the survival of an organism, say, the respiratory pigment protein cytochrome *c*, might evolve a lot more slowly than another protein that is not essential for survival. Therefore, care must be taken to appropriately choose the protein pairs to estimate the evolutionary distance.

By the same token, proteins that are found to evolve slower than other proteins can be recognized as being under a stronger natural selection than other proteins.

6.5 Evolutionary Distance Can Also be Estimated from DNA or RNA Base Substitution Rates

It should be intuitively clear that DNA or RNA sequence alignments can also be used to estimate phylogenetic distance, but such an endeavor must utilize a different metric in addition to using DNA or RNA sequences that are generally well-conserved between species. This was first adopted by Motoo Kimura (1924–1994), a profoundly insightful evolutionary geneticist working in the picturesque city of Mishima, Japan, under a grand view of Mount Fujiyama.

Kimura proposed two different models for estimating evolutionary rates; the first model is general, and the second model, simpler, applies well to eukaryotes (Kimura 1981). Because of this work, we can arm ourselves with a quantitative measure for estimating the evolutionary time difference for a pair of DNA or RNA sequences. The cost of DNA or RNA sequencing is far cheaper and is far more easily done than the task of directly sequencing proteins. The importance of this task should be clear considering what we learned in Chapter 1 for the evolutionary arguments that favored the Virogenc hypothesis for the causes of human cancer, which ultimately led to the discovery of oncogenes. This is, of course, not the only reason, for evolutionary distance estimation has been indispensable for all fields of biology and biotechnology; we will gradually appreciate how important this concept has been to the growth of biotechnology as we progress through the book.

Kimura discovered that evolutionary distance can be deduced from a pair of aligned DNA sequences in a manner analogous (but the formula is different) to the method using protein sequences (see Chapter 16, Section 6.4). Using his method, Kimura estimated the evolutionary distances, K, for the third codon positions in various homologous protein-coding DNA sequences between humans and rats, and between rats and rabbits, respectively. The corresponding divergence time, T, estimated in this manner is listed in Table 6.1 (Kimura 1981).

We see that the common precursor of the human and rodent globin, somatotropin, and insulin genes had diverged about 60 to 75 million years ago, in the Mesozoic era. Fossil records are consistent with this evolutionary timescale. Interestingly, the α-globin and the β-globin genes probably

Table 6.1 Evolutionary divergence times estimated from DNA sequence differences between pairs of related genes.

Gene pairs	Third codon divergence, K	Divergence time, T (in millions of years)
Human vs. rat growth hormone (somatotropin) genes 1 and 2	0.75	75
Human vs. rat proinsulin genes encoding A and B protein chains	0.60	60
Rabbit vs. mouse α-globin genes	0.65	65
Rabbit vs. mouse β-globin genes	0.49	49
Rabbit α- vs. rabbit β-globin genes	1.19	119

had diverged from an ancestral gene that predated the split between rodents and other mammals by about twice as long ago, that is, by at least 100 million years.

6.6 Molecular Phylogenetic Tree

A phylogenetic tree is a common device to illustrate hypothetical evolutionary relations among organisms. We will examine here one toy example of how to draw phylogenetic trees and leave it to the reader to consult textbooks on computational biology or bioinformatics for a more rigorous introduction to this extremely important area (Deonier, Tavaré, and Waterman 2005).

Consider a series of animal species, such as dogs, goats, cats, snakes, and kangaroos, who might or might not have in common certain morphological features, such as mammary glands, retractable toenails, hooves, tails, canine teeth, and hair. We can write down a matrix of animals by features, where the absence or presence of a feature can be represented by a 0 or 1, respectively (Fig. 6.5a). This matrix can also be represented as a tree diagram, in which these relationships encapsulate the degree of relatedness (or clustering property) among the entities (Fig. 6.5b). It is intuitively obvious that if such tree structures can be constructed from a matrix of traits, they can also be constructed, with more accuracy, for

(a)

TRAIT	Animals:	Dog	Cat	Goat	Snake	Kangaroo
	Mammary glands	1	1	1	0	1
	Retractable toe-nails	0	1	0	0	0
	Hooves	0	0	1	0	0
	Tail	1	1	1	1	1
	Canine teeth	1	1	0	0	0
	Hair	1	1	1	0	1

(b)

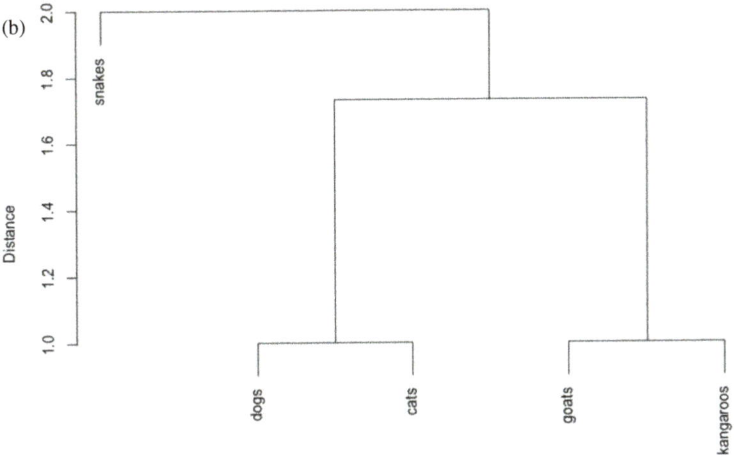

(c)

```
OmicronBA1    ---LDSKVSGNYNYLYRLFRKSNLKPFERDISTEIYQAGNKPCNGVAGFNCYFPLKSYSF    57
Delta         --NLDSKVGGNYNYRYRLFRKSNLKPFERDISTEIYQAGSKPCNGVEGFNCYFPLQSYGF    58
Alpha         ---LDSKVGGNYNYLYRLFRKSNLKPFERDISTEIYQAGSTPCNGVEGFNCYFPLQSYGF    57
Beta          ---LDSKVGGNYNYLYRLFRKSNLKPFERDISTEIYQAGSTPCNGVKGFNCYFPLQSYGF    57
Wuhan         SNNLDSKVGGNYNYLYRLFRKSNLKPFERDISTEIYQAGSTPCNGVEGFNCYFPLQSYGF    60
Gamma         SNNLDSKVGGNYNYLYRLFRKSNLKPFERDISTEIYQAGSTPCNGVKGFNCYFPLQSYGF    60
                 :::::.***** *************************..***** ********:**.*

OmicronBA1    RPTYGVGHQPYR    69
Delta         QPTNGVGYQPY-    69
Alpha         QPTYGVGYQPYR    69
Beta          QPTYGVGYQPYR    69
Wuhan         QPTNGVGYQ---    69
Gamma         QPTYGVGYQ---    69
              :** ***:*
```

(d)

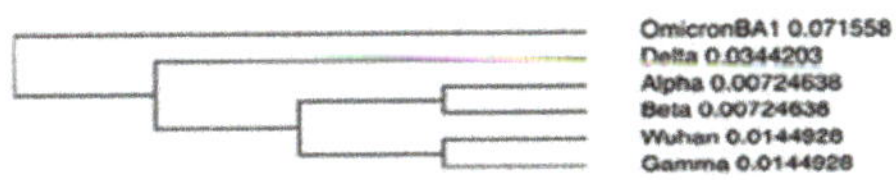

Figure 6.5 Phylogenetic tree made from traits. (a) A trait can be an enumeration of a conventional property. Here it is 0 or 1. But it can be an integer or a real number (e.g., the mean weight of femurs of adult female mammals or the evolutionary distance computed from an alignment score matrix). (b) A phylogenetic tree constructed from the matrix in (A). (c) Alignment of six variant SARS-CoV-2 Spike protein sequences overlapping a segment that binds to the human ACE2 receptor protein. The numbers on the right of the alignments are the number of amino acid residues aligned. (d) A phylogenetic tree calculated from the evolutionary distance metric inferred from the alignment in (C).

representing degrees of similarities among a set of aligned protein or DNA sequences (Fig. 6.5c, d). In this last example, we have represented the evolutionary distance among a few isolates of the COVID-19 pandemic virus SARS-CoV-2 based on the alignment of the protein (Spike protein) that is needed for the virus to bind to the human cell's ACE2 receptor protein in the lung epithelium and thus gain entry.

6.7 Significance of the Molecular Clock

The reader should do well to ponder the numbers shown in Table 6.1: these protein-coding genes have retained their ancestral resemblances in terms of their encoded protein sequences over tens to hundreds of millions of years, while their DNA sequences have been changing slowly through the random processes of mutation and divergence.

Thinking along this way, the Darwinian evolutionary framework has provided us not only with a way to rationalize the resemblances of gene and protein sequences across geological timescales but also given us the wherewithal to devise a molecular clock of evolution, using DNA/RNA or protein sequence divergence over evolutionary times. An important concept to emerge in this chapter is that random variations in inherited characteristics, as Darwin had posited, have found numerical precision in the context of the inheritance of DNA, RNA, and protein sequences.

The sequences of DNA (or RNA) in genes and the sequences of proteins are akin to phenotypic characteristics, albeit molecular, because these sequences ultimately provide function: either directly or indirectly by shaping the three-dimensional structure of proteins. These sequences may change at random by mutation. Therefore, the inheritance of variants of these DNA or protein sequences is equivalent to the inheritance of random variants in characters that Darwin had proposed.

Natural selection is the process by which certain traits or genetic variations become fixed (or, conversely, go extinct) in a population over generations based on their ability to increase (conversely, decrease) the survival and reproduction of individuals possessing these traits.

There is yet another process by which a variant can be fixed in the population — genetic drift. Genetic drift is the random fluctuation of the frequency of traits in a population due to chance events, which includes the rate of formation of the variations themselves and additional factors such as population migration, diseases, catastrophic environmental factors, and social factors such as selective mating or reproductive isolation (such as eugenics or ethnic cleansing, caste systems), warfare, or other forms of violence, among many others. Small populations are more susceptible to genetic drift, which can cause certain traits, along with their causative genetic variants, to become fixed in the population at the expense of others.

The reader, however, might also be wondering how these abstract notions of evolutionary conservation and divergence of protein and nucleic acid sequences have impacted the practice of medicine and biotechnology. More specifically, why does a practicing physician, or for that matter a biotechnologist who wishes to bring a new medicine to the market, need to be aware of these intricacies of evolutionary biology?

Natural selection and genetic drift together shape the evolution of a species by determining which traits and variations persist and become widespread in the population, and these allow for the development of new adaptations to changing environments. The persistence of some of the genetic variants that provide selective advantage under a condition may also cause certain diseases, such as the sickle cell disorder or several types of thalassemia, or several degenerative brain disorders, such as Huntington's disease. Conversely, the persistence of certain variants, perhaps with little selective impact in the population in the past, might exhibit more profound effects because of changing environments. For example, metabolic disorders associated with some genetic variations might be more common today than in the past because of changes in food and behavioral choices. In the next chapter we will explore how these abstract notions of evolution can and do profoundly impact human societies.

Chapter 7

Evolution of Coronaviruses

Nothing affected the present generation of humans on earth as much as the last coronavirus pandemic. This global catastrophe that has killed at least 6,978,672 people until the day of this writing underscores the need to appreciate the impact that the emergence of new viruses can have on humanity. The way to understand and better prepare against this menace is to understand how coronaviruses evolve. This chapter presents the background of the coronavirus pandemic of 2020 to 2022.

***

A bustling café. Candle-lit chandeliers cast a warm glow over marble tables and plush red velvet chairs. A young doctor, Edward Jenner, dressed in a silk shirt with high collars and a bowtie' sits upright, sipping tea. Darwin, an older man with a flowing beard, ever the observer, sketches a pigeon on a napkin, while Louis Pasteur, a stocky gentleman with a French accent, animated and expressive, waves a small vial of some mysterious concoction.

Jenner: (stirring his tea) Gentlemen, let us toast to viruses — though they bring misery, they are fascinating adversaries. While I subdued the pox with my vaccine, I confess I've little insight into how these infernal particles evolve.

Darwin: (placing his pencil down) Ah, Jenner, evolution drives their cleverness. Like my finches, adapting their beaks, viruses mutate and face selection. Simplicity is their strength; they adapt faster than any bird or beast.

Pasteur: Oui, Darwin, but hosts are critical! These creatures you admire are crucibles where viruses evolve, jumping species to acquire new traits. I've seen it with rabies — a terrible journey!

Jenner: How? Surely viruses, being so simple, cannot rival the complexity of your finches, Charles.

Darwin: Simplicity can be advantageous. Viruses mutate rapidly, creating diversity, while natural selection ensures only the most adaptable strains thrive.

Pasteur: Exactly! Tiny changes can alter a virus entirely — making it milder or, worse, more deadly. Coronaviruses could adapt to new hosts or evade immunity.

Jenner: And what of the immune system? Surely it challenges these invaders.

Pasteur: It is a microscopic arms race! The immune system evolves defenses, while the virus counters with new tricks — like your cowpox vaccine. Vaccines exploit the body's memory to outwit foes.

Darwin: Fascinating. These viruses form a tree of life in miniature — a viral phylogeny! Their branches diverge as they adapt to hosts or new infections. A coronavirus today might spread through mammals, birds tomorrow.

Jenner: And where might this "tree" lead? Could a virus grow so adapted as to threaten entire populations?

Pasteur: Perhaps, but vaccines are our tool to guide their evolution, altering selective pressures on viral populations. Controlled evolution, you might say.

Darwin: Controlled evolution? How curious. Could humanity truly shape the course of evolution, even at such a small scale?

Pasteur: Why not? We shape wine and cheese. Why not viruses too?

Jenner: A strange thought — mankind as the gardener of microbial life. But could such shaping ever backfire?

Pasteur: Bien sûr. Poorly done, we might create worse strains. This is why we must proceed with care. But knowledge is our ally.

Darwin: To knowledge and adaptation — not only of finches but of men and their sciences.

Jenner: And to the hope we stay ahead of these clever particles.

Pasteur: (clinking glasses) Santé! To the great battle of evolution and ingenuity!

Glasses clink, mingling with laughter and the hiss of steaming coffee pots.

7.1 Recent Viral Pandemics

Never in our living memory did we witness how profoundly evolutionary biology affects our daily lives until the coronavirus pandemic hit the world in early 2020.

Surely, the influenza pandemic of 1918 was of a similar scale, but human memory is short. Besides a few epidemiologists and historians, few remembered the impact of the 1918 "Spanish Flu" pandemic. Furthermore, at that time we did not have an understanding of how these pandemics originate; there was no vaccine against influenza; the only way to contain the spread of the pandemic was the age-old use of face masks and quarantining of the affected individuals, practiced at least since the bubonic plague epidemics of the medieval era (Earnest 2020).

Ernest Hemingway (1899–1961), the epitome of the American way of life under the big sky, already a war hero, had to quarantine for 3 months in Antibes on the French Riviera in 1926, not because of "Spanish Flu" but because of whooping cough that had afflicted his son! During the Spanish Flu, Hemingway was an ambulance driver in Italy, where he saw enough deaths to last a lifetime. Later, he wrote the following lines in a short story entitled, "A Natural History of the Dead" (Hemingway 1933):

> *"The only natural death I've ever seen, outside of loss of blood, which isn't bad, was death from Spanish influenza. In this you drown in mucus, choking, and how you know the patient's dead is at the end he turns to be a little child again, though with his manly force, and fills the sheets as full as any diaper with one vast, final, yellow cataract that flows and dribbles on after he's gone."*

I am unsure how much of that description can be generalized, but the Spanish flu had killed about 50 million people worldwide, next only to the medieval bubonic plague by the numbers killed. Russia's typhus epidemic in 1918 to 1922 killed about 2 to 3 million people. In 1957 to 1958, another influenza pandemic killed about 1 to 4 million people, and yet another in 1968 to 1969, the Hong Kong flu, killed about the same number. The HIV/AIDS epidemic, first described in 1981, which possibly had started a lot earlier and remains unresolved to this day, has so far killed over 36 million people. In 2009, the swine flu pandemic killed about half a million.

There are many more to list, but it appears that viral or bacterial pandemics have killed more people than all wars combined in the 20th century alone, even if one doesn't consider the three coronavirus pandemics of the 21st century: Severe Acute Respiratory Syndrome (SARS), Middle East Respiratory Syndrome (MERS) and Coronavirus Disease 2019 (COVID-19).

7.2 It Is Important to Surveil Coronaviruses of Animals

The first reported incidence of a coronavirus infection was in the 1920s, when an epidemic of respiratory disease of chickens, with high mortality, was noted. Since then, other groups of respiratory infections by viruses were reported for domestic animals in the 1930s and 1940s; all were found to be caused by viruses producing similar antibody profiles in the affected animal hosts (related serotypes).

Human coronaviruses were first recognized in the 1960s, when certain respiratory viruses causing common cold-like symptoms could not be cultured by methods standard at the time for culturing other respiratory viruses. Special conditions using serial passages through cultured human embryonic tracheal cells led to the identification of a unique group of viruses that under the transmission electron microscope revealed a roughly spherical virion structure with a lipid bilayer in which there were embedded club-shaped particles. This outer layer resembled a corona or the radiating aura around the heads of the divinity in medieval paintings, hence the naming of this group of viruses.

Soon thereafter coronaviruses were found to possess single-chained RNA as the genetic material. Unlike the retroviruses we discussed in Chapter 1, however, there was no trace of reverse transcription during the lifecycle of coronaviruses.

Instead, the coronavirus genomic RNA was found to serve as the direct template of its replication, via the production of a complementary RNA template upon which further rounds of RNA synthesis generate more copies of the genomic RNA to be packaged into the virions (Fig. 7.1).

The advent of rapid DNA and RNA sequencing by the late 1990s led to the sequencing of coronavirus genomes from all over the world. This

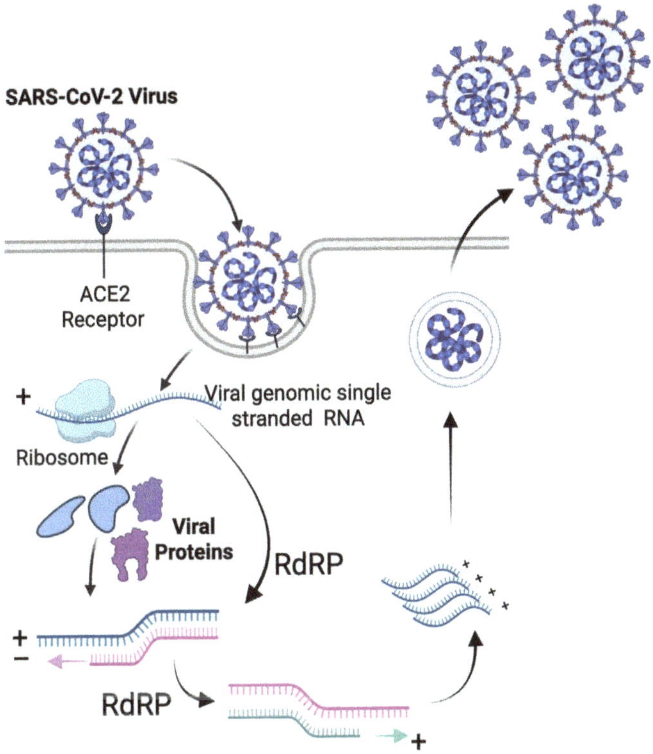

Figure 7.1 Replication of Coronaviruses involves an RNA-dependent RNA polymerase (RdRP) that can take the viral genomic RNA (which is also an mRNA) and can replicate it to the (−) strand (chain) RNA complement, and at the next step make further copies of the (+) strand (chain) viral genomic RNA. [This figure was drawn using Biorender.com under an academic license]

effort had accelerated during and immediately following the severe acute respiratory syndrome (SARS) pandemic of 2002, which started in November of that year in the Guangdong province of China.

By the end of 2003, the SARS coronavirus had spread to 28 countries globally, had infected at least 8,096 people, and had killed at least 774 (therefore, at least a 9.5% case fatality rate).

This first SARS pandemic was initially thought to have been started with the virus jumping originally from bats to Himalayan palm civets and/or raccoon dogs to humans. However, upon extensive surveillance work, it was later clear that while these animals might have been the conduits to humans, they did not constitute the actual reservoirs of the virus in which it had evolved (Graham and Baric 2010).

But what exactly were the scientists looking for in these bat viruses that could help us understand their potential for becoming future agents of human pandemics?

First, bats were thought to be the most likely natural reservoirs of coronaviruses (Li et al. 2005). How did we come to that conclusion?

About 400 bats representing six genera were trapped in their natural locations, their droppings, throat swabs, and blood were examined for the presence of coronaviruses; and the RNA genomes were sequenced. The organization of genes in one of the genomes isolated from a bat was found to be nearly identical to that of the SARS virus isolated from human patients several years earlier (Fig. 7.2) (Li et al. 2005).

This would be considered a highly unlikely chance event: think about all the possible genome organizations! There are enormous number of possibilities. Out of these nearly infinite possibilities, two viruses from different hosts as diverged as humans are from bats happened to have a very similar genome organizational pattern.

Therefore, the only reasonable conclusion was that both the human SARS virus and the one from a bat had descended from the same ancestral virus.

Significantly, however, several bat coronaviruses with widely varying genomic sequences were discovered, among which the SARS-like coronaviruses formed only one specific group.

In other words, there are multiple types of coronaviruses that exist among bats, more so than in any other animal species examined until then.

Based on this critical observation, it was proposed by Li et al. that bats might constitute a common natural reservoir of coronaviruses, with pools of many different viruses (Fig. 7.3a). If we can identify some signatures in these patterns of viral genome organization, which might allow us to recognize an imprint that is a hallmark for human transmission, then we can prepare in advance of a potential pandemic.

The SARS pandemic therefore had lent a new urgency to deciphering the evolutionary pathways within natural reservoirs of coronaviruses, especially in bats.

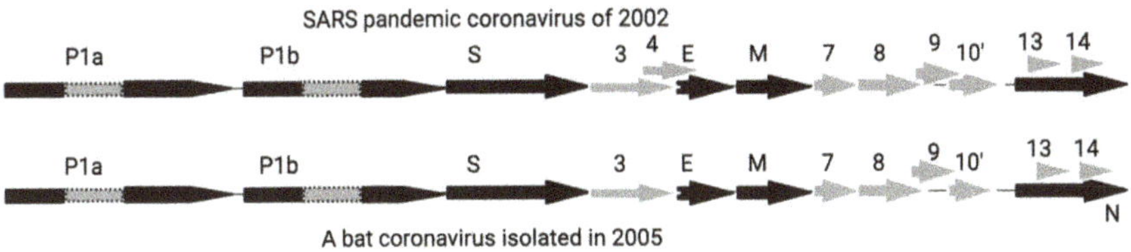

Figure 7.2 Genomic organization of the 2002 SARS virus and that of a bat virus isolated in 2005. The numbers and letters correspond to protein-coding genes that are transcribed into mRNA in known coronaviruses. A few genes are overlapping with other genes. Gray arrows denote genes with substantial sequence variations between the two viruses. Despite sequence variation, the genomic organization is very similar. The grayed regions in the middle of the P1a and P1b genes represent regions of the genes not drawn to scale. Data are taken from Li *et al.* (2005).

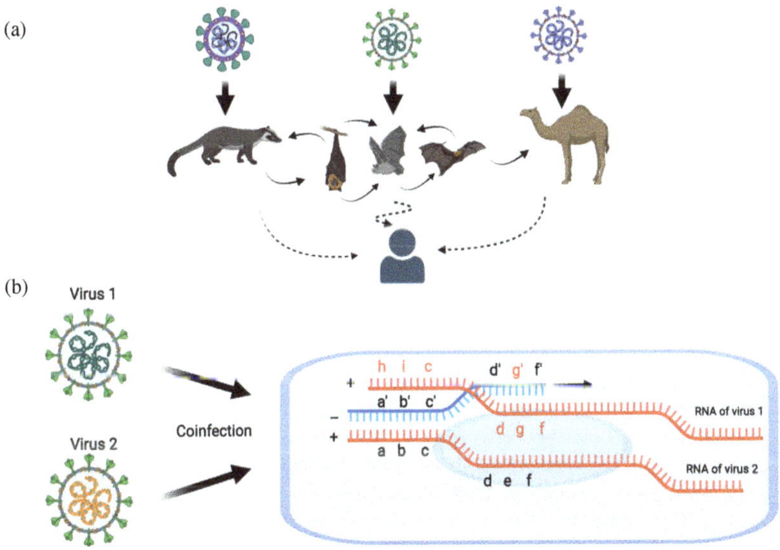

Figure 7.3 Complex population biology of coronaviruses. (a) Natural reservoir of coronavirus variants probably exists in some animals, such as bats, from which it circulates among several animal species, including civets and camels, and is also transmitted to humans. Switching of hosts probably involves mutations that allow the virus to attach to certain receptor molecules of a different host. (b) If a host cell is co-infected by two different variant viruses, a novel virus might emerge, which has acquired RNA sequences from both viruses through "template switching" during RNA synthesis. The RNA chain of one coinfecting virus might read *abcdef* and the other *hicdgf*, whereas a recombinant virus' sequence might read *abcdgf*. Complementary sequences are written with a prime (e.g., *a* and *a′* are complementary). [This figure was drawn using Biorender.com under an academic license]

7.3 The MERS Epidemic of 2012

The Middle Eastern Respiratory Syndrome Coronavirus (MERS-CoV) pandemic emerged in Saudi Arabia in 2012 and subsequently spread to 27 different countries, mostly in the Middle East and North Africa but also to the United States of America and South Korea. There were over 2,000 laboratory-confirmed cases and over 700 deaths (Ramadan and Shaib 2019). Approximately 64% of the cases occurred in males with an average age of 48 years and had a 35% case fatality rate — one of the highest human case fatality rates ever recorded from respiratory coronaviruses.

MERS-CoV was found to be associated with camels (dromedaries) in the Middle East. It is often difficult to demonstrate with confidence the path of movement of a virus among geographical locations and, as discussed above, among animal hosts. This virus obtained from a dromedary could be cultured in human embryo-derived tracheal cells, thus suggesting that these viruses could have jumped directly from camels to the human host. However, one must be open to the possibility of the opposite pathway, movement from humans to camels, as well. An extensive so-called "metagenomic" sequencing effort was carried out, in which viral RNA from nasal swabs of 79 camels yielded 93 unique genomic sequences (Sabir et al. 2016), suggesting that camels were a reservoir of this virus. This procedure is termed "meta-genomic" because individual viruses were not isolated, but their presence was inferred in samples through the nucleic acid (RNA/DNA) sequencing technique — thus, a single sample may contain more than one distinct virus.

This procedure revealed diverse genomic sequence patterns that enabled the construction of phylogenetic trees of the viral genomes based on evolutionary distances as described in previous chapters. The results were quite striking: even though these viruses resembled some of the bat coronaviruses, the lineage of MERS-CoV appeared to be almost as diverse among camels of the Middle Eastern region as they are among bats.

These results were suggestive of frequent recombination, or exchange of bits of their RNA genomes among the MERS-CoV variants (Fig. 7.3b). A coronavirus genome isolated from the droppings of a South African bat was 83% identical to the MERS-CoV genome at the nucleotide level

(Corman et al. 2014), suggesting that the bat and the camel viruses probably constituted a continuous pool of viruses across vast geographical distances.

Infection of the same animal host by multiple viruses simultaneously, as inferred from metagenomic sequencing results, could potentially produce hybrid (more appropriately, recombinant) viruses through the exchanging of parts of their respective genomes. If such a recombinant virus gains a selective advantage for infectiousness, crossing the host barrier, replicative efficiency, or a combination thereof, we might get a potentially dangerous virus for humans.

Perhaps such viruses are often generated by co-infection and recombination at the interface of different species. Based on these considerations, it was suggested that (Sabir et al. 2016):

> *". . . the prevention of transmission at the animal/human interface is likely to be the most efficient measure to contain the threat from this virus."*

These initiatives taken against the threat of this highly virulent strain led to the development of a vaccine against MERS-CoV for camels (Alharbi et al. 2017), with the hope that camels immunized against the virus would no longer provide an open field for the virus to evolve and hone its fitness to perhaps jump the species barrier to infect humans.

7.4 Where to Look for Human Host-Specific Patterns?

Are there pools of coronaviruses in bats that evolve to jump from one animal species to another? If so, is it possible to identify certain strains of coronaviruses that thus far were found only in animals but could be poised for jumping to humans? Are there tell-tale human-specific signatures in the genomes of coronaviruses that mark them as potentially infectious to humans?

More generally, we need to answer this somewhat abstract question: how do sequence patterns of amino acids in proteins evolve under selective pressure? This is because any change in the biological property of the virus must ultimately derive from changes in the sequence patterns of amino acids in their encoded proteins.

To understand how selection might favor one stretch of amino acids than another to change during evolution should be important for designing vaccination strategies by utilizing relatively more conserved regions of the viral antigens. This should protect against the emergence of future variant viruses to overcome vaccine-mediated immunity.

To answer some of these questions, international and national efforts were mounted in the early 2010s to sequence the RNA genomes of coronaviruses isolated from numerous different animals and humans. These RNA sequences and their encoded proteins were aligned to one another to measure the evolutionary distances among them.

The idea is that the closer one coronavirus is to another in terms of one of the sequence-based molecular clock metrics (*D*, defined in Chapter 6, Equation 6.2), the higher is the probability that these evolved from a common ancestral virus.

To accomplish the above task, one can compare the whole genomes of the viruses. However, among closely related viruses, one can also focus on a highly conserved virus-encoded protein, such as the "Spike" protein, which is identical to the club-shaped particles anchored on the membrane of coronaviruses seen under the transmission electron microscope.

Focusing on the Spike protein has biological relevance because this protein has evolved to bind to animal cell surface receptors to enter the cell. A change in the Spike protein sequence might imply a change in its host specificity, infectiousness, or even virulence, because the efficiency of virus multiplication depends in a complicated way on how efficiently it enters the host cells. Therefore, monitoring Spike protein sequences encoded by coronaviruses from wild and domestic animals may, in principle, provide important clues about the potential risks of one of these coronaviruses jumping to humans.

One of the crucial molecular determinants for host specificity appears to be a region of Spike, termed the Receptor Binding Domain (RBD) (Fig. 7.4) (Lan et al. 2020). This region was first recognized by aligning Spike amino acid sequences isolated from a variety of animals, such as various species of bats, civets, and humans. Alignment revealed a region of high sequence conservation, while revealing the presence of smaller stretches of amino acid

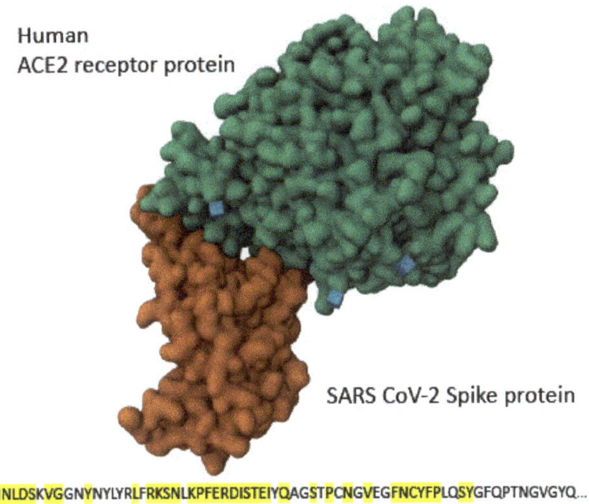

Figure 7.4 Co-crystal structure of human ACE2 receptor with SARS-CoV-2 spike protein's receptor binding domain. The stretch of amino acid residues written at the bottom (in one-letter code) corresponds to the subset of residues at the binding interface. Yellow highlighted residues are often different (sometimes absent) in Spike proteins of coronaviruses isolated from various animals. This suggests that amino acid residues at the interaction interface determine host specificity.

[Image reconstructed from the data of Lan et al. (2020), PDB id 6M0J, retrieved from PDB DOI: https://doi.org/10.2210/pdb6M0J/pdb]

residues within this largely conserved stretch. These smaller regions exhibited frequent changes in amino acid residues that appeared to encode species specificity (Fig. 7.4). It is as if, within a complex design of patterns that was largely invariant among all SARS-like coronaviruses, there were small regions of local signatures of individual species of animals from where the parent viruses were isolated. The conclusion was that these smaller regions of variable sequence patterns contributed to host specificity.

If the RBD domains of various coronaviruses encode a signature for host specificity, the host receptor that binds to the respective Spike protein must also exhibit amino acid residues that contact these signatures. This indeed was shown to be the case (Fig. 7.5). The coronaviruses thus appear to be in an evolutionary game of "chicken" in high gear with the mammalian ACE2 receptor — mutating and evolving to infect one host or another at a dizzying rate due to their short generation time.

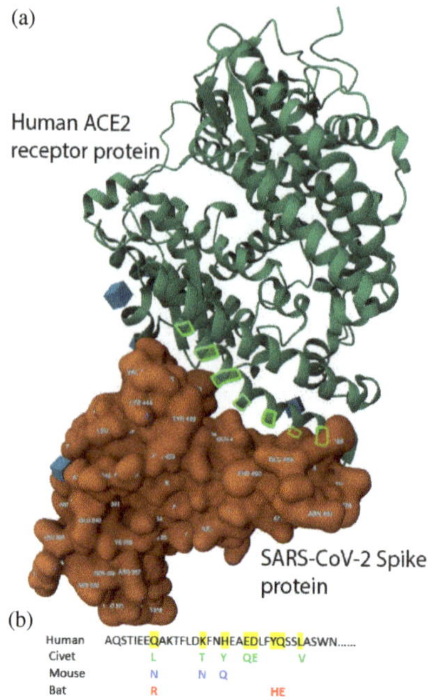

Figure 7.5 Mammalian ACE2 receptor protein has changes in amino acid sequence at the interaction surface with the Spike protein's receptor binding domain. (a) The green ribbon diagram represents the human ACE2 receptor protein in a co-crystal with the SARS-CoV-2 Spike protein (in the red space-filling model). The bright green rectangles on the helical region are residues that are different in the corresponding ACE2 proteins of civet, mouse, and bat. The blue cubes are sugar residues. (b) A segment of the helical protein sequence of the human ACE2 receptor, with corresponding changes in the amino acid residues (yellow) in civet (green), mouse (blue), and bat (red).

[Image reconstructed from the data of Lan et al. (2020), PDB id 6M0J, retrieved from PDB DOI: https://doi.org/10.2210/pdb6M0J/pdb]

Based on these results of inter-species homology of the Spike proteins, their differences, and their corresponding receptor protein sequences, it was possible to suggest (Graham and Baric 2010) that either the human SARS virus evolved from bat coronaviruses through an intermediate vector, such as a civet, or it might have evolved directly in both human and civet hosts independently. While the data could not resolve between these two possibilities, it was clear that bats were involved because the sequences were closely related to several known bat-borne coronaviruses.

7.5 The Emergence of SARS-CoV-2

As the reader can see from the foregoing rather incomplete summary of the SARS and MERS pandemics, there was ample warning for the possibility of an emerging coronavirus pandemic.

Pandemics are not rare. Ebola virus outbreaks occur nearly every year, which reached its peak in 2014 upon infecting 28,715 people and killing 11,372 in countries extending from those in central and western Africa to the United States. The Swine Flu (H1N1) pandemic of 2009 caused a minimum of 18,449 deaths globally. The Zika virus epidemic of 2015 to 2016 had infected 84,276 people in 87 countries.

The US National Security Advisory panel to President Obama had considered an emerging pandemic to be the greatest threat to US national security and had prepared a "playbook" in case of such a global pandemic. With an explicit mandate for uncovering threats of pandemics from viruses worldwide, the PREDICT (or "Predict, prevent, and detect zoonotic diseases with pandemic potential") project was launched in 2009, supported by a grant from the United States Agency for International Development (USAID) for epidemiological surveys. The PREDICT project scientists had reportedly collected over 180,000 samples from wild and domestic animal species throughout the world and had found over 1,000 novel viruses, including a novel Ebola virus strain, according to a New York Times editorial (McNeil 2019) published on October 25, 2019.

Lamentably, however, PREDICT's funding had reached the end of its 10-year initial period in 2019 and was discontinued thereafter by the US administration of that time. PREDICT was just the initial testing phase of a more ambitious Global VIROME Project which, devoid of US governmental backing, subsequently took on a privately funded mandate for continuing the work begun by the PREDICT project.

Against this background of heightened awareness of potentially emerging pandemics, the first report of a coronavirus having infected a cluster of people in Wuhan, China, came in early December of 2019.

By January 25, 2020, scientists in China had isolated and sequenced the viral RNA, based upon which they realized it was a novel coronavirus, though related but importantly different from that of the SARS or MERS versions, and generally related in its genomic organization to several previously sequenced bat viruses (Wu et al. 2020).

In late December of 2019, an important unanswered question was, did this virus have the potential to cause a widespread human pandemic?

To address this question, Wu et al. looked more closely at the Spike protein structure, which revealed that this protein's sequences were rather closely related to the previously identified SARS-CoV virus' Spike protein.

In fact, the sequence of amino acids in the region of the RBD domain (see Section 7.4 above) of this novel coronavirus had about 74% amino acid identity with respect to that of the SARS-CoV RBD domain.

It is as if a SARS-CoV-like viral genome had apparently recombined with two other broadly similar bat viruses at two genomic locations, thus producing an ancestor of this novel viral genome (Fig. 7.3).

Critical for successfully binding to the human Angiotensin Converting Enzyme or ACE2 receptor, with which SARS and SARS-like viruses enter the human cells, is a short region on the Spike's RBD domain: there were telltale signatures of amino acid changes in this novel coronavirus Spike protein that made it highly likely that this virus could efficiently bind to and infect human cells that express on its surface the human ACE2 (or hACE2) receptor protein (Fig. 7.4 and 7.5).

That fateful report, published online on January 25, 2020, concluded:

> "*This outbreak highlights the ongoing ability of viral spill-over from animals to cause severe disease in humans.* (Wu et al. 2020)."

The government of China warned the World Health Organization (WHO) about this novel coronavirus and the possibility of a pandemic in early January 2020. WHO had coined a name for this virus, SARS-CoV-2, given its similarity to the SARS-CoV virus of 2003, and the emerging pandemic the name COVID-19.

By March 16, 2020, there were over 18,000 daily new cases of COVID-19 reported globally, with over 650 deaths reported. At the time of this writing on December 24, 2021, there have been 279,029,029 reported cases of COVID-19, with 5,405,420 deaths, and the pandemic was still raging, with no country on earth having been spared.[1]

Consequences of the Darwinian evolution of coronaviruses were in full swing by March 2020 and will be described in more detail in the next chapter. The Congressional Research Service of the United States estimated that the global economic impact of COVID-19 was over $90 trillion by early 2021 — more than that of any single factor in at least a century (https://sgp.fas.org/crs/row/R46270.pdf). Such is the impact of evolution on human society, which we have uniquely experienced in our own lifetime.

---

[1]This chapter was initially written on December 25, 2021. See updates at the end of Chapter 8, Section 8.5, written on December 24, 2024.

Chapter 8

Evolution of SARS-CoV-2 Variants

circa December 2021 and postscript in December 2024

This chapter, first written during the final week of the year 2021, thick within the COVID-19 pandemic, examines the ongoing emergence of new variants of SARS-CoV-2 at that time. Two weeks after completing this chapter, the author himself was infected, even though he was twice vaccinated by that time. Thankfully, his symptoms were not life-threatening. The chapter describes how the author, sitting in a one-bedroom apartment in the mountain town of Quito, Ecuador, in which he had confined himself to write this book, was monitoring and calculating the evolutionary rates of the emerging virus variants while a cold rain fell outside. A postscript written in December 2024 evaluates his expectations and summarizes the fallout of the pandemic.

***

In a parallel universe in a quiet, book-lined study, four men sit around a low table. Albert Camus was a Nobel-winning author who was one of the founders of the philosophy of existentialism and wrote an influential novel named The Plague. Francis Crick, a physicist who discovered the Double Helix with James D. Watson, was often considered the smartest biologist ever. Jonas Salk and Albert Sabin, both medical researchers, independently developed two different vaccines for polio.

Camus: Ah, the plague returns — not rats in the streets this time, but something unseen, and global. COVID-19. A pandemic that reminds us how absurdly fragile human existence can be.

Crick: Absurd, you say, Camus. Yet science has met this absurdity with elegance. The RNA vaccines are among our finest achievements — a triumph of molecular biology.

Salk: Certainly, Francis. Though my own work with the polio vaccine was slower, what astonishes me is the speed — less than a year to develop and deploy these RNA vaccines.

Sabin: Speed, yes. But such haste has bred mistrust. Many fear these vaccines and dismiss them as rushed or experimental.

Camus: Precisely, Sabin. The politics of fear and division fuel this mistrust. In my time, we spoke of the human condition's indifference to suffering. Now, humanity itself perpetuates indifference — this time toward truth.

Crick: Truth is indeed under siege. But misinformation thrives because we scientists have failed to communicate clearly. The public doesn't understand mRNA, spike proteins, or antigen presentation. They need more than papers and jargon — they need trust.

Salk: Trust, yes, but it must be earned.

Sabin: Leadership faltered. Unclear messaging behind changing policies eroded confidence. Public health is not just about the science; it's about how that science is delivered. The reasoning behind the changing policies was kept under the belt — this lack of trust by the government in the common people's ability to judge what is good or bad policy was suicidal.

Camus: And there lies the rub: the intersection of science and the human condition. You speak of vaccines and policies, but what of the despair? The isolation? The upheaval in the human psyche? Science cannot heal loneliness or restore dignity.

Salk: No, Albert, but it can give hope. The vaccine is not merely a tool — it's a symbol. A symbol that humanity, for all its flaws, can rise to the challenge.

Crick: And yet, hope must be tempered with realism. The virus has evolved, as viruses do, challenging us to adapt. Variants emerge, requiring boosters and updates. The battle against nature is never truly won.

Sabin: Still, the question remains: how do we unite science and society? How do we foster a response that is as inclusive as it is effective?

Camus: By acknowledging the absurd, my friends. Accept the chaos, yet act as if meaning exists.

Salk: A toast: to science, to compassion, and to our shared humanity.

Crick: And to the curiosity that drives us.

Sabin: To the balance of reason and empathy.

Outside, the rain continues, a soothing rhythm against the windowpane. Inside, the conversation fades, but the ideas linger — science, society, and the enduring struggle to make sense of it all in a world defined by inspiration and absurdity.

8.1 A Measure of Natural Selection

If natural selection works on a collection of randomly generated natural variants, produced by the process of mutation and recombination of the genetic material, then it should be measurable.

The intensity of natural selection, or selection pressure,[1] was first defined by J. B. S. Haldane (Haldane 1924). We will examine this idea as adapted to the context of SARS-CoV-2 virus evolution. This has been helpful for identifying virus variants that appear to have high selective advantage and therefore are potentially more infectious (Dorp et al. 2021a).

The spread of a virus and its subsequent evolution depends on two factors: what happens when the virus infects a person, and what happens when the virus is transmitted to the next person. Within an individual, the virus undergoes mutation and recombination and thus evolves under natural selection imposed by the host's immune system, the variants competing among one another. The second process, as the virus moves from one person to another, is governed by various physical properties of the virus and how it interacts with a susceptible host.

Consider a time, in units of discrete generations, in which a variant allele (by mutation) appears in a population of the virus as it infects its

[1] See Chapter 14 for a quantitative description of natural selection and fitness.

host, before any natural selection occurs. Notice that most of the variant mutants will either have no effect on the survival of the virus (i.e., neutral mutations) or will be deleterious for its survival. This is because a third of mutant bases in the RNA genome would be in the third base position of a codon, which often will not cause a change in the encoded amino acid, or else it might cause a substitution with a synonymous amino acid (one with a similar property) (see the table of genetic codes, Fig. 6.2). Thus, these mutations would have little, if any, effect. If a nonsynonymous substitution, however, does occur, there is a good chance that protein function will be impaired, because it is always easier to destroy a structure (as dynamite does it) than to build one that functions better.

Rarely, however, a variant might have a slight advantage over the original strain, because the resultant variant protein might, perhaps, be more heat tolerant, have a slightly higher catalytic activity (if it is an enzyme, say the RdRP of coronaviruses), or, in the case of the Spike protein, perhaps bind more tightly to the hACE2 receptor to enhance entry of the viral particle into the host cells.

There are many ways such advantageous variant phenotypes could arise. Nonetheless, the frequency with which these arise is much, much less than the frequency of disadvantageous or neutral variants. However, once a variant with a slight advantage appears in the population, its clonally descendant genomes will likely produce a slightly higher number of progeny viruses of the same variant per unit time — if the variant is more infectious, if the variant replicates more efficiently, or perhaps if it avoids the host immune surveillance, and so on. A more quantitative discussion on this topic is provided in Chapter 14 (section 14.3) and Chapter 17.

8.2 Evolutionary Trajectory of SARS-CoV-2 Variants

We are familiar with the glacial speed of natural evolution, which occurs over centuries and millennia. How then did the global waves of SARS-CoV-2 variants, a few with significantly stronger selective advantage relative to the prevailing strains, occur over only a year and a half? We have seen in the past a rapid rise in drug-resistant variants of the HIV/AIDS virus as well.

SARS-like coronaviruses have a proofreading RNA-dependent RNA polymerase (RdRP), meaning that this RNA polymerase backtracks its newly synthesized stretch of RNA during replication, destroys the newly laid-down RNA chain (a few base pairs might have been accidentally incorrectly paired, which are then taken out), and the polymerase starts again (Fig. 8.1). Thus, on average every nucleotide on the RNA during replication has at least two chances of getting it right.

If the probability of an incorporation error per nucleotide is, say, 1 in 100, then the net probability of a misincorporation error is about 10^{-4} per base if each base has two chances to make it right[2]; it is 10^{-6} if it has three chances. The length of the genomic RNA in SARS-CoV-2 is about 30,000 nucleotides. If there are two rounds of synthesis (one edit per replication)

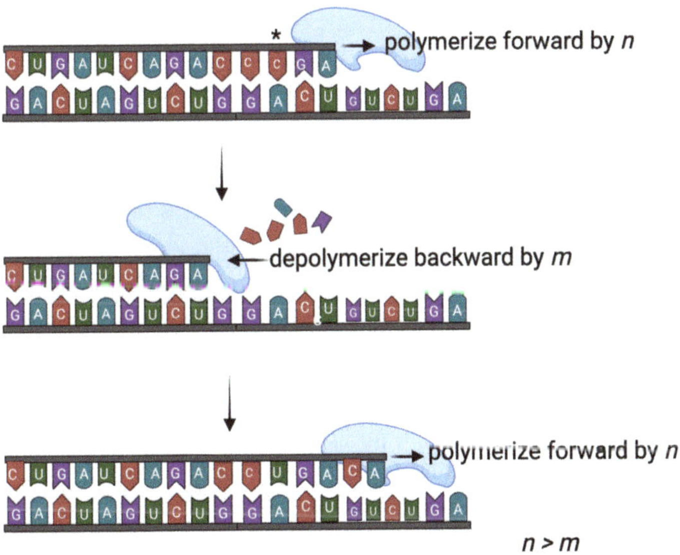

Figure 8.1 Proofreading by RNA-dependent RNA polymerase (RdRP). The RdRP enzyme polymerizes n nucleotides, backtracks to depolymerize m nucleotides, and then runs forward again while polymerizing n nucleotides on average. The asterisk indicates an error in base pairing. Because the newly laid RNA chain (top) can be depolymerized relative to the template chain (bottom), there is a chance that the erroneous base will be removed and the correct base will replace it. If the probability of an error in one run of polymerization is r per nucleotide, then the average error probability is expected to be $\sim r^2$. [This figure was drawn using Biorender.com under an academic license]

[2]This can be approximated as p^2 if the process is like two independent rounds of error-prone polymerization. In actual RdRP proofreading, the error is removed and the correct

on average per base, then there would be, on average, 0.03 base-change mutations per genome per act of replication (or 3 mutations per 100 genomes per act of replication). For three rounds of synthesis (i.e., two edits per replication), there should be only 3 mutations per 100 newly replicated genomes.

The SARS-CoV-2 RdRP appears to have an overall error rate of 10^{-4} per nucleotide (Bradley et al. 2024). By contrast, the replication process of retroviruses, including HIV/AIDS, is intrinsically error-prone, causing about 1 error per 100–1000 nucleotides per act of replication.

In this light, SARS-CoV-2 appears less susceptible to the occurrence of spontaneous mutation relative to retroviruses. Influenza viruses, with their segmented genomes, can mutate and reassort their genomes even more rapidly, producing numerous variant strains. This is one reason flu vaccines need to be redesigned every year. Coronavirus genomes appear relatively stable in comparison.

It is therefore interesting to determine to what extent mutant genomes of the virus arise within a single individual during a full course of infection. Importantly, do these mutant genomes of the virus recombine within an individual as they multiply within the infected cells? In other words, is recombination necessary for the generation of variant combinations, thus increasing the probability of the emergence of potentially advantageous variants? Let us make an estimate.

It is still unclear what the minimum infective dose of SARS-CoV-2 is for humans. From animal studies with related coronaviruses, and human volunteer studies with a coronavirus type that produces common cold symptoms in humans, an estimate of about 100 viral particles has been proposed as the minimum infective dose that produces COVID-19 clinical symptoms (Karimzadeh, Bhopal, and Tien 2021). However, it is quite likely that different strains of SARS-CoV-2 may have different minimum infective doses.

The viral load of SARS-CoV-2, that is, the number of virus particles per milliliter of blood of an infected patient, varies from about 200 on day two (asymptomatic period) to a peak of about 10^8 (median, ~$10^{6.5}$, rarely

---

base reinserted, which may not be entirely independent, but the effective error suppression is still ~p^2.

reaching 10^9) by day five (Jacot et al. 2020). With an average blood volume of 4.5 liters in a person, this translates to a sustained number of viral particles per person of about 10^{10} per day, often 10 times higher. Since many of these particles are expected to be degraded and renewed over about two weeks, one may expect at least 10^{12} progeny viruses per infectious generation within a patient.

Assuming a mutation rate per nucleotide of 10^{-6} per act of replication, with a genome of 30,000 nucleotides, one expects about 3×10^{10} mutant nucleotides distributed among 10^{12} progeny viruses. This analysis shows that even in a single infected person, every nucleotide of the genome of the virus is likely to be mutated at least once somewhere within the total virus population inside a patient.

It is difficult to predict how many mutants among this pool of variant sequences survive per generation of infection and can successfully infect a new host. If a mutant viral genome had no deleterious phenotypic effect on virus viability or infectivity, then it is likely that at least one particle per 200 in an infectious dose is, in fact, a mutant.

In other words, viruses having on average a single nucleotide mutation, many with potentially zero or low cost to the virus, should be quite common within an infectious dose. Considering that about 280 million people had been infected by the end of December 2021, it is overwhelmingly likely that all possible mutations in every nucleotide of the viral genome had already been sampled by this time.

Given this rate of mutation in the SARS-CoV-2 genome, it was surprising that until December 2020, 12 months into the pandemic, there was no evidence of any variant with a significant growth advantage over the original one. This suggests that to obtain a significant selective advantage, multiple beneficial mutations, including amino acid substitutions and deletions/insertions, might need to occur simultaneously.

8.3 Parallel Evolution of a High Fitness State

An interesting question about the probability of evolving a highly adapted virus is the following. There appear to be many more mild viruses, including cold viruses, that infect people than there are highly virulent viruses. A simple intuitive explanation is that a highly virulent virus, which might cause considerable morbidity, has less opportunity to spread due to the incapacitation of its host or due to host mortality — a dead host has little further chance of spreading the virus by coughing. But such effects are at best small, because most respiratory viruses have evolved an unusual property: the host is most infectious during the early, even asymptomatic, phase of infection. The same is also true of SARS-CoV-2. Therefore, such effects might be minimal at best.

An alternative explanation might be that viruses evolve also to maximize their numbers (i.e., their progeny viruses) within the host — that is, to maximize the viral load. This would naturally increase the chance to spread, and a variant would spread even more so if the host were nearly asymptomatic — a silent parasite!

In other words, a virus variant that has evolved to minimize the activation of the host's immune system without causing substantial damage to the host, without inhibiting its own replicative efficiency or infectivity, and in fact by enhancing these very properties, might exhibit a very high fitness. How many such possible variant states might exist for a coronavirus?

No one knows the answer to this question for sure, but let us assume that multiple such variant states are possible for SARS-CoV-2. Let us imagine that there are quite a few different combinations of mutant alleles, say, amino acid replacements, deletions, or insertions in the Spike protein as well as in a few other coronavirus-encoded proteins, which may impose such an adaptive value to the virus. What is the probability that in two independent patients' bodies such an identical state could be achieved, in parallel, twice?

Remarkably, the answer is simple. Allen Orr (1960–), a brilliant researcher of evolutionary biology at the University of Rochester, had arrived at a surprisingly precise and simple answer for parallel adaptation to the same environmental condition by molecular evolution (or parallel evolution), which is not too difficult to test in the laboratory (Orr 2005). Orr showed that if there are n possible different mutations (where n is small) such that each can provide a benefit (fitness) to the organism (i.e., if each of the n mutations singly contributes a fitness increment relative to the unmutated version), and provided that any combination of these mutations does not depress fitness, then the probability P that two different populations evolving independently (say, the viruses in two different patients or patient populations with no contact between them) to the same identical combination of mutations is given by Orr (2005):

$$P = \frac{2}{n+1} \tag{8.1}$$

If only one mutation can provide increased fitness, or $n = 1$, then the probability of attaining this fitness state by this single mutation is 100%, which it must be. Although this formula works only when the number of such mutations is small, say, about 3 or 4, we will assume for illustrative purposes without much loss of rigor that this formula will apply in our example case.

As it so happens, the omicron variant has at least eight mutations that map on a small region of the spike protein's receptor-binding domain (RBD) relative to the sequence of the original Wuhan version of the SARS-CoV-2 virus' RBD domain.[3] Besides these eight, there are over 50 changes in the omicron variant relative to the original variant, but let us assume for now that these other changes do not affect fitness.

Let us also assume for the moment that all these 8 mutations are together required for the virus' increased transmission, say, because these amino acid replacements might increase the affinity of the virus for binding to the human cell's receptor protein. Since there are 20 different amino acids of which any one of 19 others can replace the existing one, a rough realistic estimate that exactly these 8 mutations occurred by sheer chance is 1 in $(19)^8$ or 1 in 16,983,563,041.

There is one more complication here. We are speaking about replacing one amino acid by another as if a single mutation can do this replacement. That is of course not true. The replacement of one amino acid by another can only be accomplished on average by ~1.8 RNA nucleotide changes (given the table of genetic code). We will approximate this as ~2. Therefore, the actual probability is about 1 in $(2 \times 19)^8$ or ~1 in 4.3×10^{12}. The denominator is not very different from the number of virions made by a single patient during the course of a coronavirus infection (see above).

[3] (https://covariants.org/variants/21K.Omicron)

If, however, each of these precise $8 \times 2 = 16$ replacements is slightly more beneficial to the virus than other mutations such that there is a positive selection favoring each mutant variant from its precursor, then the probability that this precise combination of mutations might arise independently within two different hosts, that is, in parallel, by equation 8.1, is 0.118 or ~12%. This is surprisingly high if one doesn't consider the trajectory by which this fitness state is reached.[4]

Purists will correctly counter that Allen Orr's formula likely does not apply when the number of mutations is over about 5 or 6. Works by several researchers including our own have shown that to achieve an Omicron-like high fitness from the original Wuhan virus might be possible with only two amino acid changes, or about $3 \times 2 = 6$ nucleotide changes, if fitness is defined as the binding affinity of the RBD domain to hACE2 (e.g., S477N + Q498H mutations; Bate et al. (2022)). This would put the value of the probability of parallel evolution at around 29%.

Allen's simple formula appears counterintuitive. How can parallel evolution of such a complex set of mutations at the molecular level be so frequent?

Stephen Jay Gould (1941–2002), a Harvard paleontologist and one of the greatest popularizers of Darwinian theory of evolution in the 20th century, spoke one evening at the Massachusetts Institute of Technology in 1989. It was a packed crowd, standing room only. He had just published his latest book, *Wonderful Life*. He asked us to do a thought experiment in which we rewind the tape of life… yes, in those days we used to watch videos on tapes!

Would we see humans evolve again?

His pronouncement was definite and intuitively clear: No. Evolution proceeds through tinkering, through myriads of accidents; its products are the products of chance, the stochasticity of nature. He wrote in *Wonderful Life* (Gould 1989):

> *"Replay the tape a million times ... and I doubt that anything like Homo sapiens would ever evolve again."*

---

[4]This estimate assumes that all 16 mutations contribute independently to fitness and no intermediate combination reduces fitness. In practice, the fitness landscape may be more epistatic or rugged, making such parallel evolution less likely.

And yet, we have Allen's remarkably simple formula (Equation 8.1) that indicates otherwise.

Allen's formula was tested several times. In experimental evolution in the test tube, under a strong artificial selection, parallel evolution of substitution mutations at the amino acid level was as high as 50% for a bacterial virus named ϕX174 (Wichman and Brown 2010). Experimental evolution in the bacterium *Pseudomonas fluorescens* under several different environmental conditions also produced a high frequency of parallel evolution but the extent of parallel evolution was sensitively dependent on the structure of the population and the environmental condition (Bailey, Rodrigue, and Kassen 2015). Significantly, parallel evolution was detectable only when the two populations were under a strong selective pressure, leading to the emergence of mutations with strong adaptive properties. Mutations selected under such conditions have been termed adaptive mutations.

It is important to remember, however, that the term "adaptive" does not imply that the mutations were produced because of the selective pressure — as in the now unpopular "Lamarckian" theory of adaptation.

The mutations are produced randomly, but some individuals (bacterial cells or viruses) are simply selected in favor, leading to a relative flourishing of the variants containing those specific mutations. This is the crux of Darwinian evolution.

This effect of natural selection is well illustrated by Richard Dawkins (1941–) in his book, *The Blind Watchmaker* using an analogy (Dawkins 2015). Imagine a monkey pounding on a typewriter with a 26-letter keyboard and a key for a space. What is the probability, Dawkins asked, of the monkey typing at random the phrase, *Methinks it is like a weasel* (borrowing from Shakespeare's *Hamlet*)? Obviously, with no selection pressure for the success of intermediate arrangement of the letters, the random probability of typing that phrase would be infinitesimally small.

But suppose, as Dawkins did, the typewriter were special, such that when the correct (or nearly correct) letter or space is typed at a given place in an emerging sentence the typewriter freezes that accident for all subsequent presses of the key. In other words, there is a selection for the right letter for the right place. Then, showed Dawkins, the probability of a monkey typing that phrase is quite reasonable — something that could be achieved in about 40 generations of sentences by the cumulative selection of valid English phrases given how many monkeys are typing and the rate of typing. Note that in this process, combinations of letters that are somewhat meaningful might arise independently — these combinations (i.e., meaningful words) are akin to Lego pieces or modules. Thus, the modularity of meaningful or beneficial clusters of mutations or genes could provide a selective advantage to accelerate the rate of evolution. In other words, small modules of proteins with functional elements probably evolve independently, are selected for imparting some benefit so the organisms carrying them flourish, and these modules could be shuffled, reassorted, recombined, and fine-tuned by additional mutations for further optimization of the benefit to the organism (fitness).

Cumulative selection has seen wide support among evolutionary biologists, who have shown experimentally that artificial evolution of microorganisms can occur in this fashion (Brown et al. 2010; Hall, Agan, and Pope 2010; Khan et al. 2011; Lozovsky et al. 2009). That which is true at the molecular level for a microbe is generally found to be true for mice and men as well.

8.4 Some Evolutionary Fitness Trajectories Are Untraversable

Are all evolutionary trajectories possible in nature? Could the omicron variant have arisen from different ancestral sequences equally easily?

How many different evolutionary trajectories or paths can n different mutations — some only mildly beneficial, very rare ones strongly beneficial, and many either neutral or disadvantageous — take to arrive at a combination of mutations that together provides a advantage for survival (i.e., become fixed) in the population? The answer is that, assuming all n mutations are required and are acquired sequentially without reversion, the number of distinct shortest paths is $n!$.

One can think of numerous molecular mechanisms that might select for a combination of mutations to enhance or optimize the fitness to a particular environment. Interestingly, however, it appears that only a few of these theoretically possible trajectories are available for Darwinian selection, at least insofar as protein functional evolution is concerned (Weinreich et al. 2006).

Consider the hypothetical example that a series of 4 mutations on the Spike protein each causes a slight increase of, again hypothetically, its hACE2 binding activity to endow the virus encoding this multiply mutant protein with a significant selective advantage over the prevailing virus. There are 24 possible evolutionary trajectories (or shortest paths) by which each of these single mutations can appear together and the resulting virus will be fixed in the population. But it appears from several studies that not all 24 possible trajectories can be explored by evolution. One reason might be that some of the possible trajectories have mutant combinations so unfit that these don't survive at all. Another way of saying the same thing is that these evolutionary trajectories traverse unbridgeable valleys of (low) fitness in the fitness landscape of the protein's evolutionary space — a fitness landscape specifies the fitness value for each possible genotype or combination of mutations in this space (Fig. 8.2). A particular trajectory from the original spike protein sequence to one that has a high selective advantage can be accessible by evolution if, and

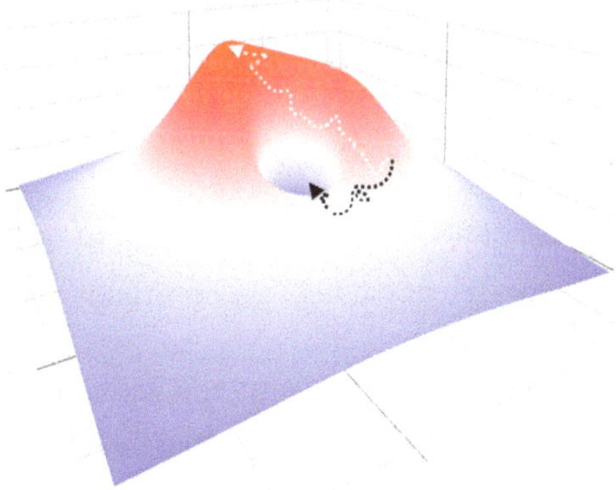

Figure 8.2 Evolutionary trajectory in a fitness landscape. The landscape might have deep valleys of low fitness and peaks of high fitness. To traverse from a suboptimal peak to an optimal peak, the trajectory might have to traverse unbridgeable low-fitness values (black trajectory). By contrast, the sequence space of biologically relevant proteins might lie in a region of local high fitness, where the trajectory (white) might be a relatively smooth ride.

only if, each unitary step in the trajectory constitutes either an increase in fitness, however mild, or has the same fitness as that of the previous step. This is precisely the "generalist" variant represented in Part I, Plate II.

Clearly, therefore, not all such combinations of mutant alleles may provide such a path. In fact, it is quite possible that no such path exists — unless multiple mutations occur in parallel, in which case a trajectory skips over a low-fitness valley along that trajectory (Fig. 8.2). However, given the low mutation rate, it is highly unlikely that simultaneously two, three, or more mutations will arise. Alternatively, the trajectory becomes more convoluted by the reversion of a mutation, and then yet another, different mutation needs to arise to skirt around the fitness valley.

Perhaps this requirement for such convoluted evolutionary trajectories prevented the early emergence of any highly adapted variant over the prevailing ones of SARS-CoV-2 and may explain the long delay between the onset of the pandemic and the appearance of the first variant with higher fitness.

Note, however, that within a host the virus may evolve significantly more rapidly than through single point mutations, because the single mutant variants might coinfect the same host cell, followed by recombination of the two genomes, giving rise to a recombinant genome in which both mutations co-occur. This may significantly accelerate the pace of evolution under certain conditions (Uecker and Hermisson, 2016).

This scenario of evolution by recombination among independently produced mutant coronavirus genomes seems likely because of the observation of many different mutant viral genomes detected by deep sequencing of circulating viral genomes in single host-patients' body fluids (Gunadi et al. 2021; Lau et al. 2021; Wang et al. 2021).

A question remains as to how the omicron variant had emerged. It has been quoted in the media that this variant might have evolved in an immunocompromised host, because in such a host a prolonged lifetime of the virus could have favored the emergence of this multiply mutated variant. While this is the prevailing hypothesis, the alternative — that intense immune selection in vaccinated hosts may drive convergent adaptation — is also plausible and merits consideration, as we discuss below.

The mutation rate is sufficiently high within any individual host, as discussed above, to have produced nearly all the mutant sequences.

Therefore, these variants together might have traversed much of the evolutionary landscape to reach a fitness peak. The mutation rate of the virus might itself have evolved to a higher value in certain lineages. Spontaneous variants with higher sequence diversity appear to correlate with substitution mutations within the gene encoding the RdRP protein, the proofreading polymerase that replicates the SARS-CoV-2 genome (Fig. 8.1). A strong natural selection would then favor the survival of a rare variant better adapted at infectious spread. Therefore, it is equally possible that the omicron variant might have evolved in an individual with rather strong immunity against the parental virus, perhaps within a vaccinated individual, by adaptive evolution.

This might indeed be a relatively desirable outcome for the world suffering under the crushing blow of the pandemic for the past two years — a variant with extremely high infectivity but low virulence may spell the end of the great pandemic by accelerating the attainment of herd immunity without causing unacceptably high morbidity. Natural selection among mostly vaccinated populations for a longer-lived and/or more infectious virus should be strong. There appears to be little equivalent selection in favor of high virulence causing higher mortality. Therefore, such a virus producing a generally milder effect but one that is highly infectious will likely become endemic in the population.

While selection plays a critical role as discussed above, stochastic processes such as drift and founder effects in large transmission events (e.g., superspreaders) may also contribute to the initial spread of some mutations.

8.5 Postscript to the Pandemic

I had written the above paragraph on December 31, 2021. That night, my wife Sumita joined me in Quito, Ecuador. Five days later, we hiked up to the base camp of one of the highest active volcanoes of the world, Mt. Cotopaxi, located at an altitude of 4864 m (15,953 feet). During our return to Quito in a bus, we sat beside a visitor from Europe who had landed the day before; he was coughing the entire trip behind his M95 mask and was in obvious distress. Two days later, we flew to one of the Galapagos Islands. Upon returning, both my wife and I tested positive for the coronavirus. That was despite both of us having been fully vaccinated and having had a booster dose two months earlier. We immediately quarantined ourselves in an apartment for ten days. Although neither of us suffered much other than some cough, low-grade fever, and the fear of being sick in a foreign country whose language we did not speak, it was not a pleasant experience.

Fast forward to December 22, 2024. I am now finishing up this book, which evidently took longer than I had anticipated because of all the other things I, as an academic, must do. Meanwhile, the pandemic is gone, although COVID-19 continues to kill some patients throughout the world. Vaccination has been wildly successful. The two people who had contributed to the development of RNA vaccines against SARS-CoV-2 — Katalin Karikó and Drew Weissman — were awarded the 2023 Nobel Prize in Physiology and Medicine. I had met Dr. Karikó at a conference in Cold Spring Harbor Laboratory, Long Island, New York, just days before her award was announced (Fig. 3.1). In her talk at the conference, she had emphasized her puzzlement in the early 2000s that RNA molecules isolated from bacteria, when exposed to mammalian cells, produced a massive hypersensitive response (by killing the cells), whereas mammalian RNA did not. This could make sense only in an evolutionary context — mammals had found a way to "cloak" their own endogenous RNA molecules. Their defensive strategy would be mounted to find and target the sources of uncloaked foreign RNAs, and thus would kill the invading bacteria or viruses and prevent further spread of infection. And yet, their own cloaked RNA molecules would not mount this defensive response.

Karikó and Weissman's work soon identified certain biochemical modifications of mammalian RNA as the wonderful cloaking mechanism that prevents mammalian mRNAs from instigating an allergic response (Karikó et al. 2005). The detailed knowledge of this cloaking process paved the way for rapidly developing and globally deploying mRNA vaccines against COVID-19. This unprecedented advance and millions of lives saved could not have been possible unless there was an evolutionary framework available to Karikó and Weissman — to think about searching for mRNA-modifying enzymes that were made by mammals but not by prokaryotes or viruses. By learning from the RNA cloaking mechanism, mRNA molecules encoding the spike protein of SARS-CoV-2 could now be cloaked (i.e., chemically modified) and be injected into humans as a vaccine to produce spike protein molecules in human muscle cells. This would induce an immune response by allowing antibodies to be formed against the virus' spike protein, and yet would not produce a massive allergenic response in the vaccinated persons in response to the foreign mRNA.

While the pandemic appears to be over for now, more than seven million deaths have been attributed to this disease (source: WHO data accessed on December 22, 2024). The virus is still circulating; within the 28 days preceding December 8, 2024, at least 4,399 cases had confirmed the presence of SARS-CoV-2 (Fig. 8.3a). The virulence of the virus has

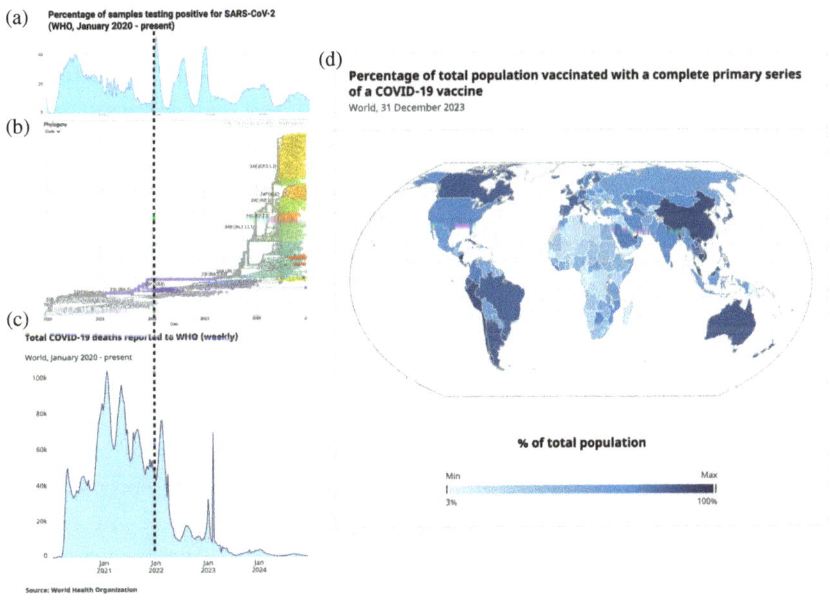

Figure 8.3 Evolutionary Dynamics of SARS-CoV-2 and Vaccination. (a) Rate of detection of SARS-CoV-2 virus in the population between January 1, 2020, and December 14, 2024. (b) Phylogenetic tree of different SARS-CoV-2 variants for the same period as (A). The main variant lineages are color-coded. (c) Weekly deaths during the same period. (d) Percentage of people who have been vaccinated by December 2024. The vertical dashed line through A–C marks December 31, when the bulk of this chapter was written. [Data courtesy of the World Health Organization, accessed on December 14, 2024. https://www.who.int/emergencies/diseases/novel-coronavirus-2019]

been falling, however, since approximately March 2022 (Fig. 8.3c). Accompanying this fall in virulence, the number of major virus variants in the infected population, by contrast, has exploded since March 2022 (Fig. 8.3b), exactly as predicted earlier in this chapter. Meanwhile, a large majority of the world population, except in Africa, has been vaccinated (Fig. 8.3d).

An intriguing possibility is that the immunity to the virus provided by the vaccine might be imposing an evolutionary selection on the virus, which is driving the emergence of new variants that are perhaps even more infectious than the original virus but are less virulent (Duerr et al. 2023). Thus, insidiously, natural selection by vaccines might be directing virus evolution to evade the vaccine-induced immunity (Lewnard et al. 2023).

The last pandemic had hardly left a family in which at least one member did not personally know someone who had succumbed to the virus. Among my own friends and relations, there were four deaths; one was younger than 55 years, yet another, only 40, survived with damaged lungs. The impact of the pandemic on the economy and society has been profound. As of this writing, disruption of the global trade infrastructure is still fueling financial, social, and political instability. On the positive side, we have discovered much about the science of virus evolution and how to combat a future pandemic. These are important lessons for the future readiness of nations, for another pandemic will certainly come.

Chapter 9

Making Proteins Fold

It is clear from previous chapters that the engines of biological organisms on earth are proteins and nucleic acids. It should also be clear that the shapes of proteins define their biological (chemical) function. Thus, a very interesting problem arises: how do proteins form their specific structures? Is it possible to predict their three-dimensional folded structures from their sequence of amino acids alone? This is an important question because if we can tell how a protein chain might fold, we could potentially engineer its function. As it turns out, this was one of the most challenging problems of biology in the 20th century. By the second decade of the 21st century, however, that challenge has been largely solved, and we will see in this chapter the nature of this challenge and how, without Darwinian evolutionary theory to guide us, we would never have found this solution.

***

In our fictional parallel universe, three pioneers of protein structure and dynamics gather at a pub around a low table. A hard rain drummed on the hazy glass panes, and the faces of our protagonists were dimly visible in the semi-darkened room. Our characters are Charles Anfinsen, the Swedish biochemist who first demonstrated that proteins could fold spontaneously; Linus Pauling, the famous American chemist and statesman who systematized chemistry by demonstrating how chemical bonds are formed and played an important role in the nuclear disarmament movement; and Cyrus Levinthal, an American physicist–molecular biologist who is considered the father of computer analysis of protein structure.

Anfinsen: My experiments with ribonuclease A showed that heating it destroys both its structure and enzymatic activity. But when I slowly cool it under the right conditions, the enzyme regains almost 100% of its activity. This demonstrated that all the information for folding into its native structure is encoded in the protein's primary sequence.

Pauling: Intriguing, but not entirely surprising. I've long said that structure determines function. I proposed this for hemoglobin decades ago, showing that sickle cell anemia is due to a structural change in the protein. Your observations align with this principle.

Anfinsen: Perhaps. But the most striking part is that the native structure, complete with its alpha helices and other features, reforms perfectly without external guidance. This suggests a remarkable intrinsic property of the sequence itself.

Pauling: Yes, yes, the alpha helix. A beautiful structure, and it underpins much of protein chemistry. But tell me, how do you know the structure was entirely lost when you heated the enzyme? Could fragments of structure have persisted, even if not detectable?

Anfinsen: Good point. I used spectroscopic methods to measure the structural content of the protein at high temperatures. The results showed no detectable structure. In contrast, the room-temperature protein showed strong signals for structures like alpha helices. You'd be familiar with those.

Pauling: Familiar? I discovered the alpha helix, and the Nobel Committee recognized its importance, though I thought my valence-bond theory deserved more acclaim. Your findings are intriguing. Even if the average structure signal is gone, individual molecules might retain small fragments. Spectroscopy might miss those.

Levinthal (interjects): If I may add, this raises another question. Proteins are polymers of amino acids, each linked by peptide bonds. These bonds allow rotation around two key angles, phi and psi, constrained by quantum mechanics. If we assume just three stable angles for each bond, then for even a small protein of 101 amino acids, there are 3^{200} or roughly 10^{95} possible conformations.

Pauling: Yes, I've done the math. 10^{95} is a staggering number. And if a bond rotates in a picosecond, it will take 10^{83} seconds to explore all conformations randomly. That's longer than the universe has existed.

Levinthal: Exactly. And yet, proteins fold in minutes or hours. This is the paradox: how does a protein navigate this astronomical space of possibilities to find its unique native state so quickly?

Pauling: Fascinating. Let's name it "Levinthal's paradox." It's reminiscent of other great paradoxes in science. Resolving it could advance our understanding of protein folding. But I doubt there's a single folding pathway for proteins. Perhaps a small number of alternate pathways exist, governed by physical laws.

Levinthal: I propose that evolution has optimized these pathways. Proteins that fold inefficiently would be selected against, leaving only those with robust and efficient folding mechanisms.

Pauling: An interesting thought. Delbrück would agree. He'd say Darwin has the last laugh here, as evolution fine-tunes even the molecular details of life.

Anfinsen: Incredible. What began as a simple experiment to restore enzyme activity now connects to principles of physics, chemistry, and biology. It's remarkable how much we can infer from something as basic as protein folding.

Pauling: Indeed. And all this over a light beer, though I'd have preferred a stout. Anfinsen, your work adds another chapter to our understanding of life.

***

9.1 Protein Engineering Is Important for Biotechnology

The reason we should care about understanding how proteins fold is because proteins are the workhorses of biology as well as of modern medicine and biotechnology.

Many proteins are enzymes and signaling proteins, which catalyze chemical reactions that are highly inefficient at the normal range of temperature and pressure that living beings generally thrive in. Some proteins are regulators of the activities of enzymes. The latter's activity is modulated by their mutual interaction; alternately, some regulatory proteins control the structure and dynamics of DNA or RNA by binding to nucleic acids. The ability to bind to or interact with another protein (or a some other molecule) defines the property of most proteins, and perhaps the most famous among these are the antibodies, which bind to foreign proteins and inactivate the latter. Some proteins constitute ion- or chemical-selective channels or gates that control the flow of molecules across cellular compartments. Other proteins are so-called structural proteins, which provide a substratum upon which other molecules can anchor.

Since the early 1970s, an industry of biotechnology has emerged for making proteins to accomplish manufacturing tasks. Globally, the entire biotechnology industry is valued at several hundred billion US dollars, providing employment to thousands of people and producing high-value products such as novel pharmaceuticals, industrial, and agricultural products. A substantial fraction of pharmaceutical and biotechnological products constitutes protein-based therapeutic products such as antibodies and other protein drugs, commercially known as "biologics" (also includes RNA and DNA products), with an estimated 2021 market value of over US $461 billion, which is expected to grow at an annual rate of about 10% for the next nine years. Synthetic biology is an approach for engineering artificial biological systems that are not found in nature. A large part of synthetic biology constitutes the engineering of novel proteins that nature has not yet invented. The estimated market value of synthetic biology applications for human use is expected to grow to over US $80 billion by 2032.

The production of industrial enzymes, with a current market size valued at approximately US $12 billion, is another goal of protein engineering. Proteases, enzymes that degrade proteins, and lipases, which break down certain fats and oils, are used in the food-processing industry and as pharmaceuticals. The Danish biotechnology company *Novozymes* produces novel proteases and lipases to make detergents to remove stains and accomplish efficient cleaning. Companies such as *DSM* and *DuPont* make amylases that break down starch, which are then used in the paper industry as well as in beverage production. DNA polymerases are enzymes involved in DNA replication and amplification. They are crucial in molecular biology techniques such as polymerase chain reaction (PCR). Companies such as *Thermo Fisher Scientific*, *New England Biolabs*, and *Takara Bio* have commercialized DNA polymerases used in research and the diagnostics industry. The widely used coronavirus detection technology during the early phases of the last pandemic, which was used for monitoring and making public health decisions globally, depended for its success on the industrial production of DNA polymerases. A vitally important emerging industry in response to climate change is the production of biofuels and carbon assimilation to reduce the impact of fossil fuels by diverting energy sources to renewable resources. These emerging industries are experimenting with novel industrial enzymes to facilitate the conversion of plants growing in fallow lands into biofuels.

The foregoing handful of examples are sufficient to demonstrate that the engineering of proteins with novel functionalities is a commercially important endeavor. Understanding the evolutionary relationships among proteins has been a key to successful protein engineering, which we will explore in this chapter.

9.2 Chemistry Constrains the Number of Possible Protein Structures

The great American physicist Richard Feynman (1918–1988) once exclaimed that he understood something when he could take it apart and put it together. He gave the example of a radio: he understood the radio if he could put together a radio that worked. What he really meant is that if we are to engineer an object, we must first understand how the object's functionality works.

In biology, structure defines function, which is especially true of proteins. If we were to design and engineer a novel protein, we would need to understand how the protein's shape or structure determines the function that we target.

The structure of a protein, a polymer of amino acids linked by peptide bonds, is primarily determined by the interaction of the amino acid side-chain residues with the solvent (generally water) and with other atoms that compose these side chains, which in turn determines the most stable values of a pair of angles, the phi and the psi angles (Fig. 9.1a), of its peptide backbone.

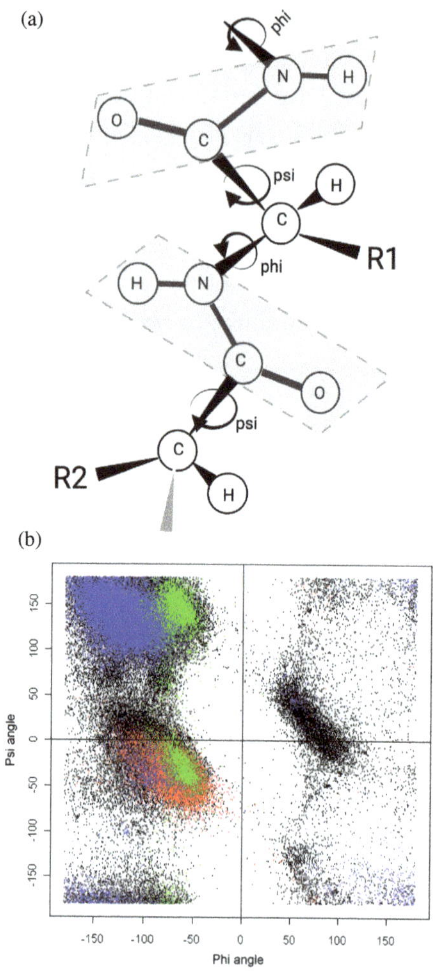

Figure 9.1 Protein structure is determined by the interaction of amino acid side chains with solvent and with one another, which fixes the polypeptide backbone geometry. (a) The phi and psi bonds can rotate and attain a stable value depending on side chain (R1, R2) properties. The shaded bonds are planar. (b) Ramachandran plot of the phi-psi angle combination of all peptide bonds of 500 naturally occurring proteins with no detectable sequence similarities. Each point represents a peptide bond. Red, alpha helix; blue, beta sheet; green, proline polymers; black, coils. [Panel B has been adapted from Song et al. (2012) under the Creative Commons Attribution License]

These two so-called dihedral angles can rotate around 360° but natural proteins tend to have a limited set of dihedral angles, which can be seen in a plot of these dihedral angle combinations of every peptide bond of a set of naturally occurring proteins (Song et al. 2012) in a form that has been named the Ramachandran plot after the Indian biophysicist G. N. Ramachandran (1922–2001), who first formulated it (Fig. 9.1b). The task of protein structure prediction primarily involves setting each of the dihedral angle values and subsequently the positions of atoms of the side chains (R1, R2, etc. in Fig. 9.1a).

9.2.1 *Proteins fold spontaneously to their lowest free energy states*

If a protein does not exist in nature — as is often the case for a protein engineer trying to make a novel protein with a heretofore unheard-of function — then we must properly fold the candidate protein and predict how its function might be influenced by its folded structural form. If it is a protein that is designed and engineered *de novo*, considering Levinthal's paradox, the task of predicting its folded structure (so-called *ab initio* folding prediction) might seem impossible. And yet, some proteins fold in microseconds to milliseconds. Anfinsen and others had the obvious idea that a protein's native structure must be at the lowest free energy state, which is a consequence of the laws of thermodynamics. However, how a protein gets to its lowest-energy, therefore the most stable, configuration remains a difficult question. An explanation of the physical mechanism as to how proteins might fold spontaneously to their lowest-energy state was first pointed out by a group of physical chemists working then at the National Institutes of Health (NIH) in the US.

9.2.2 *Protein folding pathways are constrained by chemistry*

Biman Bagchi, a brilliant physical chemist, was an undergraduate student a year ahead of me in Presidency College (now, Presidency University) in the eastern Indian city of Calcutta. In 1991, Bagchi was spending time at the NIH. One day, on a 30-minute commute to the laboratory in a public bus, he was reading a semi-autobiographical book, *What Mad Pursuit*, by Francis Crick (1916–2004). There he came across a reference to Richard Dawkins's whimsical example of evolution by the accumulation of small changes, which we have already discussed in Chapter 8 (Section 8.3): that of a monkey typing randomly and coming up with the words from *Hamlet*, "*Methinks it is like a weasel*." Bagchi suddenly realized that the solution to Levinthal's paradox might be just that. Deep in thought, Bagchi missed his bus stop that day and had to take a train back to the NIH. Following up from this thought, in a 1992 paper Bagchi and colleagues wrote:

> *"In both examples, folding proteins or writing Hamlet, biased searches are much more effective than completely random searches."*

This trio (Zwanzig, Szabo, and Bagchi 1992) of physical chemists derived a master equation that related the mean time it should take for a protein to fold to the correct combination of phi and psi angles for all peptide linkages of that protein in terms of the available thermal energy. The bias in its search process was just the energy term. An equation that they derived is given below:

$$\tau \cong (1/Nk_0)\,(1 + k_0/k_1)^N \tag{9.1}$$

Here, τ is the mean passage time (in seconds) it takes a protein starting from a random collection of bond angles (a random mixture of correct and incorrect angles, diagrammatically expressed as a random order, for example, *ccicicïiiccicicicccc*..., where *c* is the correct angular value and *i* is an incorrect value) to first get all of its bond angles correct (i.e., *ccicicïiic-ciciciccc*... → *cccccccccccccccc*....); k_0 is the rate constant with which one correct bond angle can become incorrect ($c \rightarrow i$); k_1 is the opposite rate constant, with which an incorrect bond angle becomes correct ($i \rightarrow c$); and *N* is the total number of peptide bonds. It can be shown using the Arrhenius equation that $k_0/k_1 = \nu \cdot e^{-U/kT}$, where U/kT is the energy term with *U* as the energy penalty for an incorrect bond, *T* as absolute temperature in degrees Kelvin, *k* as Boltzmann's constant, and ν as the number of possible bond angles (here it is assumed to be 3 for phi and 3 for psi, therefore a total of 6). Note that under the condition $U = 0$, that is, with no energetic penalty term for incorrect bond angles, Equation 9.1 reduces to

$$\tau = (1/Nk_0)\,(1 + \nu)^N \tag{9.2}$$

Equation 9.2 describes the unbiased (random) search process that Levinthal saw as a physical impossibility.

To examine the implications of equation 9.2, let us take a toy example. For a protein with $N + 1$ amino acids, there are N peptide bonds. Thus, for a protein with 101 amino acids, there are 100 peptide bonds, which means that $N = 100$, $\nu = 6$. Assume that the rate constant for incorrect-to-correct bond angle formation, $k_1 = 10^9$ per second, and the opposite, the rate constant for correct-to-incorrect bond angle formation, which includes the penalty term, $k_0 = 10^9 \times \nu \cdot e^{-U/kT}$ per second. For these values, Fig. 9.2 shows a plot of the mean first passage time for folding to the correct state

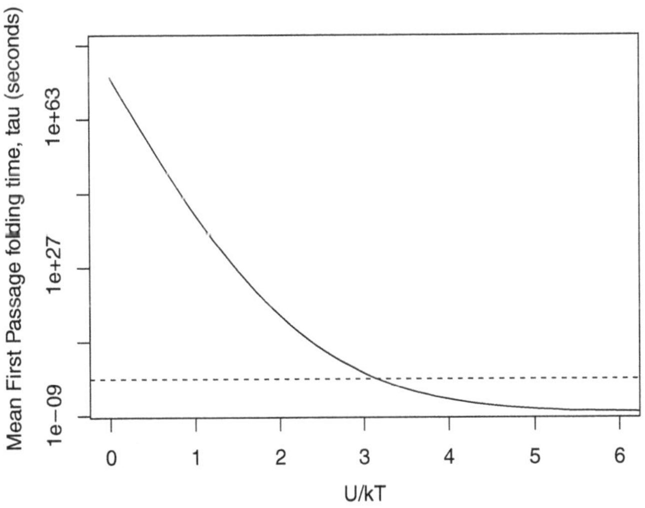

Figure 9.2 Estimated mean first passage time for correct folding of a protein of 101 amino acid residues as a function of thermal energy penalty (U/kT) for incorrect bond angles. The biologically significant domain of the mean folding first passage time (of the order of $\sim e^{-4}$ or about 18 milliseconds) corresponds to an energy term of ~3.7 in units of *U/kT*. The dashed line corresponds to 1 second, which intersects the curve at about 3.2 units of *U/kT*. [Calculated using equation 9.2]

for various values of the energy term in units of *U/kT*. It is quite clear that at *U/kT* of ~3.2 units, the protein will spontaneously fold in approximately 1 second. This is a physically reasonable condition and is easily reachable for a small protein within the temperature range used in Anfinsen's laboratory experiments.

Thus, these simple statistical thermodynamic constraints greatly reduce the severity of Levinthal's paradox. Therefore, there is no need to postulate the presence of unique pathways of folding for most proteins. Thermodynamic parameters constrain the search space of folding geometry from an astronomical number to a manageable range.

While the above considerations resolved, by a proof of principle, Levinthal's paradox, they did little for us to predict the precise structure of a protein or whether a protein will indeed fold to a unique structure at all.

Levinthal himself, however, never thought of the protein folding problem as a paradox. He clearly understood that the solution to the problem must be the existence of directed protein-folding pathways, guided by thermodynamic and kinetic rules, well-honed by natural selection for well-folded proteins. It is well known that proteins that fail to fold properly clog up the intracellular machinery by precipitating and aggregating — a phenomenon associated with a host of degenerative diseases, such Alzheimer's and Parkinson's diseases, to name but two of the most prevalent diseases of this type. It remains to discover these pathways, to avoid problems with engineered proteins that might not fold easily to functional forms, and, per chance, to cure diseases that are presumably caused by protein misfolding.

9.3 Computer Search Using Energetic or Geometric Constraints on Protein Folding Is Computationally Hard

If the search space for bond angles is physically constrained by thermodynamic parameters, could high-speed computer models using these constraints find the most probable structure of a folded protein?

9.3.1 *On the hardness of computing*

How rapidly does a computer search? This depends on two factors: the speed of the computer hardware (largely determined by its central processing unit, or CPU, and the random-access memory, or RAM) and the efficiency of the algorithm used. The former is a matter of engineering, with continuous improvements in the speed of basic operations. The latter, governed by mathematical principles, defines how long it takes to find a solution — a notion known as Computational Complexity.

Problems in the Polynomial Time (**P**) class are those where computing time grows as a polynomial function of input size (e.g., $C_t \sim ax^n$, where C_t is the computing time, a is a constant, x is the number of input units, and n is a fixed exponent). These problems are considered ractable, with computation time increasing manageably as input grows (Fig. 9.3). A broader class is Nondeterministic Polynomial time (**NP**): problems for

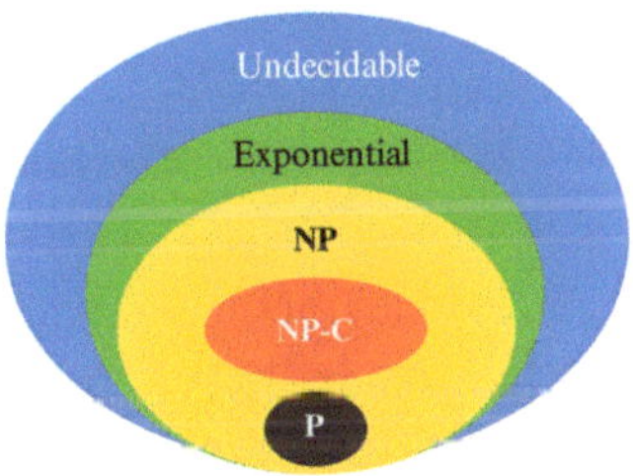

	Problem Size	Time to Compute
P	10	1 microsecond
	100	10 microseconds
NP	10	1 microsecond
	100	3.9×10^{11} Centuries

Figure 9.3 Classes of computational complexity. Problems in the Polynomial time (P) class are the most tractable: computing time grows manageably with input size. P is a subclass of the broader Nondeterministic Polynomial time (NP) class, which contains problems whose solutions, even if difficult to find, can be verified efficiently. Within NP, the NP-Complete (NP-C) problems are the hardest: no known algorithm solves large instances efficiently, yet any candidate solution can be checked quickly — which is what makes heuristic and approximation approaches viable. NP problems that are neither in P nor NP-C are NP-Hard; unlike NP-C, solutions to these cannot be verified efficiently either. Beyond NP lies the Exponential class, and further still the Undecidable class — problems for which no algorithm can ever guarantee a correct answer. The table illustrates the practical stakes: for a problem not in P, scaling from 10 to 100 input variables increases computing time from 1 microsecond to approximately 3.9×10^{11} centuries. [This figure was drawn using Biorender.com under an academic license]

which a proposed solution can be verified in polynomial time, even if finding that solution is hard. Importantly, **P** is a subset of **NP** — every problem solvable quickly can also be checked quickly — and the famous open question in computer science is whether the reverse holds (**P = NP**).

Within NP, the NP-Complete (**NP-C**) problems are the hardest: no known algorithm solves large instances efficiently, yet any proposed solution can still be verified easily. This is what makes approximations powerful — a heuristic or biologically motivated constraint can guide the search, and the resulting answer can be checked. Finding a protein's folded structure from a randomly unfolded state, or an optimal travel route visiting many locations without retracing steps, are characteristic examples. Finally, **NP-Hard** problems lie outside NP entirely: they are at least as hard as NP-Complete problems, but unlike them, proposed solutions cannot be verified efficiently either — representing the outer boundary of computational intractability.

How hard are the **NP-Hard** problems? Assume that a particular **NP** problem with $n = 100$ variables takes 1 millisecond to be solved by a present-day computer. Since the runtime for such a problem scales as $T \sim O(a^n)$, where a is an arbitrary constant (>1), assuming $a = 2$, the runtime need for solving a problem with $100n$ is 1.34×10^{2923} Centuries! That hard.

The *ab initio* search for the lowest energy folded structure of a protein unfortunately is **NP-Hard** (Fraenkel 1993). This means that it is nearly impossible to do a systematic search for and find the minimum energy configuration of a reasonably sized protein using any current computer.

9.3.2 *Does competition or collaboration accelerate progress?*

In the first decade of the 21st century, there was great excitement about using high-speed spectroscopic methods to identify protein-folding intermediates and thus constrain the search space for protein folding. High-speed parallel computers, some with custom architectures, were expected to discover all protein folding rules from the structures of these intermediates and enable computational folding of any protein. To stimulate progress, leaders in the field announced innovative competitions for the fastest computational methods to fold proteins.

One of the most influential competitions has been the Critical Assessment of Techniques for Protein Structure Prediction (CASP), first held in 1994 and now a biennial event. CASP brings together researchers worldwide to assess protein structure prediction methods, where participants predict the 3D structures of proteins based on their amino acid sequences. CASP has significantly advanced protein structure prediction by comparing methods, identifying strengths and weaknesses, and fostering improvements. Other initiatives, like the Protein Structure Prediction Center's Free Modeling (FM) category and the DREAM project's protein folding challenges, have also driven progress. These competitions have raised awareness about protein folding's importance in drug discovery, molecular biology, and biotechnology, fostering new collaborations among researchers from diverse fields.

Until the mid-2010s, emphasis was on discovering the physical and chemical rules governing polypeptide backbone geometry, influenced by interactions among amino acid side chains and water molecules. Despite challenges in determining the structure of water around proteins, progress continued, with new records set at each CASP meeting. By 2015, researchers felt that the main limitations were computational resources and better solvation models, engaging fast spectroscopic methods to experimentally determine protein folding pathway intermediates.

9.3.3 *A limited number of elementary folds*

It was soon realized that Darwinian evolution has utilized a limited number of protein-folding motifs, primarily consisting of the right-handed alpha-helix, the beta-sheet, and the beta-turn. Each of these motifs corresponds to specific clusters of phi-psi angle combinations observed in natural proteins. Rather than fixed values, these torsional angles vary within certain ranges. Natural proteins combine these elementary motifs into a limited number (~133,000–170,000) of topological domains, such as helix-turn-helix, four-helix bundle, beta barrel, and alpha-beta-alpha (Fig. 9.4), despite the theoretical possibility of numerous other structural features (Cheng et al. 2014). Kinetic studies showed that short protein regions often fold spontaneously and independently into these structures, which

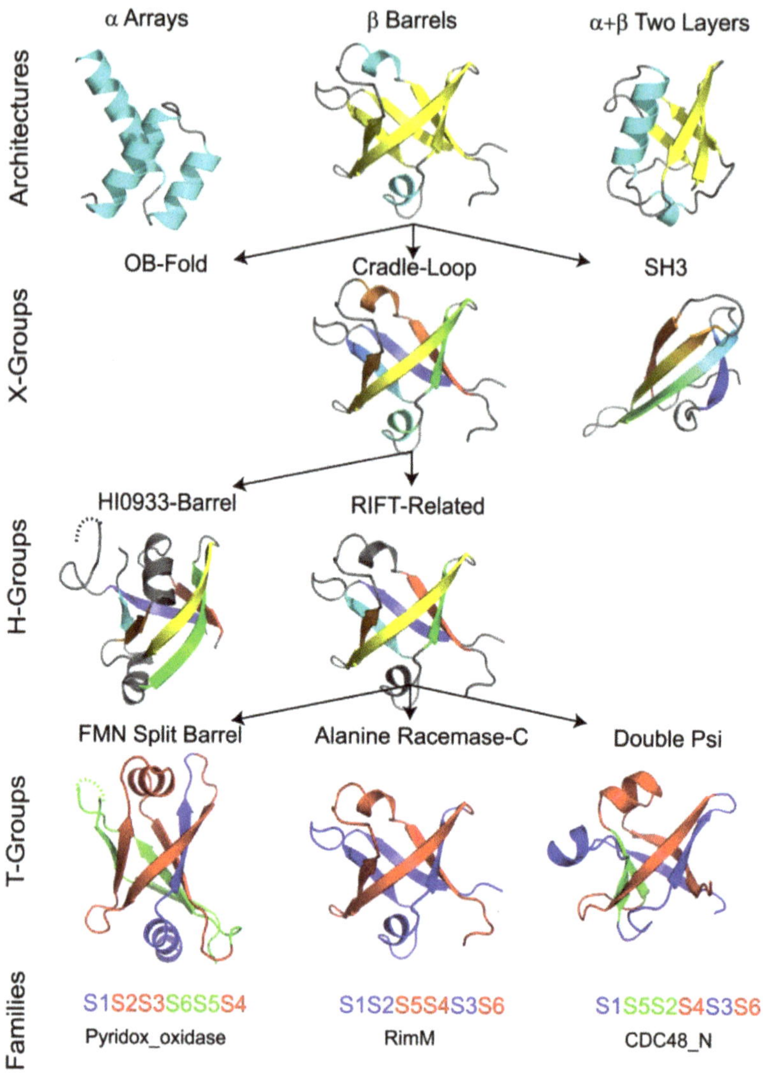

Figure 9.4 Examples of a few protein domains. Domains are artificially selected topological units that are often repeated in known crystal structures of proteins. There are several different ways that domains could be classified. Some emphasize purely topological criteria conserved over evolution, others combine topological criteria with functional properties. These domains belong to the Evolutionarily Classification of Protein Domains (ECOD) class. [Reproduced without modification from Cheng et al. (2014) under Creative Commons Attributional License agreement]

then cooperatively fold together, making protein folding a multi-staged, cooperative process. Computational *ab initio* approaches modeled these kinetically separated conformational states to form small unit folds that interact cooperatively to create higher-order structures. Despite these insights, achieving accurate *ab initio* protein folding remained challenging until the CASP meeting in December 2018.

9.4 Natural Evolution Instructs Machine Learning Algorithms to Fold Proteins

Since the advent of high-speed computing and neural networks in the 1990s, artificial intelligence (AI) has found numerous inroads into genomics-inspired biological and medical sciences. A grand success story in this field concerns the *ab initio* folding of a protein using deep learning, ushering in a new era of computationally enabled molecular biology.

To students of computer science, AI and machine learning (ML) do not require definitions, but for those readers who might be biologists or students of medicine, we should emphasize a critical distinction between the two. AI is a general term that encapsulates the idea that algorithmic computation can analyze large datasets to determine multivariate relationships among data objects. ML, on the other hand, is a more specific discipline within AI that concerns itself with learning rules from previous data to predict future data. ML has two main approaches, unsupervised and supervised ML (Fig. 9.5). The former is an approach to classify and/or cluster the data into distinct or overlapping groups based on similarities of patterns within the data using arbitrary mathematical rules that are specified by a human operator.

In the supervised approach, the data consist of a set of input and output variables (which could be nominal or quantitative). Depending on the characteristics of the output variable, it is either a classification or a regression problem.

Supervised ML requires previous data and certain instances in which subsets of data points are preassigned to specific classes. The algorithm then trains an internal model; using that trained model, the algorithm predicts the assignment of new data to the classes or predicts the values of the dependent variable.

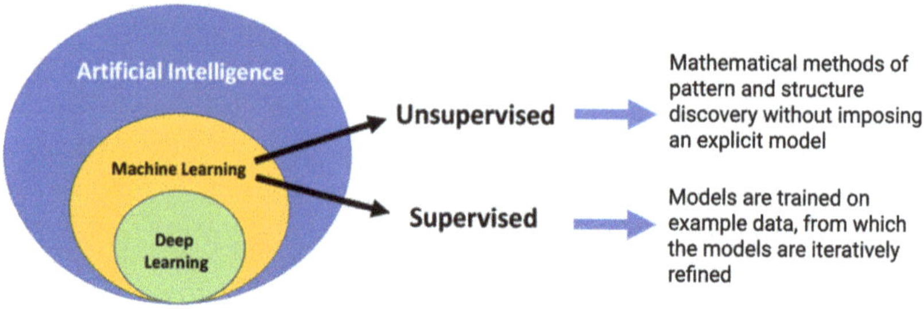

Figuure 9.5 Classes of machine learning.

Deep Learning is a specialized subset of ML, which employs neural networks (hierarchically arranged electrical logic circuits) with many layers, typically referred to as deep neural networks. These networks use feedforward connections and a key training process called backpropagation, which adjusts the weights of the network during the training process. This mechanism allows neural networks to model complex nonlinear relationships in data — relationships that are often beyond the reach of standard statistical tests.

Since Levinthal's paradox suggests there must be favored pathways of protein folding, ML might be one way to circumvent the high-complexity combinatorial search space by using as features the physical and structural properties of many proteins for which structures are solved (~174,000 proteins).

While rules of chemistry might be sufficient to dictate that all proteins must have at least some structural relatedness due to chemical constraints, it is unclear whether that is necessarily true of the universe of all possible proteins. On the other hand, the Darwinian evolutionary paradigm implies that all naturally occurring proteins, encoded by their respective genes, must embody shared structural information because their encoding genes are themselves evolutionarily related. Consequently, Deep Learning algorithms trained on the currently existing (and fast growing) compendium of protein structures experimentally determined by techniques, such X-ray crystallography and cryo-electron microscopy, has now been engaged widely to solve various sub-problems related to protein folding (Torrisi, Pollastri, and Le 2020). Most notably, Deep Neural Network was able to train the algorithm to construct and optimize a force-field potential, which was used to predict distances between pairs of atomic residues in their stable structural states within a protein (Senior et al. 2020).

A deep neural network, AlphaFold-2, made by the supercomputer DeepMind of Google, has outperformed every other competing protein-folding algorithm — the atomic coordinates of a few of its predicted structures are indistinguishable from those obtained by cryo-electron microscopy (at 0.3–0.6 nanometer resolution) of the folded proteins (Callaway 2020). These groundbreaking advances in protein structure prediction was recognized by the award of the 2024 Nobel Prize in Chemistry to three of the main contributors to this field: David Baker (1962–) (University of Washington, Seattle), Demis Hassabis (1976–), and John Jumper (1985–) (both of Google, DeepMind). Moreover, the Physics Nobel Prize in 2024 went to the two theoretical computer scientists whose work laid the foundation of neural network computing — John Hopfield (1933–) and Geoffrey Hinton (1947–).

There are now quite a few resources in the public domain for *ab initio* protein folding so as to try one's hands at the protein-folding problem (Adhikari 2020) and, per chance, to improve upon these. It is unlikely that training on existing proteins would have succeeded unless the existing proteins were evolutionarily related and therefore likely followed structural rules that have been conserved by natural selection during evolution. The problem otherwise would have remained intractable because of the immense structural space occupied by even simple proteins.

AlphaFold-2 and a slower, but on average more accurate, RoseTTAfold-2, which additionally uses a method to provide a random jitter to a few crucial atoms from their mean positions and lets them settle according to rules in statistical thermodynamics — the so-called diffusive perturbation method — both use evolutionary relationships among known proteins to train their folding models. This dependency makes it difficult for these computer models to predict proteins with no known or relatively few sequence relatives.

A solution was provided by the recent successes of large language models (LLMs) to learn the "language of folding" inherent in the sequences of naturally occurring proteins to rapidly deduce the structure of other proteins. Here again is the underlying assumption that evolutionary relatedness by descent has left footprints of an underlying rule of folding (among vastly numerous chemically possible ones) whose signatures are learnable by a language model, just as a seafarer in times past used to learn the language of the inhabitants of a strange land.

One prominent success with LLM has been the construction of ESM-2, whose prediction accuracy approaches that of AlphaFold-2 but is significantly faster (Lin et al. 2023). Yet another group has succeeded in using LLM to outperform Alphafold-2 by a millionfold (Chowdhury et al. 2022). This latter model, RGN2, performs especially well on proteins for which few evolutionary relatives are known, apparently by learning the underlying geometric relationships between atoms and their relative distances within proteins akin to the grammatical elements of a natural language (Chowdhury et al. 2022). A pity is that it is not yet feasible to decipher what geometric rules of protein-folding grammar the machine has learned, because deep neural networks act like black boxes. It is difficult, often impossible, to associate meanings in the strengths of feedback circuits that lend themselves to find an optimal solution to a complex problem, such as protein folding. There is an emerging area of machine learning — explainable AI — that needs more theoretical insights for future progress — perhaps only then we can trace back the evolutionarily conserved rules of protein folding that evidently these deep learning models have learned.

While we can now computationally fold proteins, whether the novel folded models represent the physical reality or not must depend on experimental testing. We can now make artificial antibodies far more efficient in attacking a virus protein than natural immunity can, right in our own laboratories in a collaboration with my colleague, Dr. Stefano Lonardi (University of California, Riverside). Stefano's student, Faisal Bin Ashraf, made a computer model that predicted an antibody, which my students have shown to bind more tightly than natural antibodies to the spike protein of the SARS-CoV-2 virus (Fig. 9.6), and it is at least equally, if not more, robust relative to the natural antibodies.

These are exciting times for biotechnology. Today we can start with a fully sequenced genome of a novel organism — say, a microbial genome found in the depth of the ocean or in the soil of the Brazilian

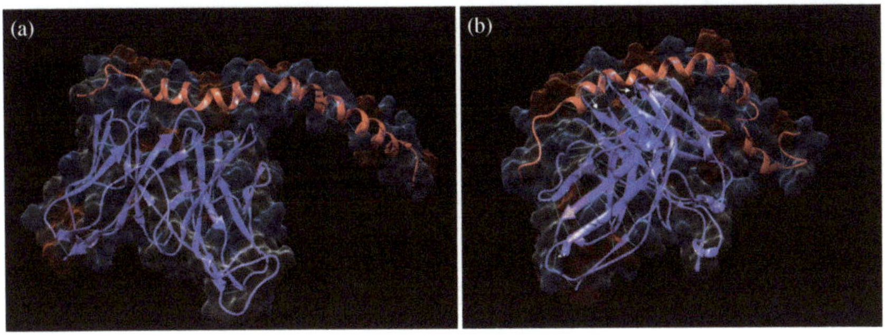

Figure 9.6 Naturally generated versus computationally evolved antibodies against the SARS-CoV-2 spike protein. (a) Complex of the natural antibody (blue) with a spike protein fragment (red). (b) Complex of an antibody evolved through a large language model and artificial computational evolution (blue) with the same spike protein fragment (red). The complex in (b) is 160-times more stable than the complex in (a). [Source: Author's laboratory]

rainforest — determine all of its encoded protein sequences using computational methods, and fold them *en masse* using AlphaFold-2 or ESM-2 or RGN2 (perhaps all three and take the best models). We can then generate computational models of interaction of a collection of small molecules with the computed protein structures and identify the best candidate small molecules that might regulate the activities of these proteins. All of this is now possible because of the marriage between artificial intelligence and the Darwinian paradigm of relatedness of protein sequences by descent.

Chapter 10

Biotechnology Meets Darwin

Here we describe how evolution conducted in a test tube can make novel reagents and drugs. The principles that guide these methods are not only fundamentally Darwinian, but they also point a way towards resolving another fundamental question of nature — how did life itself originate on Earth? Darwin was publicly silent on this latter question; evolution explains things once life starts. But ideas in molecular evolution not only help us cure diseases and allow entrepreneurs to build multibillion-dollar companies, they also imply that if molecules can evolve spontaneously in a test tube, then why not molecular assemblies that replicate — that is, life itself — evolve from molecules?

***

A dimly lit pub in a parallel universe, its wooden beams sagging slightly under the weight of ages, four figures sit around a circular table, pints (at the room temperature) in hand. The air is thick with the smell of roasted barley and ideas. A chalkboard menu offers "Primordial Soup of the Day." A. I. Oparin was a Soviet biophysicist who had independently proffered a theory of the origin of life; we had encountered J. B. S. Haldane in the past, and he had also offered his theory of how life originated from nonliving matter on Earth, and also Francis Crick, the co-discoverer of the DNA double helix; J. D. Bernal was a physicist with deep contributions to X-ray crystallography as well as to the history and philosophy of science.

Oparin: To the primordial soup! (he raises a glass) It's a shame none of us brought a beaker. Imagine the barman's face if we whipped up some amino acids right here.

Haldane: Oparin, you and your soup. You always overlook that it's not just the soup but the spice that matters — the redox gradient, the hotter and more Indian, the better it gets! How else does one coax lifeless molecules into a dance of increasing complexity?

J. D. Bernal: Spoken like a true materialist, Haldane. But tell me, do you suppose life truly began in your hypothetical oceans, or was it my beloved clays that whispered the first secrets of order into chaos?

Crick: Let's not delude ourselves. The key is not where life began but what it began with. RNA, of course! If you had your clay and soup, I had my Central Dogma to guide the way.

Oparin: Central Dogma, pah! Such arrogance, Francis. Do you forget that life is chaos before it is dogma? I laid out the primordial pathways that could lead to proteins long before RNA strutted onto the scene.

Crick: Oh, I concede your pathways, Sasha, but only as scaffolding. RNA is the true artist — it painted the first genomes and sculpted the first enzymes. Why, without RNA, your proteins would be as aimless as tourists in Cambridge without a map.

Haldane: Yet isn't it curious how protein engineering is now the darling of biotech? Evolutionary algorithms on supercomputers churn out enzymes in hours. The irony — computers mimicking blind chance to design what nature took billions of years to perfect.

Bernal: A poetic twist, indeed. And yet, while you quarrel over RNA and proteins, let me remind you that neither would be here without minerals. It is the structured rigidity of crystal lattices that provided the first templates. Even in biotech, we see the resurgence of inorganic chemistry in nanotechnology and biomaterials.

Crick: Touché, Bernal. But let us not dwell entirely on the past. The future of biotechnology lies in what I call "directed panspermia." Imagine engineering microbes to terraform distant planets.

Oparin: Ha! Spoken like a true imperialist. Always looking outward when the mysteries of abiogenesis right here on Earth are unsolved. First solve how your RNA arose, Crick, then conquer the cosmos.

Haldane: Ahh, the joy of not knowing, the thrill of discovery. Here's to evolution, in molecules and ideas!

Bernal: Indeed. And may we never stop questioning, even in a pub as fine as this.

Crick: Cheers to that. Now, someone pass me the menu — I'm curious about this "Thermophile IPA."

The conversation continues into the night, fueled by ale, curiosity, and the passion of those who dare to ask how life began.

***

A teenager hitchhiked from a suburb of Pittsburgh to Washington, D.C., to participate in a protest march against the Vietnam War. By 15, she had moved out of her home to an apartment by herself, where she paid the rent and supported herself by lying about her age to work as a waitress and a cab driver. Frances Arnold (1956–) went to Princeton University to study mechanical and aerospace engineering. Upon graduation, she went to the University of California, Berkeley, to do a PhD in chemical engineering instead, having to pick up undergraduate chemistry knowledge within a short time. Dr. Arnold ultimately came to the California Institute of Technology, where she is now the Linus Pauling Professor of Chemical Engineering, Bioengineering, and Biochemistry, and the Director of the Rosen Bioengineering Center. She won the Nobel Prize in Chemistry in 2018 for showing how to engineer proteins using Darwinian selection in a test tube.

The history of artificial Darwinian evolution of biologically interesting molecules — evolution in a test tube — did not start, however, with Frances Arnold. That distinction belongs to two different research groups, one headed by Larry Gold (1948–) of the University of Colorado at Boulder (Tuerk and Gold 1990), and the other by Jack Szostak (Ellington and Szostak 1990) whom we have already met in connection with the discovery of telomeres and who at the time was in Massachusetts General Hospital of the Harvard Medical School.

10.1 Molecules can be Artificially Evolved

For Darwinian evolution to work, there are three primary conditions: (1) there must be variants; (2) the variants as well as the primary entities must be able to replicate; (3) selection pressure can enrich the proportion of some variants through the replicative cycle. It occurred nearly simultaneously to Gold and Szostak that RNA and DNA provided an ideal opportunity to evolve these molecules in the test tube. The first two requirements for evolving RNA and DNA, that is, variants as mutant sequences, and the ability to replicate in the test tube, were well known. DNA can be replicated by the action of DNA polymerases, for example, by the famous polymerase chain reaction (PCR); RNA can also be replicated by the direct activity of RNA-dependent RNA polymerases or indirectly by the reverse transcriptase activity to make DNA followed by PCR. However, the third requirement was not at all obvious. This is because, unlike proteins, one generally did not think about DNA or RNA as having any function except indirectly as a storage of information in the form of a tape of nucleotide sequences.

10.1.1 *RNA structure can be complex enough to have enzymatic activity*

A few years earlier, it was shown by Tom Cech (1947–), at the time a colleague of Gold at Boulder, and by Sydney Altman (1939–2022) at Yale University, that RNA molecules can act as catalysts just as proteinaceous enzymes do. For their work, they received the Nobel Prize in Chemistry in 1989.

That RNA might have catalytic activity was first proposed by Francis Crick (Crick 1968) on the basis that the extra hydroxyl group present in RNA in contrast to DNA is chemically active. Crick realized that single chained RNA might fold by intramolecular base paring to complex geometrical shapes. If shapes (structures) were important for function, he speculated, then having a chemically reactive group (the extra hydroxyl group in RNA relative to DNA) might allow some RNA sequences to fold in complex shapes and, per chance, attain chemical reactivity (Fig. 10.1). Crick in 1968 and later Walter Gilbert (1932–) (Gilbert 1986) proposed that in the prebiotic atmosphere, RNA molecules might have been the first evolved entities endowed with chemical and biological function, which were able to self-replicate, and had generated a prebiotic "RNA World" upon which Darwinian selection might have worked. Later works have largely validated and refined this proposal [see, for example, (Harish and Caetano-Anollés 2012)].

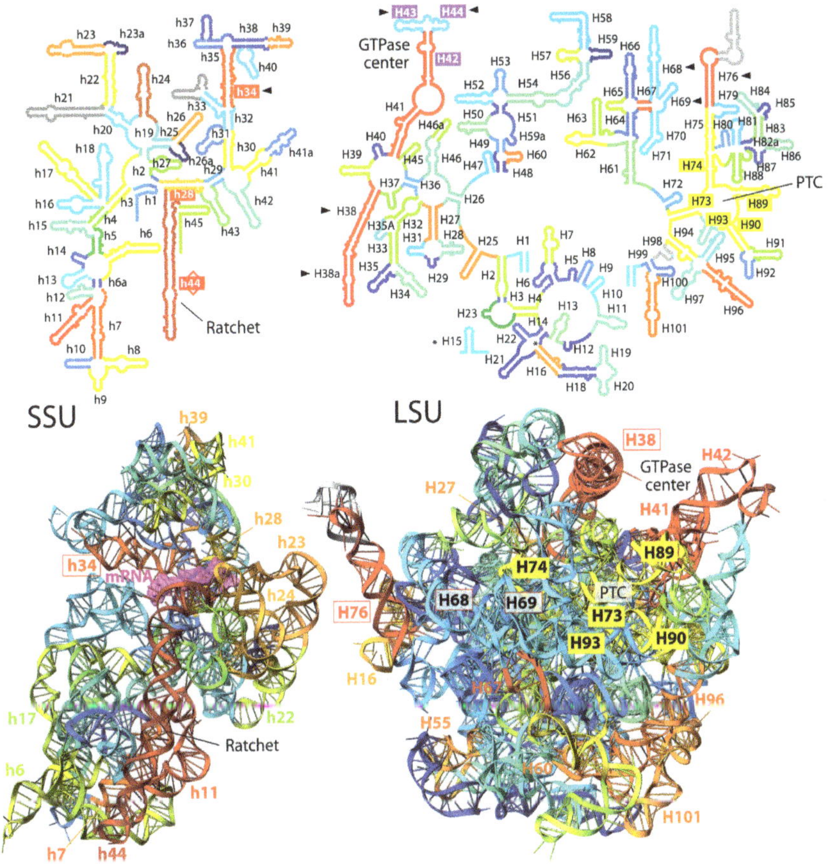

Figure 10.1 RNA can have a complex three-dimensional structure and enzymatic function. Here, the structures of two functional RNAs, the small ribosomal subunit RNA (SSU) and the large ribosomal subunit RNA (LSU), are shown. In the top two panels, a 2D representation of intra-chain base pairing is presented, and in the bottom panels, a pseudocolor rendition of the 3D structures is given. RNA helices are labeled H1, H2, etc. The RNA molecule in red is the mRNA that binds to specific regions of the ribosomal RNA. Ratchet, GTPase center and peptidyl transferase center (PTC) are three distinct enzymatically active domains. [Reproduced from Harish and Caetano-Anollés (2012) under the Creative Commons Attribution License]

By demonstrating that RNA molecules indeed have catalytic activity, Cech, working with an RNA of the protozoa *Tetrahymena*, and Altman, working with, an RNA of *E. coli*, showed that some RNA molecules have enzymatic properties that were until then thought only to be properties of proteins. In other words, RNA also has biochemical functions on which natural selection could work. Note that this is somewhat different from the more common role of nucleic acids to store information (of polypeptide chain sequence). Here, function means a chemical function, not merely an informational role.

It had independently occurred to both Gold's and Szostak's groups that technical solutions for all requirements for the test-tube evolution of nucleic acids had by then existed; all they needed to do was to find an appropriate strategy.

10.1.2 *Evolved RNA aptamers can bind small molecules or protein targets*

Szostak and his graduate student Andrew Ellington decided to evolve RNA sequences with a specific function (Ellington and Szostak 1990) (Fig. 10.2). Their natural selection was the ability of the RNA molecules to bind specifically and with high affinity to certain dyes, which were already known to bind with high selectivity and tightness to certain enzymes, resembling the cofactors of such enzymes.

Ellington and Szostak made a library of random short RNA molecules, generating up to 10^{13} unique sequences using random DNA templates. These RNA molecules were applied to dye molecules attached to inert beads in a column. RNA sequences that adhered to the beads were subsequently released by washing the columns with a high-salt solution.

The released RNA was converted back into DNA using reverse transcriptase, amplified through PCR, and transcribed back into RNA. This RNA was then reapplied to the dye columns, and the cycle of selection, replication, and transcription was repeated. After five rounds of this iterative process, starting with approximately 10^{13} initial RNA variants, only a few thousand sequences remained.

Upon sequencing, they discovered that most of these RNA sequences shared conserved motifs, showing high affinity for the dye molecules. Ellington and Szostak concluded that the RNA structural solutions to the dye-binding challenge were limited to a small number of different

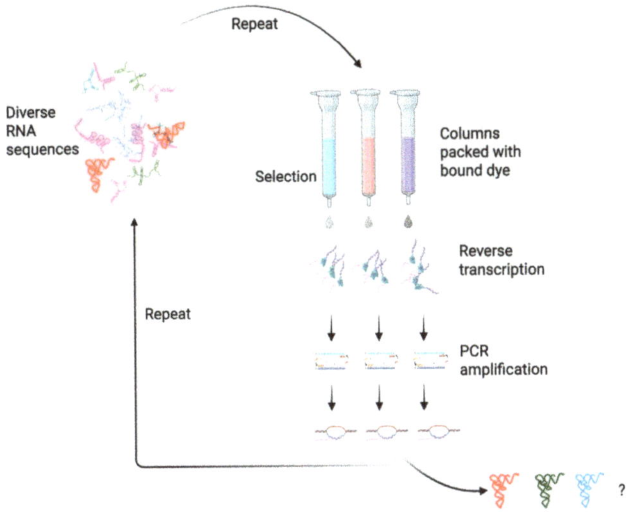

Figure 10.2 *In vitro* evolution of RNA for a novel chemical function. A diverse library of variant RNA sequences was synthesized and applied to columns packed with a dye. The columns were washed, and the bound molecules were eluted. The eluted RNA molecules were transcribed into DNA, then PCR amplified, followed by transcription back into RNA. The cycle of affinity binding to the columns with bound dye molecules, followed by amplification of the bound RNA, was repeated. The winners were identified, sequenced, and confirmed for selective binding to specific dye molecules. The procedure is reminiscent of natural selection but occurs on molecules to evolve a chemical function. Here, the function is the binding of a variant RNA to a specific dye molecule. An experiment like this was performed by Ellington and Szostak (1990). Tuerk and Gold (1990) evolved the ability of RNA molecules to bind to a protein molecule instead of a dye molecule. [This figure was drawn using Biorender.com under an academic license]

sequences. They named these short RNA molecules with specific binding properties "aptamers."

Gold and his graduate student Craig Tuerk's selection strategy (Tuerk and Gold 1990) was slightly different: They aimed to isolate RNA sequences that could tightly bind to a naturally occurring protein. Using a similar approach, they synthesized random RNA sequences and applied them to the target protein, which was immobilized on nitrocellulose filter paper. RNA molecules that did not bind were washed away, while the bound RNA molecules were stripped off, converted to DNA, and transcribed back into RNA for further enrichment.

This process of binding, selection, and amplification was repeated for six cycles. By the end of these cycles, the enriched RNA sequences displayed conserved motifs, indicating that the structural solutions for binding to the target protein were limited to a small number of RNA sequences. Gold and Tuerk named their version of this *in vitro* evolution method "SELEX" (Systematic Evolution of Ligands by Exponential Enrichment).

Gold's and Szostak's experiments paved the way for *in vitro* evolution of nucleic acid molecules with special biochemical properties: if there is a well-defined target molecule or an enzymatic activity, then it is possible to evolve molecules by brute-force selection and amplification. While in both approaches the universe of variant sequences was generated at the beginning of their experiments, it was immediately clear to researchers that it was possible to tweak their methods such that more random variations could be generated by adding a step of error-prone mutation during every replicative amplification cycle. To the best of the author's knowledge, these were the first instances of anyone attempting to artificially evolve, by natural selection, a set of molecules with a target function.

10.1.3 *RNA aptamer drug treats progressive blindness disease*

Works on SELEX and aptamers led to a significant clinical application with the development and launch of Pegaptanib (Macugen), an aptamer drug made by OSI Pharmaceuticals and Pfizer for treating wet age-related macular degeneration (AMD). AMD, often linked to diabetes and shares some vascular features with diabetic retinal diseases, is a leading cause of blindness. Pegaptanib is a synthetic aptamer designed to bind tightly to Vascular Endothelial Growth Factor A (VEGF), a blood vessel growth factor responsible for abnormal microcapillary growth and blood leakage into the retina in diabetic patients. When administered by an ophthalmologist through injection into the eye, Pegaptanib selectively inhibits VEGF activity. While it is not a cure for wet AMD, it has proven effective in delaying vision loss associated with the condition. However, the drug is no longer marketed in the United States for commercial reasons.

Currently, several other aptamer-based drugs are in clinical trials, with the aptamer drug market expected to expand. Unlike earlier RNA-based aptamers like Pegaptanib, these next-generation aptamers are typically made from DNA, chemically modified nucleic acids, or short peptides (peptide aptamers). These newer aptamers, developed using similar evolutionary selection processes (e.g., SELEX), offer advantages such as greater stability, reduced toxicity, enhanced binding properties through combinatorial diversity, and a lower risk of triggering allergic reactions.

10.2 Artificial Evolution of Proteins

When Frances Arnold started her laboratory at Caltech, it was widely believed that successful engineering of proteins with novel biochemical properties would require detailed insights as to how protein structure dictated function. Moreover, it was thought that the engineering solution for one kind of biochemical problem might be so unique that it would have rather scant lessons for another biochemical problem. In other words, there was unlikely to be a general method for engineering novel functionalities of proteins. Arnold demonstrated that one can engage the power of Darwinian evolution to rapidly evolve protein molecules, thereby enabling virtually limitless biochemical solutions.

10.2.1 *This watery life*

Life probably emerged in an aqueous environment, presumably in shallow seawater in the prebiotic atmosphere rich in organic compounds, which is why water constitutes about 60–90% of the mass of most living organisms, including those of the bristle-cone pines growing in highly arid environments. Most multicellular organisms would die if their body's water content were depleted below 60% by mass. A famous exception to this is the microscopic organism called the tardigrade, which can survive (though in an inert state) even when about 99% of its body water is depleted, which also enables them to withstand extremes of temperature, high doses of radiation, and even exposure to vacuum. Perhaps tardigrades could take the place of humans in interstellar voyages. An important consequence of water-based life is that proteins — the agents of chemical transformations that define life — are bathed in water and evolved in water. Water molecules around and inside protein molecules are necessary for the appropriate folding of proteins through hydrogen bonds and other longer-range interactions with amino acid residues. Since natural proteins have evolved in an aqueous environment and their structures are dependent on an abundance of water, most protein enzymes are insoluble in non-aqueous organic solvents. By contrast, the chemical industry, involving the synthesis of many organic and inorganic molecules, has a need to catalyze reactions embedded in organic solvents or in high ionic strength solvents (with low relative water content).

10.2.2 *Engineering subtilisin for activity in dimethyl formamide*

Arnold decided to grapple with the question of whether one can engineer protein-based enzymes that are stable as well as catalytically active in organic solvents or in solvents with high ionic strength, such as those with high concentrations of metal ions, in the presence of polar organic solvents, ionic detergents, or at high temperature.

The enzyme on which the first of these test evolutionary studies was conducted was called subtilisin E, an enzyme that degrades other proteins by hydrolysis of the peptide backbone (Fig. 10.3). Note that the same reaction — hydrolysis of proteins — can be accomplished by dissolving the protein substrates in dilute acid or alkali and boiling. The enzyme subtilisin E, isolated from the bacterium *Bacillus subtilis*, does not need an appreciable amount of acid or alkali, and can greatly speed up the hydrolysis reaction at the room temperature. Subtilisin E, however, loses its activity in the presence of the polar organic solvent dimethylformamide (DMF). Relative to the pure enzyme in aqueous solution, the enzyme exposed to an 80% solution of DMF in water retains only about one thousandth of its catalytic efficiency;[1]

---

[1] Catalytic efficiency describes how effectively an enzyme speeds up a reaction. It involves both how fast it works and how well it binds to the target molecules.

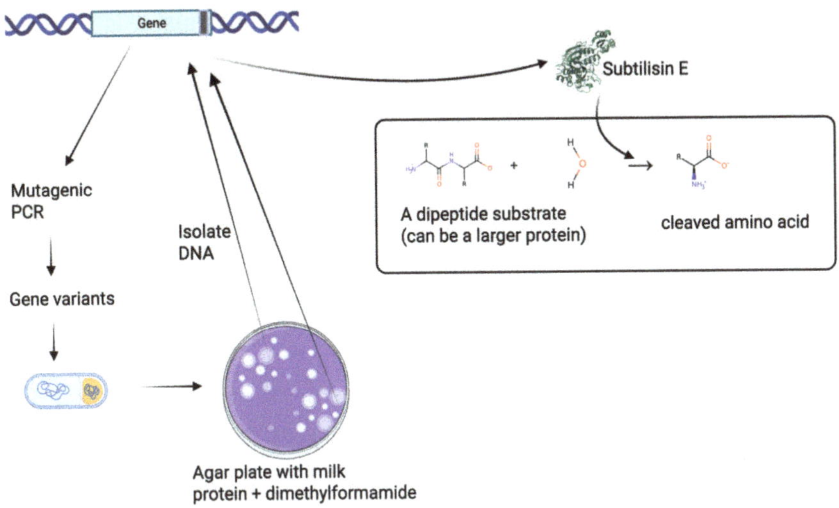

Figure 10.3 *In vitro* evolution of an enzyme with activity tolerant to dimethylformamide (DMF). Subtilisin E is an enzyme encoded by *Bacillus subtilis*. The enzyme hydrolyzes proteins and is therefore economically important. However, the natural enzyme cannot tolerate DMF, an organic solvent. Francis Arnold's laboratory produced a secreted version of the enzyme and its gene. They mutagenized the gene to generate variants, then introduced these into the bacteria, which produced variant proteins that were secreted into agar plates containing the milk protein casein and DMF (initially at low concentration). A halo around a colony was indicative of secreted subtilisin E activity around the colony. Bacterial cells from such a colony were picked, and this cycle was repeated with higher and higher concentrations of DMF, eventually evolving an enzyme adapted to high concentrations of the chemical. [This figure was drawn using Biorender.com under an academic license]

furthermore, approximately half of the original enzymatic activity in water is lost upon every 6 hours of incubation in 40% DMF. How can one engineer this enzyme to be more stable in DMF?

Starting with naturally occurring sequences of subtilisin E, students in Arnold's laboratory at Caltech decided to first make a few specific amino acid substitution mutations in the subtilisin E gene such that the encoded protein was predicted to be more stable in polar solvents. In choosing these amino acid substitution mutations, Arnold's laboratory built directly upon insights developed by the works of, among others, Australia-born protein biophysicist Brian W. Matthews (1938–) of the Institute of

Molecular Biology in Eugene, and of the British chemist Alan Fersht (1943–) of the Medical Research Council (MRC) Laboratories in Cambridge, UK, on the structural basis of thermostability in proteins.

Arnold's early work on crambin, a plant-derived toxic protein that retains its structure and activity in nearly anhydrous methanol, DMF, and glacial acetic acid, provided valuable insights. It highlighted the importance of tightly binding water molecules to both the inner and surface regions of the protein while minimizing water interaction with surface-exposed residues. This is challenging because protein folding typically results in hydrophobic residues being buried inside and hydrophilic residues exposed to the solvent. Drawing on these findings, Arnold proposed a set of fundamental rules for stabilizing proteins in polar non-aqueous solvents (Arnold 1988).

With her student K. Q. Chen, Arnold made multiple initial replacements of amino acids on the subtilisin E protein, which they thought should contribute to its stability in non-aqueous solvent (Chen and Arnold 1991). They expressed protein variants in *Bacillus subtilis*, which secreted the proteins onto agar plates (Fig. 10.3). When the plates contained casein, a milk protein, active enzymes secreted by the bacteria degraded the casein, creating halos around the colonies. The size of the halo correlated with enzyme activity: larger halos indicated higher activity, assuming equal protein production and stability.

By incorporating increasing concentrations of DMF into the agar, they identified variants that retained activity in this solvent. Some of these variants performed better than the original subtilisin E enzyme under moderate DMF conditions. One engineered enzyme, with three simultaneous amino acid replacements, demonstrated 38% higher efficiency in 85% DMF compared to the natural subtilisin E. Additionally, the effects of these three replacement mutations were found to be additive.

10.2.3 *Make random variants and select*

Encouraged that they were on the right track, the duo began to further mutagenize the subtilisin E gene,[2] hoping to generate variants that were

[2]They had introduced a fourth replacement mutation in the starting gene before random mutagenesis for reasons too complex to discuss here.

even more efficient in higher and higher concentrations of DMF (Chen and Arnold 1993). For every cycle of mutagenesis, they conducted in a test tube an error-prone PCR step — at suboptimal conditions for base-pair fidelity during DNA replication, such that base incorporation errors occurred at a relatively high frequency. The PCR-amplified DNA pieces were introduced into bacterial host cells, and the cells were spread for colony formation on agar plates containing casein and higher and higher amounts of DMF, searching for the largest halo sizes. The variant subtilisin genes of those colonies that produced the largest halos were picked (by PCR), subjected to further cycles of error-prone PCR amplification, reintroduced into the host cells, and screened for even larger halo sizes in the presence of higher and higher concentrations of DMF.

Upon several such repeated rounds of mutagenesis and selection, one colony producing a rather large halo in the presence of DMF was analyzed for its encoded subtilisin E variant sequence. This variant had six additional amino acid replacements, and the evolved enzyme had a 138-fold improvement of catalytic efficiency in the presence of 40% DMF over the original (naturally occurring) subtilisin E enzyme. Remarkably, all replacements that were important for higher activity of the enzyme in high concentrations of DMF were located on the surface of the protein, clustered around the active site and a groove that binds the substrate. It is unlikely anyone would have thought to replace these amino acid residues were they engineering this protein by deliberate rule-based tweaking (for fear of destroying the enzyme's catalytic site). A second important observation upon subsequent analysis of the mutations was that the effects of these replacements were not additive — in other words, it would have been very difficult, if not impossible, for anyone to deliberately engineer this protein with multiple replacement mutations starting from an elementary knowledge of each single mutant's activity.

The above work was the beginning of the era of artificial evolution of industrial enzymes, which Arnold led in its early years, for which she was awarded the Nobel Prize in Chemistry. We will not attempt to describe the commercial impact of this technical breakthrough, for which the reader is encouraged to read recent contributions in the field, but we will briefly describe below the conceptual breakthrough that this approach entailed.

10.3 Does Natural Selection Work Only on Smooth Slopes?

While thinking about engineering proteins, we considered randomizing every amino acid position of a given protein. For a short peptide, say, 10 amino acids long, it is difficult but possible to generate a total of 20^{10} or approximately 10^{12} different combinations of peptides. This is about the limit of what one can practically do with today's technology. Not much further improvement in diversity is feasible, not only due to technical limitations but also due to the exponential increase in the mass of such combinatorial populations of peptide variants.

Natural proteins probably have evolved from a pool of polypeptides in the prebiotic atmosphere. If we posit that all current natural proteins must have evolved from a limited number of primordial proteins (here we are agnostic about which came first — proteins or nucleic acids), using the rationale discussed earlier, the genealogy of all current proteins should be traceable to a small number, perhaps even one, primordial protein.

The important concept to arise from this is that all extant proteins, and most proteins that had ever existed in the past (some of which have become extinct), can likely be reached from any present-day protein by a series of steps in which one protein differs from another by one residue. We do not imply that the replacements had occurred in single steps, but that they can be "construed" as single replacement steps (ignoring frameshift mutations) (Fig. 10.4). We can thus trace the relationship among all proteins by lines joining the immediate neighbors of every other, which should form a network. It is immediately clear that the structure of such a network should contain all existing proteins and therefore can be thought of as the raw material upon which natural selection can work.

How large is this network? If N is the total number of protein sequences that can be derived in one replacement step from any starting protein, say, X, consisting of 100 amino acids (just as an illustrative example), and f is the fraction of protein sequences that are biologically meaningful, then $f.N$ is the number of biologically meaningful proteins on which natural selection can work by potentially improving fitness. N for a protein consisting of 100 amino acids must be a maximum of 1900, because there are 20 different amino acids and at each position of X there

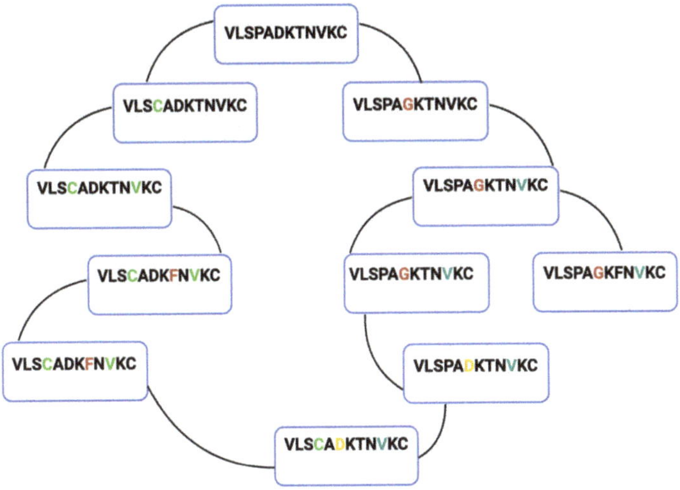

Figure 10.4 Conceptual diagram illustrating the protein space. A hypothetical protein sequence changes by one replacement per step. Each protein sequence is related to its immediate sequence variant by one replacement. Sometimes some replacements converge, forming a loop; at other times, there is no loop.

is already an amino acid assigned, which can change by replacement. In practice, however, N is less than 1900 because of the restrictions imposed by the genetic code — not all 19 amino acid replacements are possible by mutation in one step (by changing the DNA base of the first or the second codon position by a single mutational replacement).

John Maynard Smith (1920–2004), who first developed this concept of protein sequence space (Maynard Smith 1970), estimated N as ~1000. Thus, when $f.N > 1$, that is when meaningful proteins form a collection such that evolution can select one or more of them, f must be >1/1000.

In other words, the density of potentially functional proteins in any region of the currently observed protein space should be higher than 1 in 1,000. Considering the universe of all possible proteins of this length, this fraction is surprisingly large. What this means is that evolution must have selected, and continues to elaborate on, a rather limited set of proteins, given the universe of all chemical tasks necessary for accomplishing most of our terrestrial livelihood.

This is exciting news because a biotechnologist has a fair chance of success in evolving, by selection pressure alone, any protein with a conceivable chemical activity that takes place in a life-like condition. It also suggests that among closely related sequence neighbors of most naturally selected proteins, there exist other unexplored protein sequences whose biological function might be improved by the application of selection pressure on sequence variants.

To put it in a different way, the fitness landscape upon which most naturally selected proteins find themselves is likely characterized by smooth slopes of fitness that can be readily climbed by mutational exploration.

There are likely to be many alternate solutions to a biological or chemical problem; nature has merely happened to have selected a precious few by accident. Seen in these lights, the existence of the immense variety and intricacies of life does not appear to be as remarkable as one needing an intelligent engineer.

Chapter 11

Coffee with Dr. Xu: In Real Life

Dr. Julia Xu is a real-life hematologist and an Assistant Professor of Medicine at the University of Pittsburgh, who treats sickle cell disease and other non-malignant hematological disorders. She also conducts laboratory research and clinical trials on how to improve care of patients who suffer from these disorders. One late February afternoon we sat down with a cup of latte by the bay window in Tazza D'Oro near Highland Park in Pittsburgh. Outside there was a light drizzle, even a few flurries and a cold wind blew.

Author: Julia, thanks for coming out to have coffee despite your little baby at home. For the sake of time, let's jump right into it. Tell me about your specialization . . . what do you do?

Dr. Xu: I am an assistant professor of medicine at University of Pittsburgh. I am an adult hematologist, so I see patients with blood disorders, I focus mostly on red blood cell disorders and in particular sickle cell disease and other hemoglobin disorders. I see patients but I also do research, run clinical trials, I have a lab, and I also work internationally.

Author: Do you consider your role as somewhat unique among hematologists or do you see yourself as being like what other hematologists do?

Dr. Xu: My area of hematology is specialized compared to those of most general hematologists. There is a difference between a hematologist in the community versus hematologists at academic centers. Usually, at least in the US, we are dual trained in hematology and oncology for fellowship. So, when you are in the community you are usually expected to see both hematology and solid oncology patients . . . In an academic center, there are some institutions that have a focus on classical hematology, or benign hematology, which deals with non-malignant blood disorders. That would be like, high or low platelets, white blood cells, or red blood cells . . . disorders like anemia rather than disorders like leukemia, which are hematological malignancies. At an academic center you tend to subspecialize a little bit more. You can do solid oncology, malignant hematology, or classical ("benign") hematology. I am in the Section of Benign Hematology, and even within that I have a specialized focus on sickle cell disease. I also chose an academic center that has a comprehensive sickle cell program, where there is a lot of research happening, and the center can provide services, support, and specialized care for patients with sickle cell disease.

Author: Tell me about your background and how you got here.

Dr. Xu: It is certainly always a long road in medicine. Lots of years of studying and training. I always wanted to be in some sort of biomedical field, growing up. While in college I was considering going to medical school or doing a PhD. I worked in a lab on endothelial cells, looking at angiogenesis.

Author: Where was the lab?

Dr. Xu: I was an undergraduate at the University of Pennsylvania. I worked in a tissue engineering lab. We looked at the role of microenvironment, the extracellular matrix on cell-cell interaction and signaling in angiogenesis.

Author: Tell us about your undergraduate college experience.

Dr. Xu: At Penn I was in a four-year sub-matriculation program where I did my bachelor's and master's simultaneously. I took my undergraduate level classes the first two years, and the last two years were mostly graduate level classes that counted towards a master's degree in chemistry. That program — the sub-matriculation program — was called the Vagelos Scholars program at the time. That program also paid for two full time summer research internships. During the school year I also worked part-time in the lab and put together a master's project at the end.

Author: So, you were a chemistry major as an undergraduate?

Dr. Xu: I was Biochemistry major and a Spanish major. And then the master's degree was in Chemistry.

Author: That's interesting. Why Spanish?

Dr. Xu: Well, I was always fascinated with languages. I also love traveling. My first language was Mandarin Chinese . . . I was born in China. When I came to the US, I picked up English pretty quickly. When I was in high school, I enjoyed learning Spanish and liked the cultural aspects of Spanish language, music, and dance . . . and ended up studying abroad in Argentina for half a year during college.

Author: Looking back, would you comment on what your experiences in language — learning Spanish, immersion in Spanish language and culture while studying abroad — how they might have influenced you, perhaps not in choosing medicine as your career but in doing the kind of specialized medicine that you do?

Dr. Xu: Absolutely. My interest in sickle cell and red blood cell disorders really came about because of my 'global' work — when I was in medical school, I knew I was interested in hematology. But at the time I didn't know . . . what specific area I would be interested in. Then I looked for a global health project because I wanted to do research internationally, addressing health disparities, improving access, diagnosis and treatment. I ended up getting connected with my mentor whose lab was working on G6PD deficiency. [**Author's note:** Glucose 6-phosphate dehydrogenase (G6PD) deficiency is an inherited condition in which the affected individual does not make enough of this enzyme. As a result, red blood cells break down and produces hemolytic anemia.] We crafted a project looking at the epidemiology of this disease and coming up with a low-cost diagnostic method for the management of G6PD deficiency in the Dominican Republic.

Author: Why the Dominican Republic?

Dr. Xu: That is because G6PD deficiency had never been studied in the Dominican Republic. We found that it affects about 9% of the population. The inherited condition is X-linked. So mostly it affects males but sometimes it also affects even heterozygous females. It is prevalent in populations with African ancestry.

Author: I presume there must be a lot of descendants of African population in Dominican Republic?

Dr. Xu: Yes, there is a lot of racial admixtures in the Dominican Republic, especially with neighboring Haiti, there is migration between the two countries. There were also several Haitian migrants who were working in the Dominican Republic and coming to the HIV clinic where I worked.

Author: When did this internship occur?

Dr. Xu: This was between the third and the fourth year of my medical school.

Author: Tell us a little more about your pathways from college through the medical school.

Dr. Xu: After college I had thought about taking a year off to travel internationally. But I ended up getting into Columbia University for medical school. The idea of living in New York City, I was very excited about, so I postponed my travel. The other thing was that I loved Spanish, and Columbia is in Washington Heights which has many immigrants from the Dominican Republic who speak Spanish. So, I was excited to live in this neighborhood, work in the free clinic, use my Spanish in the hospital to care for patients. That was part of the reason why I chose Columbia and why I decided to go straight there after my college. It was a wonderful experience. I had a leadership role in the free clinic that was run by Columbia medical students. I regularly interpreted for patients in the hospital. Caring for patients in their native language brings a different level of interaction when you are a provider — a level of comfort (to the patient). I really enjoyed that experience. At Columbia it was the first time that I realized that my hobby of studying Spanish could be useful in my career. Furthermore, there was a strong global health presence — the Mailman School of Public Health at Columbia University is very strong — they have a lot of programs internationally. One of my mentors founded an HIV clinic in the Dominican Republic. Through him I ended up applying for an international clinical research fellowship. I took a year off between the third and the fourth year of medical school for that fellowship. That was my first exposure to global health research, and I loved it. I had opportunities to give academic talks in Spanish, for instance, to the clinic staff to explain my project to them. So, it was exciting to use Spanish in a different context.

Author: Did you think of doing research as an important part of your life early on, like while in high school, or did you think about it only later, while in late college years or in your medical school period?

Dr. Xu: When I was in high school, I volunteered in a hospital and I found that experience interesting. But when I was in college, as I mentioned I was in this four-year program where everyone was thinking of doing either MD or PhD or MD-PhD. I was planning to do an MD-PhD. After spending three years in a bioengineering lab and talking to my mentor there who was an MD-PhD, I realized that it was not really my ambition to run a large lab, but as an MD I could still do research, even basic science research, and I didn't necessarily have to have a PhD. My takeaway was that if I wanted to exclusively focus on having a lab, then getting a PhD made sense, to have the intense training needed in bench research. But if I wanted a more diverse research program, then as an MD I could still do lab research, as well as global health research, clinical research, and see patients. At the end of the day, I was more interested in having that clinical orientation.

Author: This leads naturally to my next question. You had already had a year's worth of global health experience. How did you choose the next step in your career?

Dr. Xu: Yes, I decided to go to Internal Medicine residency at Duke University largely because of my interest in sickle cell disease. During my work in the Dominican Republic, I got very interested in sickle cell disease. For instance, I heard these anecdotes . . . there was a group of nuns who ran a school, who were in touch with the founder of the HIV clinic, They had mentioned to him that they know which kid has sickle cell disease even though no one has been diagnosed — they said they could tell when the kids have pain episodes, or if they have to go to the hospital frequently. They knew there were many kids who might have been suffering from sickle cell disease but the hospitals didn't know how to diagnose them or to treat them. So, I did a side study while I was there, looking at the prevalence of sickle cell trait [**Author's note:** sickle cell trait is the condition in which the individual is heterozygous for the sickle hemoglobin mutation (hemoglobin S), and generally does not exhibit the clinical manifestation of sickle cell disease.] and other hemoglobin disorders. I found a high prevalence and got really interested in working on the global burden of these diseases. These are really devastating diseases with very few options for treatment at the time. Hydroxyurea was all there was. It was a field with a lot of need, a field where discoveries could make a lot of international impact. I was also interested because it was a disease of childhood that then becomes a chronic condition. The psychological and psychosocial aspects of the disease are really complicated — between growing up with a chronic disease, coping with chronic pain, especially in the US where those living with sickle cell disease also must deal with racial discrimination, stigma related to the opioid epidemic, many complex psychosocial issues. It is a multifaceted disease. Another reason I chose sickle cell disease is that I attended this conference — the biggest annual hematology conference — where I ended up meeting some of the hematologists who were working in the sickle cell space. Everyone I met was a bleeding heart and so passionate about their work. They really seemed to want to make a difference in their patients' lives, improve the global access to treatments; it was a medical community that worked closely with disciplines of psychology, public health, and epidemiology, to try to help improve care for patients. I just loved that environment and mentality. So, I ended up committing to sickle cell disease before I even took care of many patients with this disease in medical school. I chose my residency based on which programs were specialized in sickle cell disease. I ended up going to Duke University because they had a very strong comprehensive sickle cell center.

Author: While in Duke, did you further expand your experiences beyond treating your patients as part of your residency training?

Dr. Xu: Yes, I did. I was in another small program called the Global Health Pathway, which had combined internal medicine and global health training for four years. So, I did my internal medicine residency, followed by a master's program in global health. I then spent one year doing field research for the global health masters. To do that I applied for the Fogarty Foundation international research fellowship, I was a Fogarty Fellow for a year in Thailand working on thalassemia.

Author: Why thalassemia? What is the relationship between thalassemia and sickle cell disorder?

Dr. Xu: I am interested in all disorders of hemoglobin. I needed to find a site and a project that would be relevant to that interest and do something that would come to fruition in a year. I didn't find a suitable project in sickle cell disease. But thalassemia is genetically related as they both involve beta-globin gene mutations. You can also have one copy of the sickle hemoglobin gene and one copy of a beta-thalassemia gene, which together end up causing sickle cell disease. The severity of the disease depends on which beta-thalassemia mutation is involved. In some parts of the world there is a high prevalence of sickle-beta-thalassemia, so as a sickle cell provider, it is important to learn about thalassemia as well. Thalassemia is a rather complex set of genetic disorders. I did a population-genetic testing project diagnosing various hemoglobin disorders. [**Author's note:** this involves the use of molecular techniques in determining the exact mutational events in various mutated hemoglobin gene regions.] I also carried out a qualitative project on thalassemia screening — my entire career until that point was focused on quantitative research . . . so it was a good opportunity to learn about qualitative research.

Author: What is qualitative research?

Dr. Xu: It involves collecting and analyzing non-numerical data, like understanding attitudes and beliefs and subjective experiences. I did a lot of interviews in focused groups. You don't do statistical comparisons. Instead, you look at narratives and themes that emerge from your data, you can see how much consensus versus divergent opinions are there in the community. It is a different way of looking at data. Personal experiences are important in these narratives.

Author: What you are saying is that you do not necessarily try to find a cause-and-effect relationship but you assess what the problems are in a more complex dynamical context?

Dr. Xu: Yes, you try to understand a diversity of perspectives, then look for common and distinguishing themes in these various perspectives that will help provide key insights — for instance, I was trying to develop policies for thalassemia prevention. I was interviewing migrant workers in Thailand to try to understand what the barriers were to test for thalassemia in this population, how willing they would be to participate in screening for thalassemia and looking for insights that could inform public health policies.

Author: Fascinating. Beyond doing your project work, how did this experience enrich you?

Dr. Xu: I learned a little bit of Thai. Perhaps I would say my most fluent Thai was in ordering food. I obviously did not speak with my project participants in Thai but I learned enough Thai to get around and to provide direction to participants and to the providers in my interview groups.

Author: You must have used interpreters to do your interviews . . .

Dr. Xu: Yes, I used interpreters for the entire interviews, and then I had the help of one of my colleagues in the project who was Thai to transcribe some of the interviews into English.

Author: After your year in Thailand, what did you do?

Dr. Xu: After residency if you want to sub-specialize in internal medicine, you need to do a fellowship. By that time, I knew I wanted to focus on sickle cell disease. So, I applied to programs in hematology/oncology, specifically to programs that allowed me to focus on hematology training. I ended up going to NIH for my hematology fellowship.

Author: While at NIH you were seeing patients as well as doing research?

Dr. Xu: Yes. NIH has its own hospitals that are staffed by faculty and fellows that are training there. There we saw a great variety of patients in terms of disease severity. The patients were all on clinical trials. We had some unique groups of patients with rare diseases and we saw the entire spectrum of that disease. Some of these patients had very advanced-stage disease. Some people were getting CAR T-Cell therapy. [**Author's note:** see Chapter 5, Section 5.4] Patients with sickle cell disease getting bone marrow transplants. I took care of these patients at the NIH hospital and rotated in other hospitals in the DC area — Johns Hopkins University, George Washington University, Georgetown University.

Author: Then you moved to Pittsburgh. Tell us about your setup in your institution.

Dr. Xu: I am in the Department of Medicine in the section of Benign Hematology on my clinical side. On the research side, I have my lab in the Vascular Medicine Institute. I focus on chronic complications in sickle cell disease, particularly on anemia, on the implications of treating anemia, on the contributing factors to anemia. I have clinical trials that I have initiated or am leading, where we are trying new therapies in combination with old therapies, to try to improve anemia in sickle cell disease. I am looking at what outcomes improve once you raise the hemoglobin level. In my lab I have certain equipment with which I can look at red blood cell rheology. [**Author's note:** rheology is the study of factors affecting the deformation of matter in a liquid flow; in this case, how deformation of blood cells, such as RBCs, occurs during blood flow, due to factors such as the viscosity of blood.]

Author: So, the hope is that some of your findings will be applied in a clinical setting as biomarkers.

Dr. Xu: Exactly. Biomarkers of disease severity or disease subtype. I am doing longitudinal studies. [**Author's note:** studying the same patients over time] Looking at these biomarkers, trying to figure out which biomarkers are the best to monitor and evaluate the response to treatments, which is important with all these new treatments currently being developed for sickle cell disease. Then I have a couple of postdocs and medical students with whom I am working on some other related studies. For example, the medical student in my lab is doing a mouse study to examine the effects of a treatment for anemia on hypoxia and iron metabolism in sickle cell disease.

Author: Tell us something about the new therapies that are coming up now.

Dr. Xu: There are many new classes of drugs being developed for sickle cell disease. In the past five to ten years the field of sickle cell disease has really exploded in terms of pharmaceutical interest and drug development. It was always a neglected and forgotten disease.

Now there is practically a rat race to see which pharmaceutical company can get the next drug approved. After decades of no new drugs for sickle cell disease, the drug L-glutamine was approved in 2017. Its proposed mechanism of action is an antioxidant effect. It was shown in a clinical trial to decrease the frequency of pain episodes and acute chest syndrome, two very common complications of sickle cell disease. Then two drugs were approved in 2019, the first of which was crizanlizumab. Crizanlizumab reduces adhesion leading to decreased vaso-occlusion and frequency of pain episodes and acute chest syndrome as well.

Vaso-occlusion is the pathophysiological hallmark of sickle cell disease. RBCs become sickled and can break down or adhere to other cells or endothelium lining small blood vessels, impeding blood flow, leading to ischemia and tissue damage.

There are many other emerging drugs with different mechanisms of action, as well as second-generation versions of existing drugs. For example, there is another emerging class of drug called pyruvate kinase activators, which modifies red cell metabolism, [**Author's note:** Pyruvate kinase is an enzyme in the glycolysis pathway by which sugar is metabolized to generate ATP] to try to inhibit sickling.

When I was at NIH, I helped lead the first clinical trial of a pyruvate kinase activator in patients with sickle cell disease, and our results have led to a large clinical trial that is being conducted globally now. There is also a lot of research into inducing fetal hemoglobin expression in sickle cell disease using drugs or gene therapy. As you know hydroxyurea induces fetal hemoglobin expression — that is believed to be the primary way that hydroxyurea treats sickle cell disease.

More recently, excitingly, gene therapies have been approved. Again, different strategies have been used for gene therapies. For instance, you can insert a functional hemoglobin gene — one of the approved therapies involves inserting a non-sickling hemoglobin gene. The other gene therapy product that was approved uses the strategy of knocking out a negative regulator of the fetal hemoglobin gene, called BCL11A, using CRISPR/Cas. [**Author's note:** for more detail of gene therapies for sickle cell disease see Chapter 12, Section 12.4.]

Author: This raises the question of how impactful the new gene therapy approvals will be.

Dr. Xu: These will be very impactful to the sickle cell community, at least in terms of giving patients hope about cure and being able to imagine a life that is not governed by sickle cell disease.

Author: Who will afford it?

Dr. Xu: Yes, that is the big question. What impact it is going to make on a global scale is unclear. I think it is a good thing that money is being put into developing therapies for sickle cell disease, with initiatives such as the NIH Cure Sickle Cell initiative. The Gates Foundation has also put funding towards gene therapy for sickle cell disease. But, to treat patients with curative therapies, you need to have the infrastructure to provide complex medical care in addition to providing the standard of care. Being able to provide transfusion support for instance when you wipe out the bone marrow for gene therapy.

Most of the countries where the disease burden is high, there is not even a newborn screening program to identify patients with sickle cell disease, so how can you treat if you can't diagnose and capture the patient population? And then there are so many barriers to bringing them into care — transportation problems, geographical barriers, economic barriers . . . if you don't have access to a safe blood supply or intensive care unit or strong antibiotics to treat neutropenic infections after bone marrow transplant, it can be very dangerous.

So, I think at this point the infrastructure is a huge barrier for global adoption of gene therapy. Therapies like hydroxyurea are still needed to keep our patients alive and in good health so that one day, they may be able to consider these therapies, once curative therapies become more available worldwide.

Even in the US, I think implementation (of curative therapies) will be limited and slow. It is very cost-intensive, resource-intensive, and unclear what insurance will be covering and for which patients. And, at the end of the day patients would have to be willing to commit to the treatment, because it is not an easy process.

Author: You mean commit the time and take the risk?

Dr. Xu: Yeah . . . be willing to accept the medical risks and potential side effects that come with bone marrow transplant and gene therapy, but also commit the time. I have several patients who are young college students, who would like to get a transplant or gene therapy, but they will have to commit to taking a semester off college to go through the transplant and recovery process, so they are going to wait a couple of years until they are done with school for this — it is one of those life-altering, all-consuming endeavors to seek this kind of curative treatment.

Author: I think I have covered all the points I needed to ask you, except one. At this stage in your life, how do you manage your work-life balance issues?

Dr. Xu: Ahh . . . I am still working it out (smiles). I think that is the biggest ongoing challenge for me. In academia, and especially during medical, graduate, and postdoctoral training, you are working hard, running as fast as you can, sprinting to the next goal, the next level of training, the next grant application cycle . . . you are just constantly sprinting, sprinting, sprinting, usually ignoring other non-work but important aspects of life.

Author: So how do you balance them? What makes one think that ok, it is a balance?

Dr. Xu: You must work at it. Once you are out of the training and you're an attending physician or an assistant professor, you are still stuck in — at least I was — stuck in the sprinting mentality. I think it is counter-productive to stick to that mentality once you are done with the bench-marks that people have set for you. When you are in the training track, you must meet certain milestones, and a certain amount of time commitment is expected, and these expectations are laid out for you. But once you are a faculty, the path is not so clear and you are the one who is driving your-self forward. If you drive too fast you will burn out. So, you must be really mindful of how much to take on and at what pace you want to grow your career so that it is sustainable, because essentially you are going from run-ning a sprint to running a marathon.

Author: That's an insightful analogy.

Dr. Xu: In this marathon, you must pay attention to your support systems, self-care, and all those other things you didn't really think about when you were sprinting. You must pick up new skills, like learning to say, no. Picking your priorities, trying to not get too diluted in terms of your research focus and your time. Trying to always keep on track and focus on the big picture. For me currently the goal is to get sequential NIH funding, but that may come at the expense of interesting collaborations that I would like to work on but are peripheral to my research focus. I only have limited time, so I must make sure that I focus on my own goals.

Author: What is your advice to young students who would like to go to medical school now?

Dr. Xu: In terms of work-life balance or in general?

Author: In general. How to choose one's career path?

Dr. Xu: I think the most important thing is the exposure. I think it is hard to know what is going to trigger your interest, ignite your passion, without doing it. If you say, yes, I want to be a doctor but have not spent time with a patient, with someone who is physically or emotionally suffering — it is hard to intuit what it would feel like to do that for the rest of your life. So, it was helpful for me to have done research in a lab and volunteering in a hospital early on.

I also had some personal exposure to medicine because my parents were both doctors in China. So, I was familiar with the medical profession more than others might have been. Once you know that you are interested in a career path, I think then the most important thing is, number one, aggressively pursue those interests, don't wait passively for things to come to you, because the more you are looking for new opportunities, talking to mentors, asking people about their experiences, how they got to their next steps, the more you'll have an idea about how to navigate this new world, what this world entails, and what you need to get to the next step.

For me, every time I got to the next step on my career path, I talked to people ahead of me about what to anticipate in the following steps. Eventually it gave me foresight and a sense of where I'd want to be five years or ten years down the road. And that perspective really helped me move my career forward at a rapid pace.

My mentors were extremely valuable. Not just mentors for research but also peer mentors who have gone through the process, even just one or two years before me. Or maybe someone who is applying for the next step at the same time as me. Figure out what their experience has been and learn from them. And then the second thing is to be open to new opportunities.

Taking research as an example, some people fixate on, say, wanting to do cardiology research and that's all they want to work on, and they pass on to a good research opportunity in another field. But you really must look at what those research opportunities are going to offer you in terms of skill sets.

What you are working on as a medical student is unlikely to be what your focus is as faculty. What you are looking for is to gain research exposures and experiences and demonstrate your commitment in terms of having some sort of output. It is really showing that you are taking something seriously and can follow through, that you can take initiative and think independently, because that's what it takes to be a successful researcher.

Along the way, there are also setbacks, and these are important to experience, too. You know, the less glamorous side of science — projects that don't work out, mentoring relationships that don't work out. It happens. If someone is passionate about pursuing a career path, it is important not to let those negative experiences dictate your path. No matter what career path you choose, it is almost guaranteed that you will face some sort of adversity, so you sometimes must just push past it, because you will never know when you will find success right around the corner. And eventually if you keep at it, you will find success, and you will look back and see that those failures or challenges are just a part of the scientific process.

Author: On that note, thank you so much for chatting with me. It has been wonderful to know your story and thanks for your insights. It is still raining outside but it is late and the café is closing.

Chapter 12

A Brave New World of Genome Engineering

The second decade of the twenty-first century has become identified with breathtaking progress in our ability to engineer the genomes of nearly all organisms. The key to this ability can be traced to two important advances in the three preceding decades. The first was the realization that organisms have evolved a highly conserved mechanism to repair their chromosomal DNA if the DNA becomes damaged, such as developing breaks or other chemical lesions, including oxidative modification or chemical crosslinking of the DNA bases. The second was that the essence of a form of adaptive immunity in bacteria could be harnessed to induce targeted breakage and repair of chromosomes in nearly any organism — a matter of chemistry. The singular player in these breakthroughs is the charged word, CRISPR.[1] In this chapter, we will review these advances and their implications for mitigating human diseases.

***

In our parallel universe, we find ourselves on a yacht, its sails catching the bright sunshine on a calm sea. A young woman, Rosalind Franklin, the X-ray crystallographer who provided the first evidence that DNA contains a helix, is at the helm, expertly steering the vessel. A stocky man with bushy eyebrows and a crisp Cambridge accent with a slight South African lilt, Sydney Brenner lounges on a deck chair, a glass of wine in hand. Brenner was famous for discovering the very first three-letter genetic

[1] Clustered Regularly Interspaced Short Palindromic Repeats

code, for identifying the messenger RNA, and for establishing a genetic model for understanding the brain. A bespectacled man with a hearing aid sits at a makeshift workstation, tinkering with a small, mysterious device. He is Thomas Alva Edison, the prolific self-taught American inventor and businessman.

Franklin: It's a pity, you know. I spent so much time trying to discover the structure of DNA, and here we are, rewriting it with CRISPR/Cas9 like it's a mere blueprint.

Brenner: Rosalind don't be modest. Your X-ray diffraction images were the Rosetta Stone for this entire field. Without your work, we might still be groping around in the dark. CRISPR is just the next chapter in the book you started.

Edison: And what a chapter it is! A technology that edits life itself — what could be more ingenious? If I had access to CRISPR in my day, I'd have bred glowing plants to light the streets or mosquitoes that delivered vaccines.

Franklin: A classic Edison idea — practical, yes, but potentially fraught with unintended consequences. Have you considered what happens when those glowing plants disrupt ecosystems? Or when your engineered mosquitoes evolve?

Brenner: Edison was always the optimist, Rosalind. But you have a point. Genome engineering isn't just about what we *can* do; it's about what we *should* do. Take medicine, for instance. We're curing genetic diseases, yes, but we're also opening the door to designer babies. Blue eyes, high IQ — where does it stop?

Edison: Nonsense. Progress is unstoppable. If society wants to enhance itself, why not? In my time, people feared electric light but look at the world now. The same will happen with CRISPR. Resistance will give way to acceptance.

Franklin: But this isn't electricity, Thomas. This is life. Altering genomes affects not just individuals but generations. What about off-target effects, ecological imbalances, reduction of genetic variations, or worse, bioweapons? You can't ignore the ethical dimensions.

Brenner: Precisely. And then there's environmental engineering. Sure, we can create drought-resistant crops or carbon-sequestering super-algae, but what if those organisms spread uncontrollably? Or what if monopolies arise around engineered life, driving small farmers into ruin?

Edison: Problems like these have solutions. Regulation, oversight, innovation — these are the tools of progress. Do you know how many times people warned me my inventions would destroy jobs or disrupt society?

Franklin: Yet they were right in some ways, weren't they? Progress always comes with trade-offs. What worries me is that CRISPR has the potential for irreversibility. Once a gene drive is released, you can't simply recall it, like one of your phonographs.

Brenner: The genie is out of the bottle, and it's a clever one. But we need to ensure we're asking the right questions — scientific, ethical, and social — before we start granting wishes.

Edison: You scientists and your caution. But fine, Rosalind, Sydney — what would you have us do? Delay progress while we ponder these "questions"? The world doesn't wait.

Franklin: Not delay, Thomas. Guide. Shape. Ensure progress serves humanity, not just those who can afford it. We're the navigators of this ship, not passengers.

Brenner: Well said, Captain Franklin. But tell me, what's your stance on using CRISPR to engineer longer lifespans? Would you sail us all into immortality?

Franklin: Immortality is overrated. The beauty of life is its fleeting nature. But if CRISPR can give us more years to live wisely, then perhaps it's worth exploring. Just don't expect me to pilot that ship.

Edison: Speaking of exploration, I've just engineered a little gadget that might sequence DNA in real-time! Now, if only I could patent CRISPR itself…

Brenner: Sorry, Thomas. Feng Zhang, Jennifer Doudna, Emmanuelle Charpentier beat you to it. But you're welcome to join the fight over who owns life itself.

Franklin: Let's hope we're wise enough not to try owning it. Life is not an invention — it's a responsibility.

***

12.1 Bacteria Often Become Immune to Invading Viruses

Not only humans and other mammals but even bacteria have a form of adaptive immune system that is remarkably conserved across many bacterial genera, pointing to an ancient evolutionary origin. Unlike intrinsic or constitutive immunity, which provides a general defense against any foreign DNA, adaptive immunity is activated when a bacterial cell encounters a foreign DNA sequence it had previously encountered as an invader.

Bacteria exhibit at least two primary forms of adaptive immunity. The first, known as superinfection immunity, occurs when a virus previously integrated into the bacterial genome prevents reinfection by similar viruses. The second, more relevant to us, involves the ability of bacteria to acquire and retain a "memory" of past infections by foreign genetic material, enabling a targeted immune response against future invasions of the same kind. This second form of bacterial adaptive immunity we will now discuss.

12.1.1 *A mechanism of bacterial adaptive immunity*

A bacterial cell might by chance survive an infection by a virus (bacteriophage or phage). While the specific reasons for survival might be varied and irrelevant for us, part of the process is that the bacterium employs enzymes to cut the invading DNA into small fragments, which are then inserted into a specific location in its chromosome. These fragments serve as molecular memories, marking past infections.

The DNA fragments, typically 20 to 58 base pairs long, are arranged as tandem arrays downstream of a transcriptional promoter. Between each foreign DNA fragment lies a host-derived "spacer" sequence, 21 to 40 base pairs long, featuring inverted sequence repeats. These spacer sequences frame the tandem arrays (Fig. 12.1).

A promoter element drives the transcription of these arrays into long RNA molecules, which are then processed into short RNA segments. Each processed RNA segment therefore contains a spacer sequence and a unique invader-derived sequence. These short RNAs recognize and target

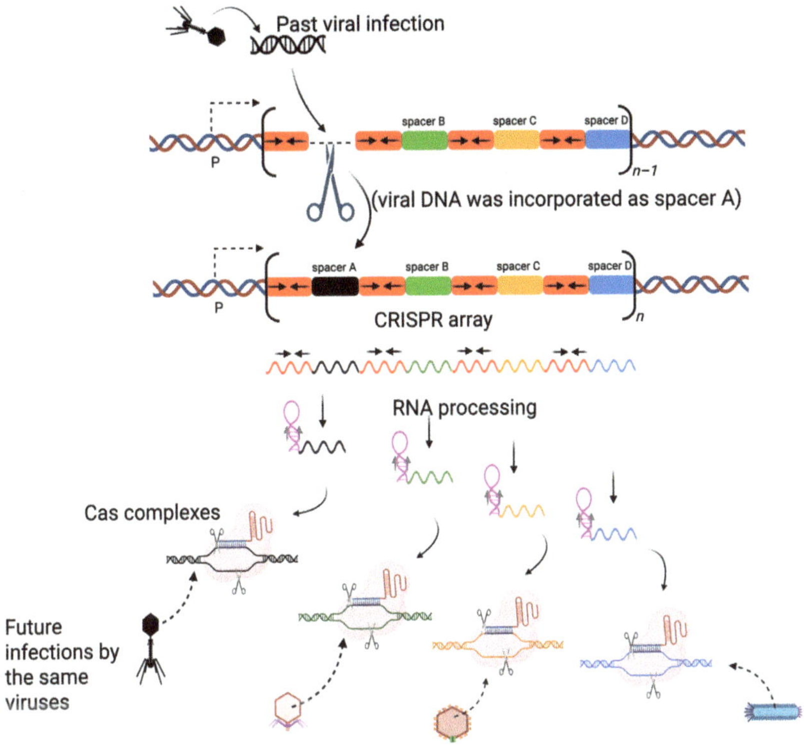

Figure 12.1 Bacterial immunity to infection by foreign DNA. See text for explanation. P is a transcriptional promoter element. Head-to-head arrows denote inverted repeats of DNA (a palindrome). RNA transcribed from such a sequence can fold into a hairpin loop. The scissors at the top of the diagram depict the site of DNA cleavage, where a new DNA segment originating in a virus (or an infectious plasmid) is inserted, which becomes the "spacer A." *The scissors at the bottom ("Cas complex") show the* sites of DNA cleavages within the DNA of the infectious entities (dotted arrows) carried by subsequent infection by the respective viruses. [This figure was drawn using Biorender.com under an academic license]

any potentially matching invading DNA during subsequent infections, enabling the bacterium to chop and destroy the intruding DNA.

Let us now see how Darwinian thinking facilitated the discovery of bacterial adaptive immunity and its translation to the powerful engineering tool, CRISPR.

12.1.2 *Discovery of bacterial adaptive immunity depended on Darwin's notion of conserved traits*

The relevant DNA repeats in the *E. coli* genome were found accidentally in 1986 by Yoshizumi Ishino (1959–) at the time in the laboratory of Atsuo Nakata (1930–) at Osaka University, Japan. They were studying a gene encoding an enzyme called alkaline phosphatase in *Escherichia coli* (Ishino et al. 1987). They stumbled upon a mysterious stretch of repeated DNA sequence on the *E. coli* chromosome. It was not possible at that time to understand the significance of these repeated sequences. Soon, more such repeats of DNA sequences were found in several pathogenic bacteria of medical importance, including *Mycobacterium tuberculosis*, *Shigella dysenteriae,* etc. The repeated DNA sequences exhibited considerable variability (or "polymorphism") among independent isolates from a variety of sources. These variable sequences acted like a barcode, which became a useful tool for identifying pathological varieties of bacteria. The method is similar to that employed in forensics to link DNA to individual persons.

In another decade, DNA sequencing techniques improved sufficiently to enable the sequencing of many more bacterial genomes. By gazing at these DNA sequences, it was soon realized that most, if not all, bacteria — both archaea and about half of eubacteria — had this arrangement of repeated DNA in their genomes. In fact, these are the most pervasive repeated DNA sequences in all of prokaryotes. Francisco Mojica (1963–) and his group at the University of Alicante in Spain realized for the first time (Mojica et al. 2000) that something must be important about these repeats if the arrangement has been conserved in both archaebacteria and eubacteria, which had diverged from each other at least 3.6 billion years ago (Battistuzzi, Feijao, and Hedges 2004). Evolutionary conservation over 3.6 billion years must signify something!

Mojica's group had initially missed the real biological significance of these repeats — mistakenly thinking at the time that the repeats might have something to do with chromosomal DNA replication and partitioning of the newly replicated DNA into daughter cells. Soon, however, they realized that these repeats represented a form of bacterial adaptive immunity (Mojica et al. 2005). This latter conjecture was based on the finding that the unique short pieces of DNA stuffed between the tandem repeat DNA elements bore the most similarity (or sequence homology) with DNA of bacterial viruses and plasmids that are transmitted among prokaryotes. Mojica's group came to this realization by painstakingly searching for sequence alignment of the unique DNA fragments using evolutionary biology procedures that were, in principle, like those described in Chapter 6. Moreover, they pointed out that those bacteria that harbor a certain viral or infectious plasmid DNA fragment in their repeats generally did not support an infection by the corresponding group of viruses or plasmids. They reasoned that the unique sequence fragments clustered in between the repeat elements might serve as memory elements of virus or plasmid infections in prior generations. The repeated elements were termed Clustered Regularly Interspaced Short Palindromic Repeats, or CRISPR (Jansen et al. 2002). Presciently, Mojica and colleagues wrote in their 2005 publication (Mojica et al. 2005):

> *"CRISPR loci can be visualized as mosaics of a repeated unit, separated by sequences at some time present elsewhere in the cell." ... "The incorporation of a given spacer in a CRISPR locus must not be devoid of functional significance. Although our current knowledge remains limited, multiple observations suggest that CRISPR could be involved in conferring specific immunity against foreign DNA..."*

While the proposal for these DNA repeats as footprints of past invaders was reasonable, and the notion of adaptive immunity tantalizing, it was not immediately obvious how this immunity worked.

12.1.3 *Evolutionary conservation suggested a molecular mechanism of adaptive immunity*

When the time is just right, apparently unrelated observations with little known significance tend to coalesce and begin to reinforce one another to weave a cohesive narrative. This occurred in the late 1990s about CRISPR. Soon enough, Mojica's group and other research teams realized that near the CRISPR repeats on prokaryotic chromosomes were located a few interesting genes. The latter's encoded protein sequences resembled the sequences of proteins already known to repair damage to the integrity of DNA double helices. The original guess concerning the function of CRISPR DNA sequences was that these had something to do with DNA replication and chromosome segregation. In prokaryotes, genes with related function are often clustered together on the genome.[2] Both Mojica's group and a group of computational evolutionary biologists led by Eugene Koonin (1956–) at the National Center for Biotechnology Information (NCBI) in Bethesda, Maryland, in the US, had independently speculated (Makarova et al. 2002) that these genes were somehow involved with the function of the mysterious CRISPR repeats. These genes were collectively termed CRISPR-associated or *Cas* genes.

The idea that selection pressure on prokaryotic genomes is strong enough to cluster together functionally related genes on the chromosome was rather compelling to Koonin. Viewed in this way, that *Cas* genes were related to DNA repair no longer made sense to him and his colleagues once the 2005 reports of CRISPR DNA's relationship to adaptive immunity were published (Bolotin et al. 2005; Mojica et al. 2005; Pourcel, Salvignol, and Vergnaud 2005). Based on the evolutionary conservation of Cas protein sequences — especially noting the presence of nucleic acid cleavage domains shared between several Cas proteins and other previously recognized DNA cleavage enzymes — Koonin and colleagues proposed a two-step mechanism:

(1) With the help of some Cas proteins viral and plasmid DNA molecules are chopped into small (~20–40 base pairs) fragments and are incorporated within the CRISPR arrays in the genome and are transcribed into noncoding RNA, and (2) when a future infection occurs, the complementary RNA transcripts bind to the invader's transcripts by base pairing, and the duplex RNA molecules are degraded by other Cas enzymes (Makarova et al. 2006).

In biology, theory alone may go far but not far enough. The Koonin team's proposal for step (1) above was largely correct. However, the proposal based on computationally predicted protein sequence homology for an RNA-degradation model of immunity as defined in step (2) turned out to be a false lead.

---

[2] This is usually not true for eukaryotes.

An important factor for this failure of evolutionary theory might not have been that the theory itself was at fault, but due to a bias in thinking imposed by the prevailing dogma. Only five years before, a novel case of gene regulation based on RNA–RNA complementary interaction, termed RNA interference (RNAi) (Torri et al. 2022), was discovered. In 2002, RNAi was called the breakthrough of the year (Couzin 2002). There was excitement at the time about a novel role of small RNA molecules as regulators of gene expression in all eukaryotes. Their potential role in diseases such as cancer, and their use as tools for studying gene function, had caught the imagination of most biologists (Bartel 2018). By 2004, the molecular mechanisms behind RNAi were being intensively studied in many laboratories, and their potential therapeutic applications for curing human diseases were already palpable (Pederson 2004). Koonin and colleagues were evidently swayed by the idea that the CRISPR/Cas system might be a prokaryotic equivalent of RNAi — so much so that they ignored their own past finding that the Cas proteins had more sequence similarity to well-known DNA-cleaving enzymes than to any known RNA-cleaving enzyme (Makarova et al. 2002).

As is usual in science, the most useful theories are those that can be readily tested. If the test result (of course, reliably conducted) is found to be different from the expectation, the theory is refuted, and one is free to look for a better explanation. The proposals by Koonin's group on CRISPR/Cas as the prokaryotic RNAi-equivalent had that attractive character — their proposal of the function of each of the Cas proteins was immediately testable by the isolation of the respective Cas enzymes and by interrogating their enzymatic activities in the test tube. Biochemists got into the act.

We will not go into a detailed discussion on how the functions of the Cas enzymes were discovered, for which the reader can refer to recent accounts (Nidhi et al. 2021; Verosloff et al. 2022; Wang, Pausch, and Doudna 2022; Wang, La Russa, and Qi 2016). We will simply point out that Cas enzymes were found to act on DNA-RNA hybrids formed between a chain of an invader's DNA and the short noncoding RNA synthesized off the CRISPR arrays. The resulting duplex is acted upon by a Cas enzyme (the most famous is Cas9) that cleaves the DNA (not the RNA) chains at two positions adjacent to the hybrid — one cleavage occurs on the same chain that hybridizes to the RNA and the second occurs on the opposite chain — such that the duplex DNA is broken (Jinek et al. 2012) (Fig. 12.1). One may imagine that a broken DNA belonging to the invader is then subjected to other sequence non-specific DNA degrading enzymes (which can cleave any DNA irrespective of its nucleotide sequence). The invading DNA is destroyed.

12.2 Bacterial Adaptive Immunity Suggested a Way to Rapidly Engineer the Genome

12.2.1 *DNA breaks are hotspots of DNA recombination*

Way back in the early 1920s, Herman Muller (1890–1967) had discovered that X-rays could break chromosomes in fruit fly and thus produce gene mutations. Since then, many had demonstrated as well as utilized the chromosome-breaking properties of ionizing radiations, most notable among them was Barbara McClintock (1902–1992). McClintock demonstrated that the broken chromosomes were highly reactive — the ends of the broken chromosomes tended to attach to other chromosomes willy-nilly and cause further breaks. In other words, for a stable genome, breaks should be avoided. When it became clear that chromosomes were DNA duplexes, it was natural to assume that DNA double-strand breaks should make the DNA genomes unstable. Shortly after the DNA double helix structure was proposed, the structure was used to explain how two viral genetic materials, from the co-infection of a host bacterial cell by two related but genetically marked strains of a bacterial virus, could potentially recombine and make a single genetic material that contained parts of both infecting virus' genomes (Levinthal 1954). This was termed a break-join model of genetic recombination, in which two DNA helices broke and joined reciprocally, thus producing a recombinant molecule (Fig. 12.2).

Many people were not comfortable with the break-join model as a natural way to recombine chromosomes because it was not clear how the two ends of the broken DNA could ever find their partner DNA molecules at precise positions such that chromosomal integrity would be preserved. Breaking DNA also seemed to be a good first step to destroying the DNA, as shown by the high reactivity of broken chromosomal ends. In response, alternate models of DNA recombination were proposed, which

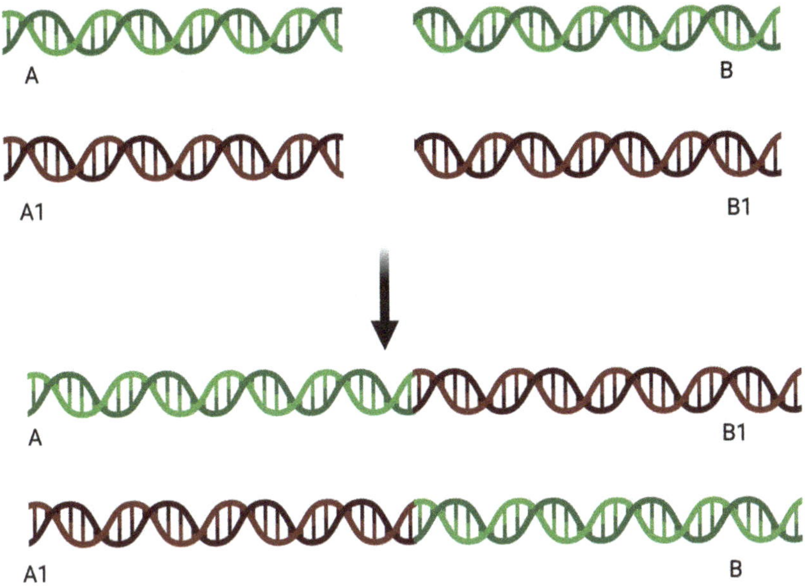

Figure 12.2 The break-join model of DNA recombination. The chromosomal DNA molecules break at precisely the same location in both chromosomes, then they simply change partners and join reciprocally. Thus originally intact molecule AB joins with A1B1 to form A1B and AB1. [This figure was drawn using Biorender.com under an academic license]

postulated that recombination — more accurately, homologous recombination, i.e., recombining two chromosomes over regions of DNA sequence homology — was instead initiated by DNA single-strand breaks, followed by the swapping of complementary single chains (Holliday 1964; Liu and West 2004; Meselson and Radding 1975).

These above models were intellectually elegant and compelling. However, no one had at the time demonstrated that DNA recombination indeed could be initiated by single-strand breaks in any eukaryote or even in any prokaryote except possibly for recombination between plasmid DNAs. This was of immense practical concern because if we were to genetically engineer eukaryotic chromosomes, one way to do that would be to increase the frequency of recombination between an endogenous chromosome with a test tube-modified DNA molecule introduced from outside the cell. If incorporation of the introduced DNA was to occur faithfully at a precise chromosome location, then one could manipulate genes at will — perhaps cure cancer by targeting an active oncogene to inactivate it or replace a disease-causing mutant gene by converting it back into the normal copy.

How could one achieve this feat, which required the accurate targeting of a piece of DNA into a diploid human genome of six billion DNA base pairs? Is it possible to activate just the target DNA sequences somehow such that only that special site becomes a hotspot of homologous DNA recombination? If only one knew how to deliver a precise single-strand DNA break at the target locus — perhaps then that locus would act as a reactive site where homologous gene replacement would occur.

Framed in the above context, it appeared shocking to most people when two groups of researchers nearly simultaneously demonstrated that in baker's yeast, one could target an exogenous DNA piece into the chromosome if the introduced DNA had a double strand break within the region of sequence homology (Orr-Weaver, Szostak, and Rothstein 1981; Scherer and Davis 1979). This unexpected observation merited a complete rethinking of how homologous recombination occurred in eukaryotes. A new model was proposed in which a DNA double strand break, not a single-strand interruption, was the initiator of homologous recombination (Szostak et al. 1983) (Fig. 12.3a, c).

I had the fortune to be in the laboratory of Franklin Stahl in Eugene, Oregon, shortly after he, with Jack Szostak, Rodney Rothstein, and Terry Orr-Weaver, had proposed that natural recombination during meiotic cell division might be triggered by DNA double-strand breaks. It was an interesting time in this field because few believed that chromosomes could survive such a drastic abuse as the breaking of both chains of a DNA helix — how would the broken ends know where to find each other to heal the break? If left unhealed, they could certainly lead to loss of chromosomal fragments, as was previously shown decades ago by classical works with X-ray-induced chromosome breaks in fruit fly and maize. I remember heated debates in meeting rooms and hallways, even on ski trips in the nearby Three Sisters Wilderness, which Frank and Mary Stahl used to arrange. Very soon, however, many of the features of this new model were tested and verified by several different laboratories working with baker's yeast as a genetic model, chiefly in the laboratories of Jack Szostak, Rodney Rothstein, James Haber, Tom Petes, and Frank Stahl.

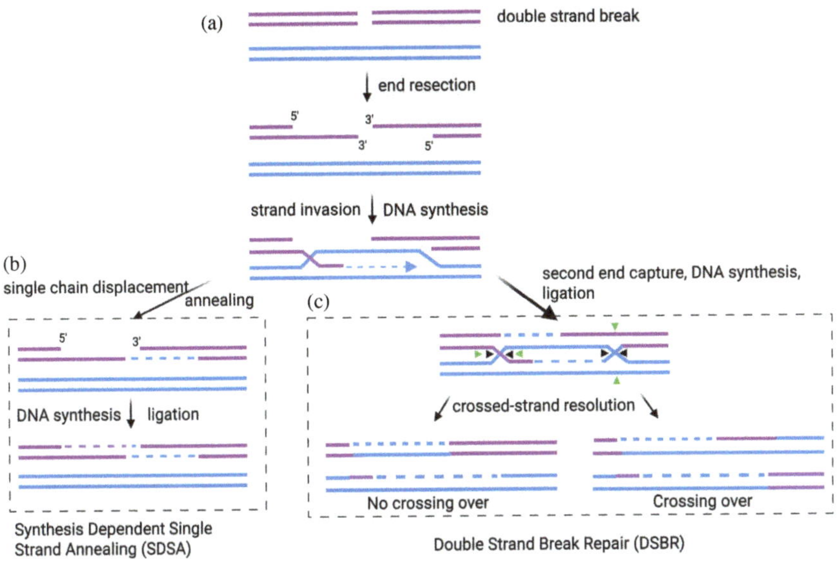

Figure 12.3 Double strand break repair models of DNA recombination. (a) Homologous recombination is initiated by a DNA double strand break in the purple duplex (the gap in the center is the site where the break occurred). This is followed by degradation of DNA single strands radiating from the site of the break on opposite strands. These single DNA strands then search for DNA sequence homology, aided by enzymes, and when they find it, the single strands invade the blue duplex, DNA replication is initiated at the ends of the invading purple strands. (b) In the synthesis dependent single strand annealing (SDSA) mechanism, the purple strands upon replication are withdrawn, followed by annealing of the broken chains over the regions of DNA sequence complementarity, then healed by further DNA synthesis and chain joining (ligation). (c) In the double strand break repair (DSBR) mechanism, the invading purple chains complete their repair synthesis and ligation but remain twisted on the blue duplex, forming two cross-strand junctions. These junctions can be resolved in two different ways—vertically or horizontally at each crossed strand. If at both locations the resolutions are in the same sense (either both vertical or both horizontal) then the original two duplexes are reformed (No crossing over occurs). By contrast, if the two junctions resolve in opposite senses (one vertical and one horizontal), crossover occurs in which one end of the purple DNA is now linked to the the other end of the blue DNA reciprocally on both double helices. [This figure was drawn using Biorender.com under an academic license]

If mouse and yeast are evolutionarily related, then that which is true of yeast should also hold for mouse. These experimental results with yeast lent confidence to researchers who were experimenting with targeted genetic manipulation of higher organisms, such as mouse and human cells, chief among them were Mario Capecchi at the University of Utah and Oliver Smithies at the University of Wisconsin (Capecchi 2005; Folger et al. 1984; Folger et al. 1985; Gregg and Smithies 1986; Smithies 1986; Smithies and Powers 1986; Thomas and Capecchi 1986a, 1986b; Thomas et al. 1986). For their pioneering work on targeted gene replacement in mouse and human cells, Capecchi and Smithies won the Nobel Prize in Physiology and Medicine in 2007.

Yet another singularly odd idea occurred to some of the workers in this field. The ability to heal DNA breaks in various organisms implied that it might even be possible to introduce a DNA break on the chromosomal target site to provoke the insertion of a DNA from outside precisely at that break-site (Fig. 12.4c). This idea was of critical importance,

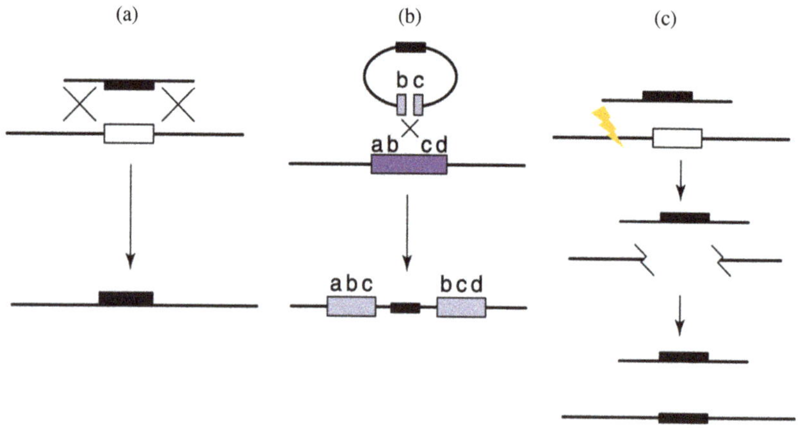

Figure 12.4 Strategies for targeted replacement of genes. The line with the filled (black) box is a piece of exogenously added DNA. The line with the open (empty, white) box is the endogenous chromosomal DNA. The box is the region of DNA to be replaced (i.e., it is the gene to be replaced). In (a), the crosses represent what must occur to replace the white box with the black. In (b), a piece of exogenous DNA with a short region spanning b and c is present at the ends of the DNA. This leads to a duplication of this region, with an insertion of the black box between the duplication (abc — black box — bcd). In (c), the endogenous chromosomal DNA is broken within a region overlapping the added endogenous DNA with black box. Gapping of the chromosome following the break is repaired off the added DNA with the black box. This results in a loss of the white box and copying of the black-boxed region into the chromosomal region that had originally carried the white box — a gene replacement by a DNA break in the targeted locus. Gene targeting by CRISPR/Cas uses this method.

probably first articulated clearly by Frank Stahl, on the assumption that these DNA break sites could act like lightning rods or SOS radio beams indicative of the need to heal the broken DNA — perhaps repair enzymes would somehow be attracted to these sites, and these would accomplish the repair by homologous recombination (Fig. 12.4c).

Several groups began to experiment with stimulating DNA recombination by delivering a targeted DNA double-strand break on chromosomes of living organisms. This was achieved by engaging naturally occurring or engineered DNA-cleaving enzymes (Chiurazzi et al. 1996; Puchta, Dujon, and Hohn 1996; Ray et al. 1988; Rudin and Haber 1988). Subsequently, enzymes were made to order — endowed with a DNA sequence recognition domain (a "reading-head") and a separate DNA cleavage domain (Fig. 12.5 top) (Kim et al. 1996). These modular enzymes allowed efficient delivery of DNA double strand breaks on the chromosome at precisely defined target DNA sequences, which then became hotspots of recombination (Carroll 2011; Joung and Sander 2013). A major problem, however, was that these protein engineering tasks were rather difficult, time-consuming, and expensive. There didn't appear to be a simple formula for recognizing just about any given DNA sequence stretch by a specific protein sequence at a reasonably low cost.

12.2.2 *Make CRISPR work in the test tube*

A potential solution to the above problem occurred nearly simultaneously to a young MIT biologist, Feng Zhang (1981–), and to a duo, Jennifer Doudna (1964–) and her collaborator, Emmanuelle Charpentier (1968–) then visiting Doudna's laboratory at the University of California at Berkeley. Both reasoned that if Cas9 could use a modular RNA as the "reading-head" to recognize a target double-stranded DNA sequence, then it could deliver a site-specific cleavage at any site complementary to the "reading-head" sequence (see Section 12.1.3) (Fig. 12.5 bottom, and Fig. 12.1).

Doudna and Charpentier redesigned the modular RNA "reading-head", now known as the "guide RNA" (Fig. 12.5 bottom) (Jinek et al. 2012), which they then used to deliver a targeted cleavage on the human chromosome. Predictably, upon cleavage the chromosomal locus became a hotspot of homologus recombination (Cho et al. 2013; Cong et al. 2013;

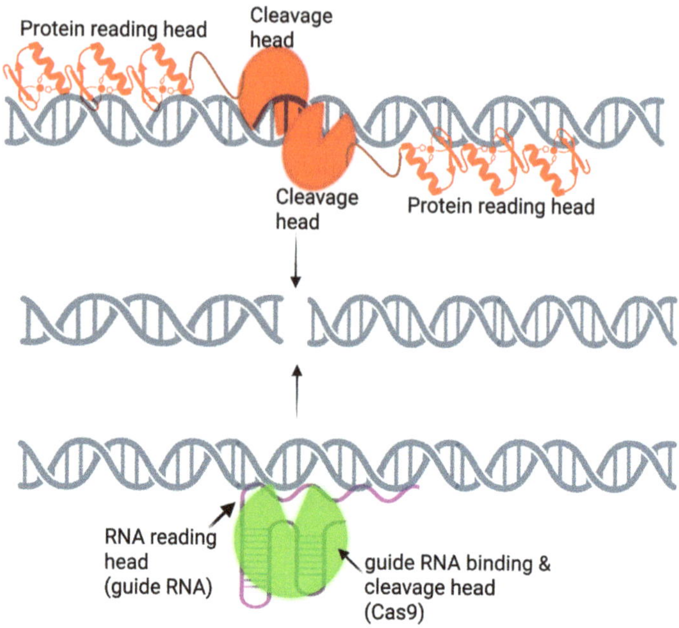

Figure 12.5 Breaking a DNA duplex at a specific target site. The target site of the DNA duplex is an arbitrary sequence chosen by the genome engineer. She then either engineers two sets of zinc finger proteins in red (top) with three zinc fingers each), one for the left side of the target and the other for the right side of the target. Each of the two zinc finger proteins has a DNA cleavage head domain borrowed from another protein. When the two red zinc finge proteins wrap around the target DNA, the two cleavage heads are precisely oriented. This complex then 'bites' down on the target sequence and cleaves it, forming the broken DNA (middle). In this case, the DNA sequence reading heads are the two sets of zinc finger protein that recognize the DNA sequences flanking the target sequence. Alternately, the genome engineer would design a guide RNA with a tail (purple) that is exactly complementary in sequence to one strand of the intact duplex (bottom diagram). The guide RNA is bound to the Cas9 enzyme (green) through its folded conformation. The cas9 enzyme has its own two cleavage heads that makes two cuts to the two DNA strands at the target site. In this strategy, the DNA sequence reading head is part of the guide RNA and the DNA cleavage head is the Cas9 enzyme. [This figure was drawn using Biorender.com under an academic license]

Jinek et al. 2013; Mali et al. 2013). The broken DNA on the chromosome could now heal either by an error-prone DNA repair mechanism known as Non-Homologous End Joining (NHEJ) (Fig. 12.6), or, in the presence of an intact homologous DNA duplex, by faithfully repairing the gap and

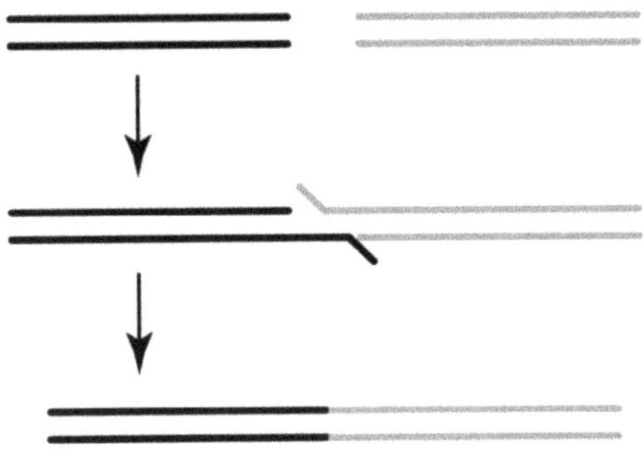

Figure 12.6 Repair of DNA doubles strand breaks by non-homologous end joining. The broken DNA ends are processed in the nucleus to reveal short single chained ends. Unrelated DNA sequences anneal over stretches of microhomology (~2–10 base pairs), the unannealed flaps, if any, are removed by enzymes, and the DNA phosphodiester bonds are sealed.

thus potentially allowing the insertion of exogenous DNA into the genome (Fig. 12.4c).

The designing and making of a guide RNA is simple, rapid, and inexpensive relative to engineering specialized protein "reading-heads." It immediately became possible to target nearly any DNA nucleotide in the human genome for genetic engineering. Soon afterwards, multiple genes could be simultaneously targeted by a battery of guide RNA molecules, allowing massively parallel genetic modification of human cells (Cheng et al. 2023; Zou et al. 2022).

12.3 Genome Manipulation in Human Health

Can human genes be engineered or manipulated using CRISPR/Cas in living humans to cure people of previously incurable diseases?

Indeed, clinical trials for curing inherited genetic disorders such as sickle cell anemia and beta-thalassemia using CRISPR/Cas enzymes have already been completed, and two separate therapies have been approved.

In both diseases, two copies of a hemoglobin gene in the patient (one of each inherited from the mother and the father) are mutated. Missing at least one normal copy of the hemoglobin gene in the adult, the patient's red blood cells become abnormal, with a complex array of pathological properties of these cells associated with a reduced capacity to carry oxygen to the tissues. Such patients frequently suffer from severe pain, blockage of microcapillaries in tissues resulting in tissue damage, often accompanied by organ failure and ultimately death.

Interestingly, these patients would generally have been normal in early childhood because a different form of the hemoglobin protein is synthesized in fetal and early childhood stages (Fig. 12.7). As the person grows, the gene encoding fetal hemoglobin becomes repressed for RNA transcription, while the adult form of the hemoglobin genes (those with the mutations) become actively transcribed, making the mutant form of the adult hemoglobin loaded in the RBCs, leading to the disease.

Could CRISPR/Cas be utilized to disable the gene responsible for encoding a negative regulator (a repressor protein, called BCL11A) of the

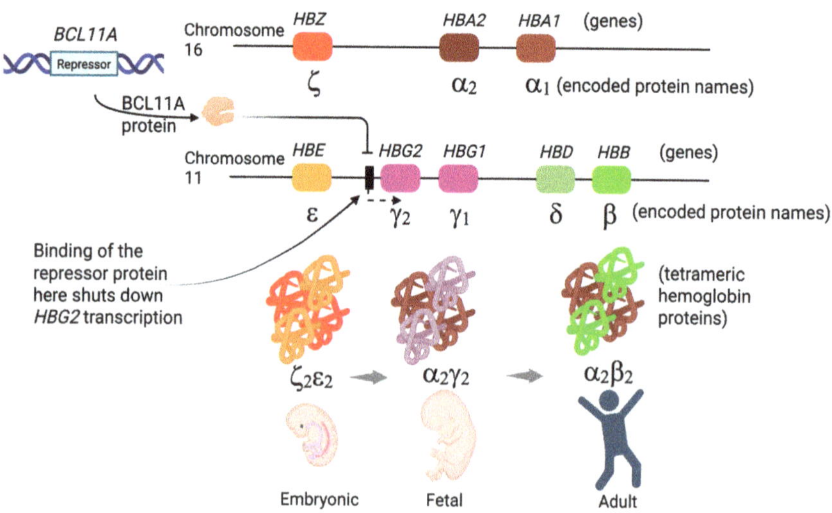

Figure 12.7 Hemoglobin genes. Two clusters of hemoglobin genes (on chromosomes 11 and 16) occur in the human genome. Human embryos only express *HBE* and *HBZ* genes, making the fetal hemoglobin containing two subunits of each protein. At the fetus stage, these two genes are repressed and *HBA1/HBA2* and *HBG1/HBG2* genes are expressed. A few years after birth, the adult form of hemoglobin (a tetramer with 2 subunits of α two of β) are expressed because the *HBG* genes are repressed by a set of repressor proteins (one of which is BCL11A) whereas the *HBB* gene (green box) is activated. As it turns out, the sickle cell disease mutation is in the *HBB* gene. [This figure was drawn using Biorender.com under an academic license]

fetal hemoglobin gene (Fig. 12.7)? Could this make possible the activation of the fetal hemoglobin gene to make the fetal hemoglobin in the adult stage, such that the fetal hemoglobin would outcompete the defective adult hemoglobin proteins?

As if in a science fiction story, bone marrow cells were collected from patients, stem cells that would ultimately produce RBCs were isolated, and CRISPR/Cas9 was introduced into these cells along with guide RNAs complementary to a DNA sequence near the *BCL11A* repressor gene. This DNA sequence is essential for activity of this gene in cells that make RBCs. CRISPR guide RNA binding at the target site caused a DNA break at that site, followed by healing of the broken chromosome by non-homologous end joining (see Fig. 12.6). Since healing of the chromosome in this manner would frequently cause DNA nucleotide deletion mutations, the BCL11A repressor protein would not be made. As a consequence, the fetal hemoglobin gene would turn back on and would be loaded in the RBCs.

This indeed was the case: the *BCL11A* gene was inactivated in some blood-forming stem cells of these patients (Fig. 12.8). These genetically modified stem cells were then reintroduced into the patient's bone marrow. The engineered cells colonized the marrow and eventually formed a subpopulation of RBC-forming stem cells. Since the BCL11A repressor protein was not produced, it was no longer able to repress the fetal hemoglobin gene in these cells. Astonishingly, the strategy worked like a charm (Frangoul et al. 2021) (Fig. 12.8). The clinical trial was successful. At the time of this writing nearly 30 patients with inherited sickle cell disorder were apparently leading disease-free lives.

The 30-year arc that spanned the accidental discovery of CRISPR arrays while studying unrelated biology of *E. coli* and the massively parallel engineering of the human genome was made possible by numerous concepts with essential roots in the Darwinian evolutionary theory, as discussed in Section 12.1 above. As a result, it has now become possible for the first time to engineer complex human genomes by design (Gostimskaya 2022).

Clearly, the early successes of these ground-breaking clinical trials to cure patients with otherwise incurable diseases by genome manipulation promise an outstanding future potential for human health. They also raise numerous uncomfortable questions. Who can afford these expensive and intricate treatments? Of the estimated 10 million sufferers of sickle cell

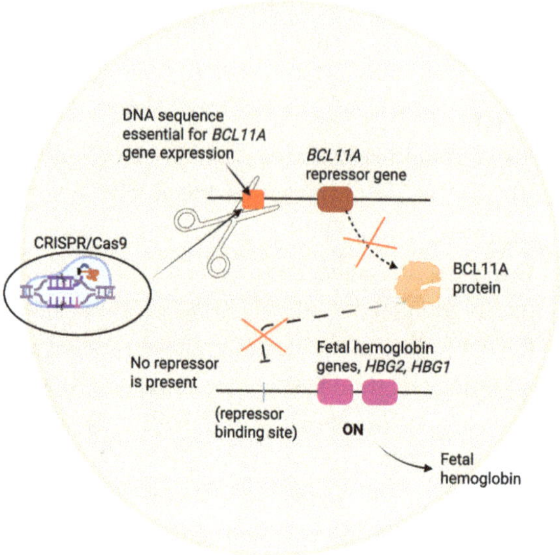

Figure 12.8 A curative therapy for sickle cell disease. This involved genetically inactivating the BCL11A gene, which encodes a repressor protein that binds to and inactivates the expression of the fetal *HBG1* genes. The process uses the DNA-cleaving reagent CRISPR/Cas9 to cleave and thus mutate the DNA sequences (red box) near the *BCL11A* gene. Once mutated, *BCL11A* cannot be expressed in the cells that make RBCs. This genetic surgery is done on stem cells that make red blood cells. Such stem cells were isolated from several RNA complementary to the DNA sequence near the *BCL11A* gene. Cleavage by Cas9 at the target DNA sequence resulted in a mutation (by NHEJ) (Fig. 12.5), such that *BCL11A* no longer can be turned on; therefore, the fetal hemoglobin genes were no longer repressed. These engineered stem cells were then introduced into the bone marrow of the corresponding patient donors. These engineered stem cells repopulated the bone marrow within a few months with genetically modified red blood cells, which allowed the fetal hemoglobin protein expression in these cells. All patients developed normal blood cells and did not have symptoms more than one year after the treatment. This therapy was approved for the treatment of additional patients in December 2023. [This figure was drawn using Biorender.com under an academic license]

disorder, the majority live in Africa and among impoverished communities in India. Few of these patients can afford to spend even one dollar a day for medical treatment. These remarkable advances in medical science will likely leave these deserving patients behind. Questions of fairness and ethical responsibility of scientists and medical researchers are now in the forefront of public discussion. Even more controversial is the ability to modify the germline of individuals by targeted editing of genes. The consequence will be that the succeeding generations of progeny would inherit the genetically modified alleles, raising the specter of "designer babies" by affluent individuals — even to the possibility of cloning themselves. The impact of such practices on human society might be incalculably disruptive.

Darwinian evolution has come full circle. Man evolved can now evolve man — a new technology ahead of its time as ethics and policy trail behind our technical capabilities (Coller 2019). Unregulated research on making CRISPR/Cas edited human genomes raises frightening possibilities reminiscent of gross violations of human rights in the past (Association 1992; Pūras 2017). Controversial work on apparently heritable gene editing in three human subjects (all conducted in the test tube prior to *in vitro* fertilization), allegedly successful, have met with public outcry and universal condemnations (Cyranoski 2019). Science is a double-edged sword. How it is used must be informed by knowledge as well as humanity.

12.4 Cheating Evolution to Eradicate Disease-Carrying Pests

Since the relatively rapid emergence of insecticide-resistant insect pests in the late 20th century, ecologists and entomologists have realized that it might be difficult, if not impossible, to fully eradicate a harmful species of insect by using chemical or biological toxins.

Mosquitoes are obligate hosts in the spread of mosquito-borne parasitic diseases such as malaria, and mosquito-borne viruses such as Zika, West Nile, Dengue, Yellow Fever, and Chikungunya. Mosquito-borne diseases together constitute a significant proportion of all infectious diseases, and contribute to nearly half the global deaths due to infection (Chala and Hamde 2021).

Many viruses of significant public health concern are flaviviruses with +ve strand (chain) RNA genomes, meaning that their RNA genomes can encode proteins directly, and the RNA genome replicates using an encoded RdRP without the need for reverse transcription. When a mosquito feeds on the blood of a previously infected host, the virus present in the host's blood enters the mosquito's gut. From there, the virus replicates within the mosquito, spreading to its salivary glands. It is then transmitted to a new host when the mosquito bites again. The mosquito is not merely a passive carrier of the virus; rather, it serves as a vector that actively participates in the virus's life cycle.

Can one eliminate mosquito-borne viral diseases by population engineering of mosquitoes? Below we provide a brief account of these emerging attempts to cheat evolution in its own game.

12.4.1 *The principle of population collapse*

If there is a single universal consequence of Darwinian evolution, it is that species are regularly extinguished. There might be many reasons for the population collapse of a species, including a sudden change in the environment, the emergence of an efficient pathogen, a change in its population sex ratio, or selectively killing a proportion of sperm or eggs that carry a particular biological factor. We will examine here whether it is possible to reduce the number of disease-carrying mosquitoes, and perhaps drive these harmful insects to extinction, without affecting the harmless varieties.

As it so happens, nature already does this for certain insects (Werren, Baldo, and Clark 2008). An intracellular parasitic bacterium called *Wolbachia sp.* infects and colonizes an insect's germline. Nearly half of all insect species can be infested with this bacterium.

In mosquitoes, this insidious bacterium is transmitted only through the female's germline — the cytoplasm of the infected mosquito's egg contains the bacteria. If an infested female mates with a male that has already been colonized by the same strain of *Wolbachia* the progeny survives. If an infested female mosquito mates with an uninfested male, their progeny also survives, and the bacteria colonize the germlines of the progeny males and females both. By contrast, if an infested male mates with an uninfested female or with a female infested with a different strain of *Wolbachia*, the progeny dies (Fig. 12.9a–c).[3] The reader can intuitively

[3]This is not the only mode of *Wolbachia*'s evolutionary strategy (see, Warren, Baldo and Clark, 2008). However, this is the mode observed with its mosquito hosts.

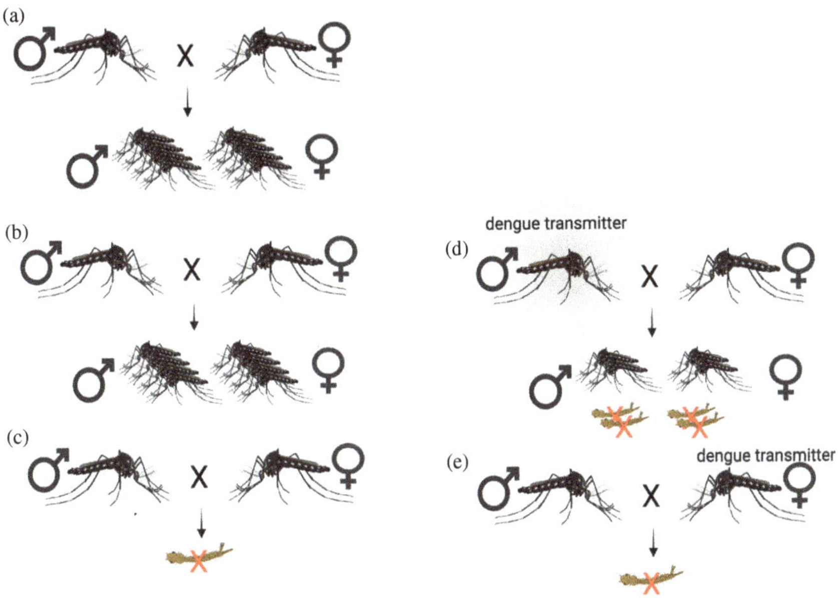

Figure 12.9 Controlling dengue-carrying mosquitos by the intracellular parasitic bacteria *Wolbachia*. The bacterial species colonizes mosquito germline cells. (a) When a *Wolbachia*-infested (green) male mosquito mates with an infested female (green), the progeny is newly infested with *Wolbachia* (green). (b) When a *Wolbachia* infested female mates with an uninfested male (black), again the progeny becomes infested with *Wolbachia*. (c) However, when an infested male mates with an uninfested female (black) (or a female infested with a different strain of *Wolbachia*), none of the progeny survives. (d) *Wolbachia*-infested mosquitos that are also carriers of the dengue virus are less fit (produce fewer progeny) than those mosquitos that are not carriers of the virus (as in B). Thus, few progeny mosquitoes carrying dengue virus survives off a mating between an uninfested dengue transmitter male and an infested female. The few surviving progeny are now *Wolbachia* infested. (e) Mating between infested male and a dengue transmitter female leaves no surviving progeny. [This figure was drawn using Biorender.com under an academic license]

see that, due to this selective death of the progeny of uninfested females in the presence of males with prior bacterial infestation, the proportion of insects with bacteria would increase with the number of generations. Thus, the bacterial parasite ensures its own propagation into the next generation of insects by killing the progeny that do not contain the same bacterial genes. This is, in principle, the same phenomenon as male lions killing the cubs sired by different males — ensuring the propagation of their own genes.

Imagine now that one releases some male mosquitoes with *Wolbachia* infestation in a particular region. Mosquitoes harboring *Wolbachia* should rapidly spread by killing the fraction of uninfested progeny.

Could this strategy be used to control mosquitoes harboring disease-causing viruses or protozoa (e.g., the malarial parasite)?

In fact, engineering of *Wolbachia* strains specifically to kill viruses might not even be necessary because a particular strain of *Wolbachia* by itself reduces the transmission probability of dengue virus by the mosquito *Aedes eagypti* (Aliota et al. 2016) (Fig. 12.9d,e). Provoked by these observations, a field trial spanning about 80 square kilometer area was conducted in Brazil (Pinto et al. 2021), which had released a large number of *Wolbachia*-infested male mosquitoes (reared in cages) harboring no dengue viruses. Within eleven months of deployment, nearly 80% of mosquitoes trapped from the deployed zones had *Wolbachia* infestation whereas none was noted earlier. Importantly, after three years, slightly under 70% fewer incidence of dengue and slightly over 50% fewer incidence of chikungunya (another virus-caused disease) incidents were observed in those localities relative to previous years. These results have encouraged further expansion of the trial (Lenharo 2023). A similar experiment is under way in Malaysia, where *Wolbachia*-infested mosquitoes of both sexes were released. Within three years, over 90% of mosquitoes caught from the trial area were infected with *Wolbachia* (Cheong et al. 2023), again demonstrating that population engineering of mosquitoes is eminently feasible.

12.4.2 *Mutagenic CRISPR amplification*

While the above approach appears promising, it is yet unclear how effective it will be for the eradication of dengue, because the effect of *Wolbachia* on virus transmission is small. Moreover, no such effects have been reported for malaria or other mosquito-borne diseases except dengue and chikungunya, reducing the general promise of this approach. To solve these problems, *Wolbachia* strains could potentially be engineered in the future to express CRISPR guide RNA molecules complementary to a virus or parasite gene, and simultaneously the appropriate type of Cas9 enzyme capable of utilizing the guide RNA to destroy the target pathogen's nucleic acids. In fact, certain Cas9 enzymes, for example, one encoded by the bacterium *Campylobacter jejuni*, do recognize and cleave even RNA molecules that are bound by guide RNAs (Dugar et al. 2018).

Lamentably, it has not yet been possible, to the author's knowledge, to introduce foreign DNA into *Wolbachia* — a requirement that must be met before successful genetic engineering of this bacterium with CRISPR is possible. Recently, a bacteriophage that infects *Wolbachia* has been identified, and its prophage DNA has been detected in the genome of certain *Wolbachia* strains (Kent and Bordenstein 2010), raising the possibility that this or a similar phage might be the conduit for introducing Cas9 and guide RNA genes into *Wolbachia*.

As an alternative approach (popularly called as "synthetic biology"), an engineered gene drive known as "Mutagenic CRISPR Amplification" has been developed (Fig. 12.10). In principle, this strategy increases the probability of transmission of a mutant gene whose wild-type counterpart is essential for the host to carry a human-transmissible virus. A host insect carrying the mutant allele cannot carry the virus (Esvelt et al. 2014). As a result of this clever strategy, the proportion of insects without the wild-type allele of the gene increases exponentially in the population. Laboratory experiments have shown this to be feasible in principle (Gantz and Bier 2015).

If insects harboring this type "mutagenic chain reaction" cassette against a virus-hosting gene of the insect are let loose in the wild, it is possible that the population of virus-hosting insects will collapse. Very soon, the synthetic gene will dominate the population of virus-carrying hosts even if it were first a small fraction of the population. Soon, this gene drive will lead to the "artificial evolution" of a host that is no longer a vector of the virus. A similar strategy has been proposed to eliminate the population of an insect, like a potentially disease-carrying insect, by not just eliminating the virus-hosting gene but the whole host organism. Imagine that the haploid gametes of a male mosquito similarly turn on a gene (***A***) harbored in one of its chromosomes (except the X chromosome) encoding the CRISPR/Cas enzyme that can destroy the X chromosome. ***A*** is now the X chromosome "shredder" allele encoding a CRISPR/Cas)

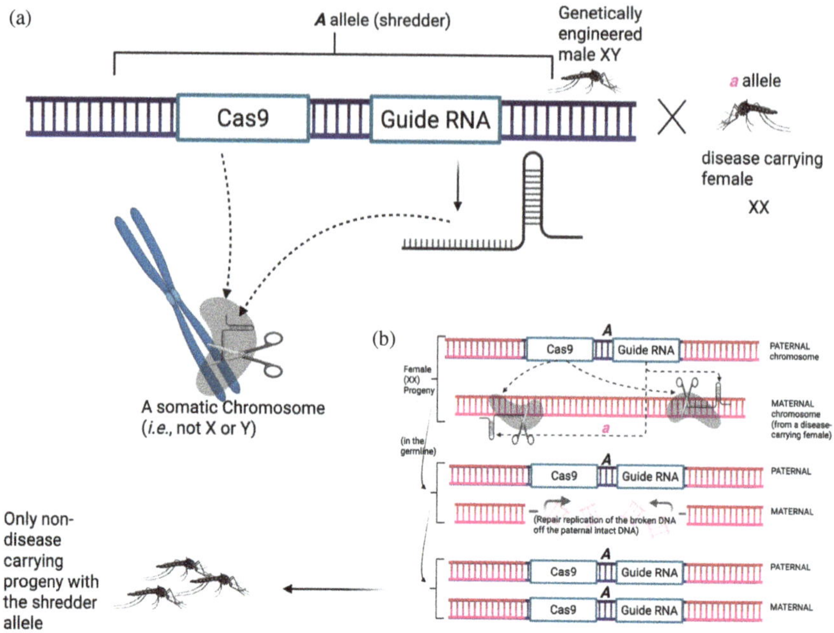

Figure 12.10 Mutagenic CRISPR Amplification. A male mosquito is engineered with *Cas9* and *guide RNA* genes in a gene that is essential for carrying a disease-causing organism. Let us call the allele of the gene that can support the disease-causing organism as "*a*". This gene is disrupted (mutated) by the insertion of the *Cas9* and *guide RNA* genes. The guide RNA is complementary to two short sequences of the *a* allele, which are not present in the mutant allele. Note that there could be two different guide RNAs, one for the left side and another for the right side of the *a* allele of the gene. Call this mutant allele "*A*". This allele must be in any chromosome except X or Y. In the germline of male and female progeny of a mating between an engineered male and a potentially disease-carrying female (having the *a* allele), the *Cas9* and *guide RNA* genes will be expressed in the germline, Cas9 protein will deliver two blows to the *a* allele obtained from the mother mosquito, and the gap in the chromosome will be repaired off the intact paternally-derived engineered chromosome carrying the *A* allele — in short, duplicating the 'gene shredder' allele. This will preferentially (exclusively) transmit the *A* allele to the progeny, which therefore cannot support the disease-causing organism. Thus, both male and female progeny will inherit the engineered allele. This will be passed on to the next generation of mosquitoes, initiating similar rounds of events, effectively eliminating the disease-supporting *A* allele from the population after a few generations of random mating. [This figure was drawn using Biorender.com under an academic license]

whereas *a* is the non-shredding (normal) allele. The guide RNA of this "shredder allele" is complementary to some target sequences that are present only on the X chromosome. Again, expression of the CRISPR/Cas enzyme occurs only in the haploid gametes of these mosquitoes. This will result in surviving sperm carrying only the Y chromosome; all X-carrying gametes die because the X chromosome is destroyed. There will be no females in the progeny of the first generation. Approximately half the first-generation males will also harbor the X-shredder allele (***A***), which will then mate with more females, generating more male-only progeny. In this way, the sex ratio will be distorted, and ultimately the whole population will collapse because there will be no more females left. The species of mosquitoes, along with the disease it normally carries, will be extinguished if sufficient engineered male mosquitoes were released. The population dynamics of the X-shredder allele are explained in more detail in Chapter 13 (Section 13.3).

Should we do this kind of population engineering of organisms? Do we understand the complex ecological relationships that each species has in nature, to evaluate the risk posed by the deliberate extinction of a species?

Part 3

I Dolci

Chapter 13

Laws of Growth and Death

Darwinian evolution applies only to individuals living within populations, because natural selection is manifested through competition among multiple individuals for higher and higher fecundity. The growth of populations of organisms affects the evolutionary processes in two different ways. First, the larger the population, the more and diverse variants will be produced by chance, and therefore the better is the chance of a favorable, even rare, variant to arise. Second, the larger the population of rare variants, the better is the chance that a fitter variant would be selected, i.e., would leave a higher clutch of progeny relative to the other variants in the population. Therefore, it is important to understand how populations grow. This chapter examines the mathematics of growth, death, and extinction. Without mathematical precision, we are left with faith in scientists' assertions; therefore, we must endeavor to quantify.

13.1 Growth Law

Let N_t be the number of individuals in a population at time t. The rate at which the number N changes with time is proportional to the number present at that time. The greater the value of N at a time, the greater is the rate of growth in the next instant of time. In other words:

$$\frac{dN}{dt} \propto N. \tag{13.1}$$

$\Rightarrow \frac{dN}{dt} = k.N$, where k, the growth rate, is the proportionality constant.

$$
\begin{aligned}
&\Rightarrow \frac{1}{N} dN = k \cdot dt \\
&\Rightarrow \int_{N_0}^{N_t} \frac{1}{N} dN = k \cdot \int_0^t dt \\
&\Rightarrow \text{In}\, N_t - \text{In}\, N_0 = k \cdot (t - 0) = kt \\
&\Rightarrow \text{In} \frac{N_t}{N_0} = kt \\
&\Rightarrow N_t = N_0 \cdot e^{kt}
\end{aligned}
\tag{13.2}
$$

This is the simple exponential growth equation, where the value of k, the growth rate constant (in units of time), can be easily evaluated from the time it takes to double the population (i.e., $k = \frac{1}{t} \text{In} \frac{N_t}{N_0} = \frac{1}{t_{double}} \cdot \text{In}\, 2$). The equation can also be used if time t is measured in generation time rather than chronological time, without too much error, so long as we consider large populations over large intervals of generation time such that the individual generation time $t_{double} \ll t$, the range of evolutionary time.

Below we explore some of the limitations of this simple growth Equation (13.2). First, let us write down Equation 13.1 with one fewer assumption.

Let $f(N)$ represent the rate of growth relative to the current population size, N. So, for example, $f(N) = 0.4$ means that the population is growing at a 4% rate when the population size has N individuals. In that case, we can rewrite the general form of the growth Equation (13.1) as:

$$
\frac{1}{N} \frac{dN}{dt} \propto f(N). \tag{13.3}
$$

This is a general form of 13.1 where $f(\cdot)$ is a function that defines a dependence of the population number on potentially other parameter(s).

In its simplest form, $f(N) = C$, a positive constant. In this case, the solution to Equation 13.3 yields the equivalent exponential growth Equation in (13.2), providing for a limitless growth in N, where α is the proportionality constant (always assumed to be positive), such that $k = \propto C$:

$$
N_t = N_0 e^{\propto C.t} \tag{13.4}
$$

The above condition is obviously biologically unrealistic for any organism unless the population density is low. Sooner or later the organism's population density will reach a level high enough that it will run out of resources needed for further growth. Therefore, a more reasonable case is one where $f(N)$ is a linear function, where A is a dimensionless parameter, of the form:

$$f(N) = 1 - \frac{N}{A}. \qquad (13.5)$$

It is easy to see that when numerically $N = A$, the population ceases to grow and reaches a steady state (Fig. 13.1). Thus, the parameter A represents the upper bound for the number of organisms in a population, and therefore is called the *carrying capacity* of the environment. It is important to realize that if a mutant arises such that it has a gene combination that allows this variant to set a numerically higher value of A, which may be interpreted as its ability to utilize the resources more efficiently relative to all others in the population, then its proportion with respect to the remaining individuals in the population will keep increasing until the total number of organisms reaches the new, higher value of A.

Let us now set the following condition, where B is yet another dimensionless parameter:

$$f(N) = \left(1 - \frac{N}{A}\right)\left(\frac{N}{B} - 1\right) \qquad (13.6)$$

This definition of the function allows us to explore two important conditions.

Condition for population expansion: We see that for $0 < B < N < A$, the population increases (i.e., $dN/dt > 0$, by Equation 13.3).

Condition for population extinction: When $N \leq B$ and $N < A$, then the population decreases ($dN/dt < 0$) (by Equation 13.3) (Fig. 13.1). In other words, if the population of an organism falls to the numerical value of the parameter B, the population will go extinct even if the carrying capacity, A, is not exhausted. The parameter B is therefore called the *critical point*

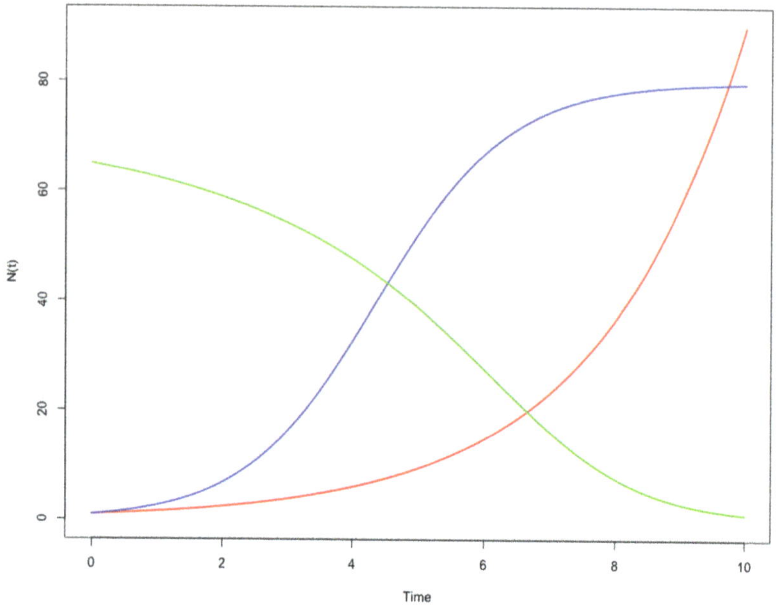

Figure 13.1 Behavior of growth laws. According to equation 13.3, the condition of $f(N) = C$ leads to an exponential growth (red). When $f(N) = 1 - \frac{N}{A}$, the population reaches a saturation when $N = A$ (blue). The population will collapse from an initial high value if $f(N) = \left(1 - \frac{N}{A}\right)\left(\frac{N}{B} - 1\right)$, when $N = B < A$ (green).

of the population. This condition is important to understand how or when a population of organisms can go extinct, which is repeatedly observed in the history of life on earth. It will also be important to understand the principle of synthetic "gene drives" to push a population of disease-carrying insects to extinction while not affecting the harmless population of that insect species.

13.2 Death Law

Let us now examine the question of death. The so-called Gompertz law holds that the risk of death in a population is defined mathematically as:

$$p_x = \alpha e^{\beta x} \tag{13.7}$$

where p_x is the probability of death at age x, and α and β are parameters that can be estimated from population studies.

The above law can be made more realistic by adding an age-independent term (λ) to the Gompertz law equation:

$$p_x = \lambda + \alpha e^{\beta x} \tag{13.8}$$

Plots (Fig. 13.2) of the function in Equation 13.8, for a few chosen values of α and β make the following points. First, as age increases the death rate (i.e., the probability of death) accelerates (the slope of the curve rises). Second, the rates of rise of the age-dependent death rate can be flattened by tuning the parameters α and β. As discussed in Chapter 4 (Equation 4.1), α is the "initial vulnerability to death" at birth. In Fig. 13.2, this is easily seen as two different initial probabilities of death at age zero for two different values of α. One could imagine that, given perfect prenatal and neonatal care in an imaginary country, and given perfect delivery and perfect protection against infectious diseases, the initial risk of death at birth might be related to genes that affect neonatal viability. The parameter β is more complex. From Fig. 13.2 we observe that β is related to the rate of rise (the slope) of the age-dependent risk of death as age progresses. We have already discussed in Chapter 4 that β is dependent on gene variants that determine susceptibility to cancer and various chronic diseases, as well as on age-dependent environmental factors, such as workplace-related exposure to hazards, prolonged disasters such as war, and poverty, among others.

Moreover, the age-independent factors (captured by the parameter, λ in Equation 13.8) that affect the risk of death are also obviously related to many environmental factors, such as affluence, exposure to war, exposure to infectious agents, vaccination rates, nutrition, and health-care availability. But some of these factors might also be related to genetic factors underlying the biological basis of organismal aging, including genetic susceptibility to chronic diseases.

By comparing the theoretical model with the actual values of global mortality rates per age collected by the United Nations (dotted lines in Fig. 13.2), we see that the value of α lies somewhere between 0.01 and 0.02,

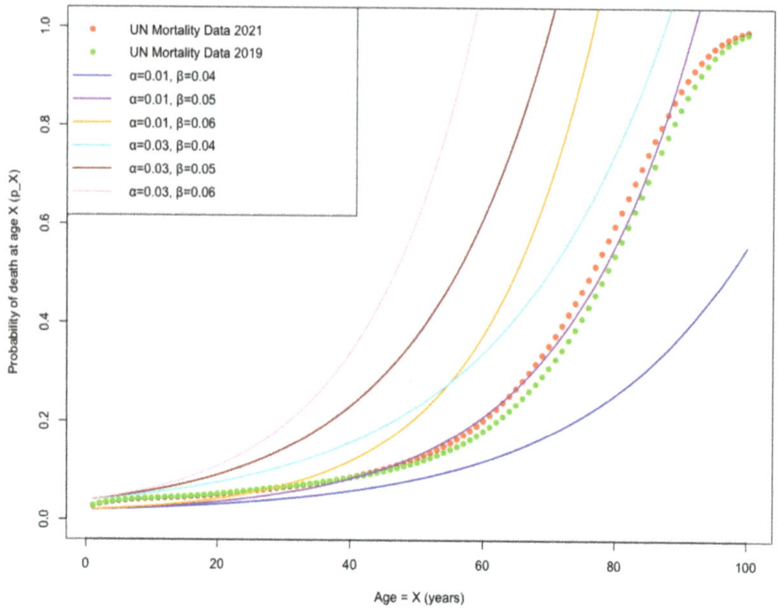

Figure 13.2 Plots of the probability of death at a certain age by Gompertz law. The curves were produced according to Equation 13.8. For each of two values of α we plot curves for three values of β. We assume a value of $\lambda = 0.01$, or 1% risk of age-independent death from all causes. The dotted curves are actual mortality data in years 2019 (green) and 2021 (red). Note the likely effect of COVID-19 on mortality rates beyond 45 years in 2021 relative to 2019 [Data source: United Nations Department of Economic and Social Affairs, Populations Division (https://population.un.org/wpp/downloads)]

and that of β is approximately 0.05. We also see a significant departure of the theoretical curve from the observed data beyond the age of 85 years — the risk of death is less than that expected by the model. What could this be due to? Might there be "longevity gene variants" that a minority of the population harbors, which make them last longer (a smaller value of the parameter β for this small segment of the population) than most others?

13.3 Distortion of Sex Ratio Can Extinguish a Population

Near the end of Chapter 12, we discussed how a mosquito population bearing disease-carrying organisms can be extinguished by changing the sex ratio using a genetic tool known as the CRISPR/Cas9 "X-shredder" gene cassette. A male diploid organism (in this case, a genetically engineered male mosquito with an X and a Y sex chromosome) turns on this gene in its germline (i.e., in the cells that make haploid gametes). Once turned on, the gene destroys ("shreds") the X chromosome-bearing gametes, thus biasing the progeny to males only. To gain better intuition of how this strategy can collapse the population of mosquitoes, let us first model the basic principle of how altering sex ratios can drive a population to extinction.

Let M_t and F_t be the numbers of males and females at generation t, respectively. Let R be the average number of offspring per female per generation; let a be the fraction of male births (the sex ratio) in the population. Assume the organism is diploid with two genetically distinct sexes, male and female.

The total number of offspring (O_t) in each generation is determined by the number of females and their reproductive rate (R):

$$O_t = R.F_t$$

Let the sex ratio at birth be biased toward males with probability α, where α ranges from 0.5 (no bias) to 1.0 (representing strong male bias). Therefore, the number of males and females in the next generation is:

$$M_{t+1} = \alpha.O_t = \alpha.R.F_t$$

$$F_{t+1} = (1 - \alpha).O_t = (1 - \alpha).R.F_t$$

Obviously, the population goes extinct when $F_t < 1$ because females are required for reproduction.

Thus, if $(1 - \alpha)R > 1$, the female population grows, ensuring population stability. If $(1 - \alpha)R \leq 1$, the number of females declines over generations, leading to extinction. Avoiding extinction therefore requires:

$$(1-\alpha)R > 1, \text{ or } \alpha < 1 - \frac{1}{R}$$

This defines a critical male birth fraction, α_{crit} for a given R. If the actual male birth fraction exceeds α_{crit} the population will collapse.

For example, if a female on average produces two surviving offspring, i.e., $R = 2$, then:

$$\alpha_{crit} = 1 - \frac{1}{2} = 0.5$$

Thus, if more than 50% of births are male *and* R is small (say, ~2), the number of females declines over time, leading to population collapse. This matches the intuitive idea that an even sex ratio (50% male, 50% female) is necessary for stable population maintenance. Higher male biases (e.g., $\alpha = 0.7$, 0.8 or 0.9) for a given R may drive extinction more rapidly because fewer females remain to reproduce. Figure 13.3 estimates the number of generations required for a starting population of 600 million humans to go extinct under various sex ratios, assuming that on average a couple produces ~2 children.

The basic principles illustrated in Figure 13.3 apply also to mosquitoes. For *Aedes aegypti* (the dengue virus carrier mosquito), a female lays approximately 100–1000 eggs in her lifetime, of which only about 5% survive in the wild. Thus, R ~5–50. Modeling shows that an appreciable probability of population collapse within 50 generations requires a strong sex ratio bias: typically, $\alpha \sim 0.9$, meaning that only ~10% of the progeny are female.

The clutch size of *Aedes aegypti*, the primary mosquito vector for dengue, Zika, chikungunya, and yellow fever viruses, varies depending on environmental factors and female health. A female typically lays about 100–200 eggs per batch, up to five batches over her lifetime, with only 1%–10% of the eggs reaching adulthood (resulting in 1–20 surviving offspring per female).

To help us understand how the mosquito population can be driven to extinction, let us make a "discrete generation" model in which time is measured in generation times so that each time step corresponds to one complete generation. With this discrete model, it is easier to understand the complex population dynamics than using the type of growth equations used earlier in this chapter.

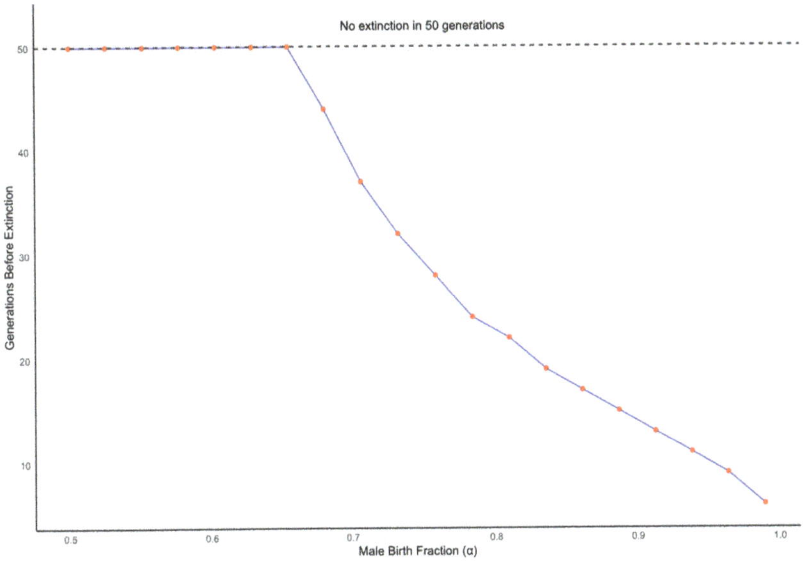

Figure 13.3 The effect of sex ratio on population extinction. The number of generations after which a population of 300 million will go extinct for various values of the sex ratio (fraction of males in the population), assuming on average two births per mating. A sex ratio of 0.5 means an equal number of males and females. A higher sex ratio for males implies fewer females are available to sustain the population. For humans, 50 generations constitute approximately 1,250 years.

Imagine introducing male mosquitoes carrying the X-shredder gene cassette into a natural population. As before, the X-shredder gene cassette selectively destroys the X chromosomes during sperm development, so males carrying the shredder produce predominantly Y-bearing sperm. Consequently, all surviving offspring are male, and all female embryos fail to develop.

To model this, let us use a discrete-time model where each time step corresponds to one full generation. We assume that each female mates once per generation and produces, on average, b surviving offspring. The outcome depends on the type of male she mates with:

- If she mates with a wild-type male, the offspring are 50% female and 50% wild-type male.

- If she mates with a gene-drive male (X-shredder), all surviving offspring are gene-drive males.

Let M_n be the number of wild-type males at generation n; G_n be the number of gene-drive males; F_n, the number of females. Each female mates randomly with a male, with mating probabilities proportional to the male subpopulation sizes.

How does the population grow from one generation to the next? To see this, we write the discrete equations that update the new population numbers for this mosquito population at a new generation as:

$$F_{n+1} = \frac{b}{2} \cdot \frac{M_n}{M_n + G_n} \cdot F_n \tag{13.9}$$

$$M_{n+1} = \frac{b}{2} \cdot \frac{M_n}{M_n + G_n} \cdot F_n \tag{13.10}$$

$$G_{n+1} = b . \frac{G_n}{M_n + G_n} . F_n \tag{13.11}$$

The critical reader will protest, however, that the conditions for deriving equations 13.9–13.11 are not realistic, because in nature there must be limits to the resources that can support the growth of mosquito populations. Therefore, to incorporate resource limitations, we assume a carrying capacity K for the total mosquito population in their current environment.

Let the density-dependent survival factor $S = \frac{2}{b} \cdot \frac{K}{N_n}$, where $N_n = F_n + M_n + G_n$.

Applying S rescales the offspring numbers due to competition:

$$F'_{n+1} = \frac{b}{2} \cdot \frac{M_n}{M_n + G_n} . F_n . S = \frac{M_n}{M_n + G_n} . F_n . (K / N_n) \tag{13.12}$$

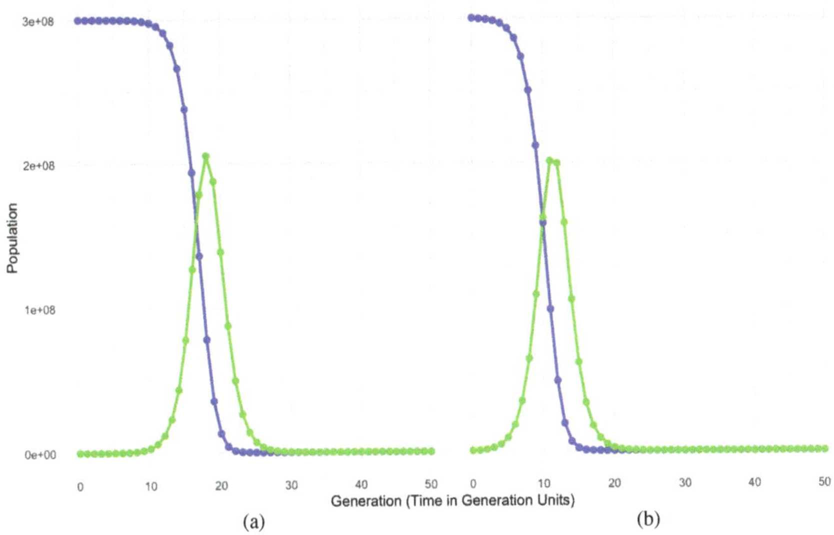

Figure 13.4 Population dynamics of X-shredder mosquitoes. In a population of 600 million (N) mosquitoes, (a) 3,000 or (b) 300,000 male mosquitoes carrying an X-shredder locus are introduced. Carrying capacity (K) of this environment is 600 million. Blue, female mosquito population. Green, X-shredder male mosquito population. The initial number of X-shredder is too small to discern in this scale. Dynamics of wild type males are not shown (they follow the females).

$$M'_{n+1} = \frac{b}{2}.\frac{M_n}{M_n + G_n}.F_n.S = \frac{M_n}{M_n + G_n}.F_n.(K/N_n) \tag{13.13}$$

$$G'_{n+1} = b.\frac{G_n}{M_n + G_n}.F_n.S = 2\frac{G_n}{M_n + G_n}.F_n.(K/N_n) \tag{13.14}$$

Notice that after adjusting with S, the updated mosquito numbers become independent of the brood size, b, due to cancellation of this term in the above three equations.

Suppose that we start with 300 million males and 300 million females (total N = 600 million), and release 3,000 gene-drive males. Simulations (Fig. 13.4) show that extinction occurs within about a year and a half (approximately 35–40 generations, assuming a 14-day generation time) even if the number of introduced gene-drive males varies by two orders of magnitude. The extinction speed is only modestly sensitive to the number of released gene-drive mosquitoes.

Of course, our simple model does not account for real-world complexities such as asynchronous breeding cycles, seasonal variations, changes in carrying capacity, differential survival of engineered mosquitoes, or the potential evolution of resistance to the X-shredder system — all factors worth exploring further.

Chapter 14

Fitness and Selection Pressure

Here we define evolutionary fitness and selection pressure in precise mathematical form. We examine a few properties that emerge from these definitions. Specifically, we examine how selection pressure favors an initially rare but more fit variant that eventually takes over the population.

14.1 Evolutionary Fitness Defined

One fundamental assumption of Darwinian evolutionary theory is that organisms that are more fit produce more offspring, which in turn leave further offspring, relative to other less adapted individuals. Thus, the relative number of viable offspring produced by individuals in a population is a crucial measure of evolutionary fitness. Below, we explore this fundamental assumption of evolutionary theory in quantitative terms, and we explore the explanatory power of these mathematically precise concepts.

The fitness, W, of a genotype (say, a genetic variant in the population, one which arose due to a random genetic process) is defined as the ratio of its abundance in the next generation to its current abundance. Thus, if n_t is the abundance of individuals of a particular genotype at the tth generation, and n_{t+1} is the abundance of individuals with that same genotype in the $(t + 1)$th generation, then the fitness of that genotype is given by:

$$W = \frac{n_{t+1}}{n_t} \qquad (14.1)$$

This implies that $n_{t+1} = n_t.W$.

For multiple generations of growth, the abundance after t generations can be expressed as (n_0 is the initial number of that genotype in the population):

$$n_t = n_0.W^t \tag{14.2}$$

It is easy to see that for fitness $W > 1$, the number of individuals of this genotype will increase over time as measured by the number of generations. Conversely, for $W < 1$, the number of individuals of that genotype will decrease after each generation due to negative selection ($W = 1$ defines an absence of selection on that genotype). Note that the abundence of individuals or their genotypes in a population can be affected by migration, drift, sampling, and density-dependence. We ignore these complexities here for our restricted definition of fitness.

14.2 Definition of Selection Coefficient

The strength of natural selection (also called the selection pressure or selection coefficient, s) on a variant genotype is defined as the ratio of the difference between the fitness of an individual genotype (W_v) and the average fitness ($\hat{W}$) of all genotypes in a population to the average fitness:

$$s_v = \frac{W_v - \hat{W}}{\hat{W}} \tag{14.3}$$

Let us briefly explore what Equation 14.3 means. If the population with a typical genotype grows 2% per year, but the portion of the population with the variant genotype v can grow at an absolute rate 6 percentage points higher, i.e., 8% per year, then $W_v = 1.08$ and $\hat{W} = 1.02$; so, $s_v = (1.08 - 1.02)/1.02 = 0.06$. In other words, $s_v = 0.06$ means the excess growth rate is about 6%.

For two variants v and a, the relative selection coefficient s is the difference:

$$s = s_v - s_a = \frac{W_v - W_a}{\hat{W}} \tag{14.4}$$

The above equation compares the selection on the two variants. Note that generally W_a is the fitness of a comparator variant, which is usually the most abundant (or the wild-type) variant in the population. Under this condition, W_a is approximately equal to the average fitness of all genotypes in the population, $\hat{W}$ (see Box 14.1 for details).

Let us now examine how a population with these two different variants behaves under selection. In short, we will let these two variants grow in the same population, and over time we will examine how their relative abundances fare when they compete against each other. Throughout, we will assume the simple exponential growth model derived earlier in Chapter 13 (Equation 13.2).

Proceeding as before, at time $t = 0$, let the number of individuals of variant a be $n(a)_0 >> n(v)_0$, the number of individuals of variant v. This is reasonable, because a newly arising genetic variant is likely to be initially very rare. After t generations (assuming no death), using Equation 13.2, we get for the two variants after t generations of growth:

$$n(a)_t = n(a)_0\, e^{k_a.t}$$

$$n(v)_t = n(v)_0\, e^{k_v.t}$$

Now we want to examine how growth rate relates to fitness. Note that we have defined time as the generation time. In other words,

$$n(a)_{t+1} = \left(n(a)_t\right).e^{k_a.(t+1)}$$

and

$$n(v)_{t+1} = \left(n(v)_t\right).e^{k_v.(t+1)}$$

Therefore, we can rewrite (according to Equation 14.2) as:

$$n(a)_{t+1} = (n(a)_t).W_a$$

and

$$n(v)_{t+1} = (n(v)_t).W_v$$

Thus, $e^{k_a} = W_a$ and $e^{k_v} = W_v$, and so we rewrite:

$$n(a)_t = n(a)_0 \cdot W_a^t$$

and

$$n(v)_t = n(v)_0 \cdot W_v^t$$

Now we want to find an expression relating the relative proportion of the two variant populations following t generations of growth given their relative fitness:

$$G = \frac{n(v)_t}{n(a)_t} = \frac{n(v)_0 \, e^{k_v . t}}{n(a)_0 \, e^{k_a . t}} = \frac{n(v)_0}{n(a)_0} . e^{(k_v - k_a)t}$$

Using Equation 14.4, we can rewrite (see Box 14.1 for a full derivation):

$$G = \frac{n(v)_0}{n(a)_0} . e^{(k_v - k_a)t} = \frac{n(v)_0}{n(a)_0} . e^{s.t} \tag{14.5}$$

The above equation shows how the selection coefficient s affects the growth of the variant v relative to that of variant a over generation time. Note that whereas this model assumes no death, no population density dependence, and assumes a purely exponential growth, the essential point — that competition between two variants may cause an exponential rise in the number of the fitter variant over the generations — will remain largely unaffected even when these assumptions are not made.

In Figure 14.1, a plot of the ratio of the number of variant v relative to that of variant a in the population as a function of the number of generations of growth is shown. In this figure, initially at generation zero ($t = 0$) the proportion of variant v was one thousandth that of variant a. We see that with a mild selection coefficient of only 0.01, corresponding to 1% selective growth advantage of the initially minor variant (v) over that of the originally major variant (a), the former takes over the population after

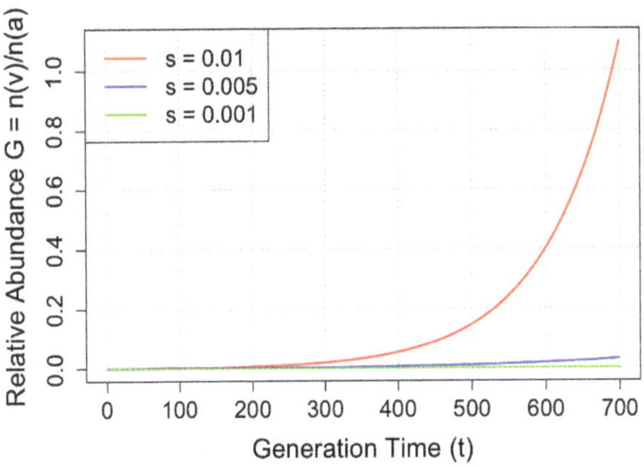

Figure 14.1 A plot of the ratio of the number of individuals of two variants in a population against generation time according to Equation 14.5. G denotes the ratio of the number of individuals of variant 2 (v) to that of the variant 1 (a) in the population. Initial ratio at $t = 0$ was $n(v){:}n(a) = 1{:}1000$. The selection coefficient, $s = 0.01$ corresponds to 1% growth advantage of variant 2 over variant 1. Results with two more values of s are also plotted.

about 700 generations of growth. For *Escherichia coli* bacteria, which live in our gut, that many generations could be less than 13 days. If it were a population of human blood cells that divide approximately once every day, 700 generations would be a little less than 2 years. If a rare (1 in a thousand) variant (mutant) blood cell arises by spontaneous random mutation, which has 1% growth advantage (on its way to becoming cancerous) over the normal blood cells, in a simplified model of exponentially-expanding cell population this variant will comprise over half of all blood cells in 2 years.

In other words, even a relatively modest selection coefficient can force the "evolution" of a population from a state in which the fitter variant was initially a minor fraction to one in which the fitter variant becomes overwhelming.

14.3 The Intensity of Natural Selection

Haldane defined (Haldane 1957) a metric for expressing the strength or intensity of natural selection, I, as follows (see Box 14.1):

Box 14.1

We had previously established:

$$G = \frac{n(v)_0}{n(a)_0} . e^{(k_v - k_a)t},$$

$$W_v = e^{k_v} \Rightarrow k_v = \ln W_v$$

$$W_a = e^{k_a} \Rightarrow k_a = \ln W_a$$

Therefore, $k_v - k_a = \ln W_v - \ln W_a = \ln\left(\frac{W_v}{W_a}\right)$

Thus, $G = \frac{n(v)_0}{n(a)_0} \cdot \left(\frac{W_v}{W_a}\right)^t$

It was shown earlier (Equation 14.4), $s = s_v - s_a = \frac{W_v - W_a}{\hat{W}}$

Rewriting, $W_v - W_a = s \cdot \hat{W}$

To proceed to the next step, we will need to use an approximation of the expression, $\ln\left(\frac{W_v}{W_a}\right)$. Let us see how we can achieve this.

Recall the Taylor series expansion of $\ln(1 + \varepsilon)$ near $\varepsilon = 0$:

$\ln(1+\varepsilon) \approx \varepsilon - \frac{\varepsilon^2}{2} + \cdots$ for small ε.

Let us write, $W_v = W_a + \delta$, where $\delta << 1$.

Then, $\ln\left(\frac{W_v}{W_a}\right) = \ln\left(1 + \frac{\delta}{W_a}\right) = \frac{\delta}{W_a} - \frac{1}{2} \cdot \left(\frac{\delta}{W_a}\right)^2 + \cdots \approx \frac{\delta}{W_a}$

$\Rightarrow \ln\left(\frac{W_v}{W_a}\right) \approx \frac{\delta}{W_a} = \frac{W_v - W_a}{W_a} \approx \frac{W_v - W_a}{\hat{W}}$, when $W_a \approx \hat{W}$, i.e., the variant a is the predominant type in the population and is close to the population average for small fitness difference.

We therefore write, $k_v - k_a = \ln\left(\frac{W_v}{W_a}\right) \approx \frac{W_v - W_a}{\hat{W}} = s$

$$I = \ln(1 + s_0) \approx s_0, \quad \text{for} \quad s_0 \ll 1 \tag{14.6}$$

To understand this, note that natural selection acts on phenotypes, but the measurable outcome is the change in allele frequencies over time. The frequency of an advantageous allele increases as organisms carrying it succeed relative to others. The selection coefficient, s_0, measures the degree to which the fitness of the variant exceeds the average.

For example, assume 99% of individuals infected with a rare variant virus successfully transmit the virus (success probability = 0.99). By contrast, 95% of individuals infected with any virus (population average) transmit successfully (success probability = 0.95). Then the relative fitness of the variant is:

$\frac{W_0}{\hat{W}} = \frac{0.99}{0.95} = \sim 1.0421$ (according to the definition in Equation 14.3).

Then the selection coefficient,

$$s_0 = \frac{W_0 - \hat{W}}{\hat{W}} = \frac{W_0}{\hat{W}} - 1 = 1.0421 - 1 = 0.0421.$$

By Haldane's definition (Equation 14.6), the intensity of selection on the variant becomes, $I = \text{In}(1 + 0.0421) \approx 0.04$ or about 4%.

This logic holds for any other quantitative phenotype, not just on the growth phenotype.

For example, assume that a variant in a human gene enhances neonatal survival by an additional 1 in 10,000 births (0.0001), whereas the prevailing neonatal mortality rate is 3.7 per 1000 births (0.0037). The latter figure corresponds to the US neonatal mortality rate for 2019. Thus, the mean neonatal survival frequency, $\hat{W}$, is = 1 – 0.0037 = 0.9963, and the enhanced survival frequency of the variant, W_v, is 1 – (0.0037 – 0.0001) = 0.9964. Therefore, the intensity of selection (as the effect on the survival frequency, not growth) of this variant is:

$I = \text{In}\,\frac{0.9964}{0.9963} \approx 0.0001$ or ~0.01%, if the variant is not associated with any other effect on growth, development, or fecundity.

The intensity of most natural selection on genes is thought to be similarly feeble. Variants with mild effects take many generations to establish within populations. Generation time thus plays the deciding factor for when a variant genotype might replace another genotype, accounting for the glacial slowness of the evolution of most genetic traits. Viruses, because of their extremely short generation time relative to that of humans, on the other hand, can evolve much faster, as we saw above in the high value of the intensity of natural selection to which they might be subjected.

Chapter 15

Mutation and Cancer

Here we explore the algebraic consequences of a few mathematical models of cancer and the evolution of cancer by gene mutation. We are interested in this pursuit because it provides deep insights into how cancer evolves through somatic mutations, which enable the cancerous cells to thrive, flourish, and outcompete normal cells. We explore how a classic experimental strategy attempts to understand whether drug-resistant cancer might arise because of the drug itself or independently of the drug — a crucial question in the treatment of those cancer types that cannot be treated by surgery or radiotherapy alone.

15.1 An Estimate of the Minimum Number of Cancer-Causing Mutations

Can a single mutation confer sufficient growth advantage to cause cancer? We will show that this is unlikely. We proceed according to Armitage and Doll (1954b).

Let us now estimate the number of cells that acquire exactly r mutations required for malignant transformation over a period of time t. We assume that each of the r mutations occurs independently of the others, and that each occurs at a constant, time-independent rate per cell per unit time. Let the probability per unit time of the ith mutation be denoted by p_i and assume that mutation events are rare and occur in a fixed order along a lineage of clonally dividing cells. Under this model, the probability that a single cell lineage acquires all r specified mutations by time t is approximately:

$$(p_1 t) \cdot (p_2 t) \cdot (p_3 t) \ldots (p_{r-1} t) \cdot (p_r t) = (p_1 p_2 p_3 \ldots p_r) \cdot t^r$$

This expression represents the rate at which a unique cell lineage accumulates all r mutations, assuming each step is independent and rare, and that the mutation order is arbitrary.

However, since there are $r!$ different possible orders in which these r independent mutations can occur, we must divide by $r!$ to account for this overcounting. Therefore, the expected number of cell lineages (or cells) in a population that acquire all r required mutations by time t is:

$$N(t) = \frac{(p_1 p_2 p_3 \ldots p_r) \cdot t^r}{r!} = K \cdot t^r \tag{15.1}$$

where $K = \frac{(p_1 p_2 p_3 \ldots p_r)}{r!}$ is a constant that encapsulates the mutation rates.

This model implies that the number of malignant cells is proportional to t^r, allowing us to estimate the number of cancer-causing mutations r from population-level data if we can observe how the incidence of cancer increases with age.

Now we want to find a way to relate this number, N, to something that is observable, such that we can estimate its value from observation. If we believe that the existence of even one cancer cell in a person is sufficient to ultimately cause a diagnosable disease incidence, then N, the number of cancer cells, is equivalent to the number of cancer patients diagnosed as having cancer in a population. Dividing both sides of Equation (15.1) by the total number of people in the population, we get:

$$\begin{aligned} &\frac{N}{N_{total}} = f = \frac{K}{N_{total}} \cdot t^r \\ &\rightarrow \log f = \log \frac{K}{N_{total}} + r \log t = C + \log t \end{aligned} \tag{15.2}$$

It is important to note that equating N, the number of cells that have acquired all r required mutations, with the number of patients diagnosed with cancer is a modeling idealization. This simplification rests on the assumption that the emergence of even a single malignant cell in a person is sufficient to lead, with high probability, to a clinically diagnosable cancer. In biological reality, this may not always hold true: not every transformed cell necessarily progresses to a detectable tumor, due to factors such as immune surveillance, tissue microenvironment constraints, or stochastic loss of viability. Nonetheless, this assumption serves as a useful approximation for estimating the number of mutational events (r) that typically precede the onset of cancer, particularly when using population-level data on cancer incidence as a proxy for transformation events. As such, it provides a tractable framework for linking the molecular evolution of cancer within cells to observable epidemiological patterns.

Note that in this model, the variable t represents the age of the patient, interpreted as the time elapsed since birth (with $t = 0$ at birth) until the acquisition of the final, or the rth mutation that leads to malignant transformation. Accordingly, if one plots (as predicted by Equation 15.2) the logarithm of the frequency of cancer incidence against the logarithm of age (Fig. 15.1), the resulting graph should yield a straight line, with the slope equal to r — the number of mutations required for cancer to develop.

At the time of the original Armitage and Doll analysis (1954b), comprehensive data on cancer incidence by age were lacking, so they used cancer mortality as a proxy for incidence. Based on this approximation, they estimated $r \approx 6$ (Armitage and Doll 1954b), suggesting that on average six mutations were required for a cancer to become fatal.

Today, with access to more complete epidemiological data, we can revisit this estimate. For example, in 2021, approximately 1.9 million new cancer cases were diagnosed in the United States (Siegel et al. 2022). Given the total population roughly 300 million, the incidence frequency is approximately $\frac{1.9\times10^6}{3\times10^8} \approx 0.0063$, or $\log_{10} f \approx -2.2$.

Comparing this to the plot in Fig. 15.1, which shows a linear relationship between $\log_{10} f$ and $\log_{10} t$, the slope is consistent with a lower estimate of $r \approx 3$.

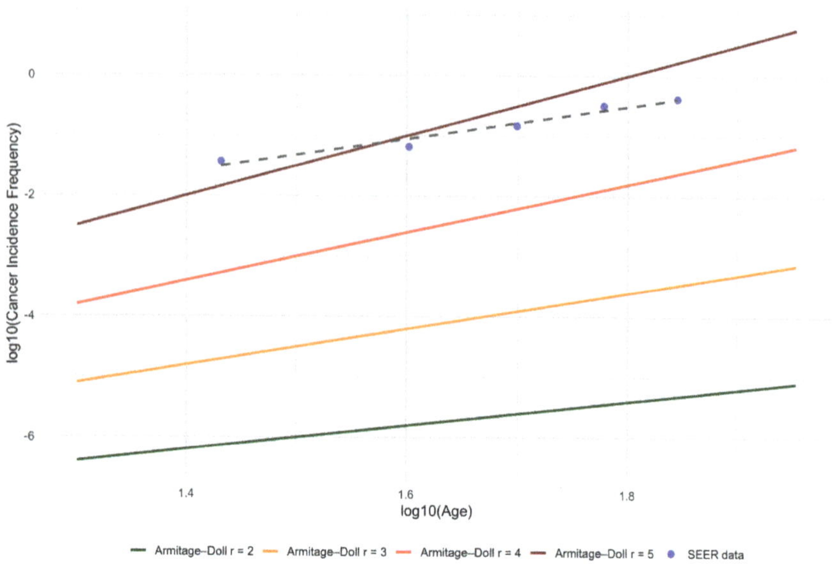

Figure 15.1. An estimate of the number of cancer-causing mutations present in cancer cells from population data. We assume the population of the US is 300,000,000, to estimate the frequency of cancer at each age from the 1921 US data on cancer incidence (see text). Log_{10} of age to cancer incidence (t) is plotted (according to Eq 15.2) against the $\log_{10}$ of the frequency of cancer (f), for various values of the assumed number of mutations present [r = 2 (green), 3 (yellow), 4 (red), 5 (brown)] in cancer-producing clones. $\text{Log}(f) = 0$ implies that everyone has cancer at birth; $\log(t) = 2.0$ is 100 years since birth. Likewise, $\log(t) = 1.8$ is about 63 years. The reader will see that the most realistic estimate of r is 3 (yellow line) on average, because $r = 4$ is too high and $r = 2$ is too low, given the frequency of cancer incidence ($\log_{10}(0.097) \cong -1.2$), providing t ($\sim\log_{10} 1.8$) = 60.1 years. When actual cancer incidence data from years 2018–2022 (ages between 27 and 80) in the USA from the SEER database (https://seer.cancer.gov/) were plotted (blue circles), the line of best fit (black dashed line) has a slope of 2.47. This estimate, however, should not be interpreted as the exact literal number of driver mutations in all cancers.

15.2 Is Drug-Resistant Cancer Caused by the Drug?

This is a fundamental question in oncology: can a drug used to treat cancer inadvertently contribute to the emergence of drug-resistant cancer cells, thereby undermining its own effectiveness? This concern arises from the biological properties of cancer cells and the evolutionary dynamics within tumors.

Cancer cells are characterized by their ability to proliferate indefinitely. This includes insensitivity to growth-inhibitory signals, independence from external growth factors, resistance to apoptosis (programmed cell death), evasion of immune surveillance, escape from cellular senescence, and the capacity to invade distant tissues and establish metastases. These cells often behave like parasites — entirely derived from the patient's own tissues, yet capable of overriding the body's regulatory systems.

Because of their rapid proliferation, cancer cells are typically treated with cytotoxic drugs that preferentially kill dividing cells. However, these drugs also harm healthy proliferating cells — such as those in the bone marrow and intestinal lining — leading to significant side effects. Moreover, many cancer drugs are themselves mutagenic, damaging DNA and thereby increasing genetic variation within the tumor cell population. This raises a critical concern: does treatment with such agents accelerate the evolution of resistant subclones within the tumor?

Natural selection acts on the resulting variation. Cancer cells that acquire mutations conferring resistance to the drug may survive and proliferate, giving rise to a drug-resistant tumor. Clinically, this often manifests as a relapse: the initial treatment appears effective, but the cancer returns months or years later, now unresponsive to the original therapy. This leads to a troubling question: did the drug cause the emergence of resistance, or did it simply select for preexisting resistant cells?

This dilemma parallels a problem addressed in a classic experiment by Luria and Delbrück in 1943 (Kendal and Frost 1988; Luria and Delbrück 1943), which distinguished whether resistance in bacteria arose spontaneously or was induced by exposure to a virus. Their results supported the Darwinian view: resistance-conferring mutations occurred randomly before exposure, not in response to it.

A similar logic can be applied to cancer. Do resistant cells preexist in the tumor before drug treatment, or are they induced by the treatment itself? Consider a controlled experiment with multiple cancer cell cultures derived from a single patient who has not yet received treatment. Assume that mutations conferring drug resistance occur spontaneously at a small constant probability p during cell division. If each culture starts with a small number of identical cells – ideally one, but in practice perhaps

~50 — and grows for a defined time, then the number of resistant cells per culture can be modeled using the Poisson distribution. The probability of exactly k resistant mutations occurring is:

$$p_k = \frac{\mu^k e^{-\mu}}{k!}$$

where μ is the expected number of resistance-conferring events per culture. In particular, the probability of no resistant cells in a culture is $p_0 = e^{-\mu}$, so the probability of at least one resistant cell is $(1 - p_0) = (1 - e^{-\mu})$.

If mutations arise spontaneously, only a small fraction of cultures will contain resistant cells, and some of those will show large numbers of resistant cells due to early mutational "jackpot" events (Fig. 15.2b, right panel). For example, if $\mu = 0.001$, then approximately 1 in 1,000 cultures will contain resistant cells, with the rest showing none.

By contrast, if drug resistance is induced by the drug — rather than arising spontaneously — then most cultures should exhibit some level of resistance, because every cell has a small but nonzero probability of becoming resistant when exposed to the drug (Fig. 15.2b, left panel). Given that each culture contains millions of cells, even a low induction rate would yield a high probability of at least one resistant cell per culture.

Although rigorous Luria–Delbrück-style fluctuation tests are difficult to perform in human cancer cells, recent studies suggest that drug exposure can induce non-genetic forms of resistance. For example, in some cases of skin cancer, treatment with the drug vemurafenib has been observed to trigger transient transcriptional changes that render cells drug-resistant (Shaffer *et al.*, 2017). When the drug is withdrawn, these cells revert to their original gene expression profile and regain sensitivity — indicating a physiological, not genetic, adaptation.

Such findings imply that at least some cases of drug resistance may be reversible, and that combination therapies or epigenetic modulators might restore drug sensitivity. However, in many other cancers, resistance arises from stable genetic mutations — such as point mutations in drug targets or efflux pump activation — which persist even after drug withdrawal. These represent the classical, irreversible cases of drug resistance, consistent with Darwinian selection acting on spontaneous mutation.

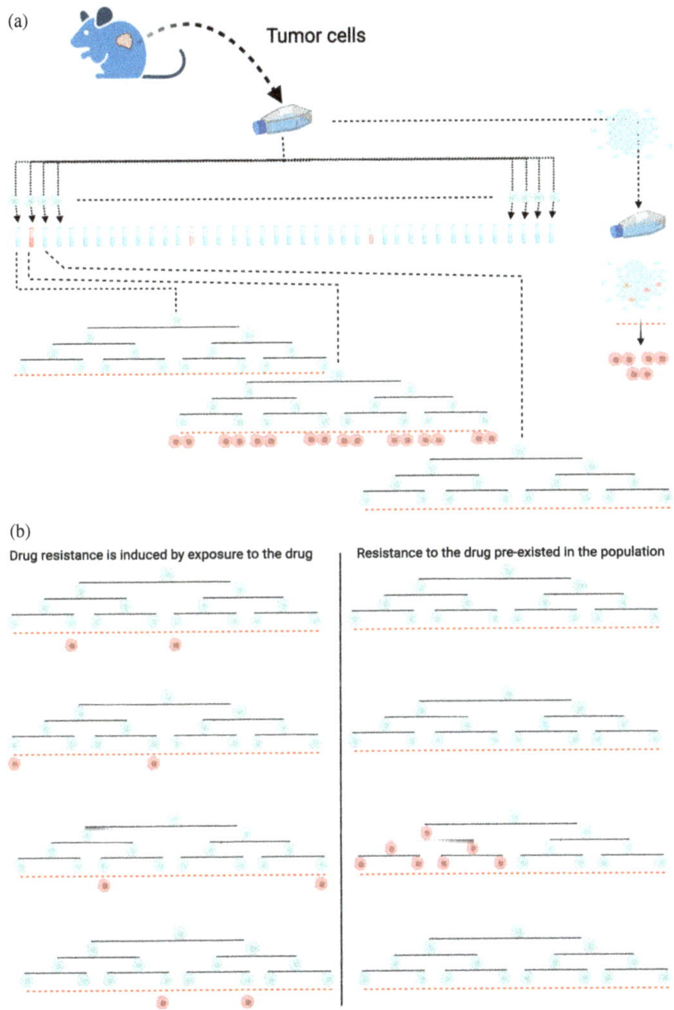

Figure 15.2 Modified Luria-Delbrück fluctuation analysis on cancer cells. (a) Tumor cells were isolated, and multiple cultures with ~50 starter cells were initiated (and a single bulk culture was also started. Following unselected growth without selection, an anticancer drug was added to the culture. Drug resistant cancer cells survive. (b) Expectations from the two models of drug resistance evolution. If the drug induced the resistance, then most cultures would have a few resistant cells; if resistant cells were pre-existent in the culture before drug addition, there would be a rare culture with many drug-resistant cells but most cultures would have none. Shafer et al. (2017) has reported such an experiment and results showed that the drug (vemurafenib) appeared to induce a transient drug-tolerant state. This contradicts the original Luria-Delbrück paradigm of the independent origin of variants from selection pressure. [This figure was drawn using Biorender.com under an academic license]

15.3 The Rate of Evolution of Cancer Cells?

Understanding the speed at which cancer cells evolve is critical to grasping both tumor progression and therapeutic resistance. A foundational contribution to this question was made by a group at the Imperial Cancer Research Fund in the UK, led by Walter Bodmer (1936–), a student of the eminent evolutionary geneticist R.A. Fisher (1890–1962). In a landmark paper (Tomlinson, Novelli, and Bodmer 1996), they introduced a quantitative framework to estimate how the presence of a mutator mutation — a mutation that increases the mutation rate itself — can accelerate cancer evolution. The following is a summary of their reasoning.

Bodmer and colleagues assumed that cancer develops in discrete stages, with significant time intervals between them — a view supported by substantial prior evidence. The model begins with an initial mutation in a normal cell, often in an oncogene, transforming it into a precancerous state. The key question is: How much faster does a precancerous cell lineage carrying a mutator mutation evolve, compared to one without it?

Let m_1 denote the average probability of a mutation occurring per gene per cell generation in a standard precancerous cell, and m_2 denote the same in a cell that already harbors a mutator mutation, such that $m_2 > m_1$. Suppose we randomly sample one such precancerous cell without a mutator mutation; the probability it acquires an additional mutation in the next generation is m_1. Thus, the probability that it does *not* mutate is $(1 - m_1)$.

If instead we consider a population of N_1 such cells, and assume that mutational events occur independently, the probability that *none* of these cells undergo a mutation in the generation is $(1-m_1)^{N_1}$. Therefore, the probability that *at least one of them acquires a new mutation is*:

$$p_c = 1-(1-m_1)^{N_1} \tag{15.3}$$

Now consider a parallel scenario involving N_2 precancerous cells that already carry a mutator mutation. The probability that at least one of them undergoes an additional mutation in the next generation is:

$$p_m = 1-(1-m_2)^{N_2} \tag{15.4}$$

Empirical estimates suggest that m_1, the spontaneous mutation rate in normal human cells, lies between 10^{-6} and 10^{-8} per gene per generation. For precancerous or cancerous cells with mutator mutations, m_2 can be much higher — on the order of 10^{-3} – 10^{-4} per gene per generation.

Assuming conservative values of $m_1 = 10^{-8}$, and $m_2 = 10^{-4}$, Bodmer and his colleagues showed that:

$$\frac{p_m}{p_c} = \frac{1-(1-m_2)^{N_2}}{1-(1-m_1)^{N_1}} = \frac{1-(1-10^{-4})^{N_2}}{1-(1-10^{-8})^{N_1}} \approx 1, \quad \text{if} \quad N_1 \geq 10^4 N_2$$

This implies that unless the number of cells with mutator mutations exceeds 1 in 10,000 of the total precancerous cell population, the overall mutation burden in the tumor will not appear elevated. However, this balance can shift dramatically due to evolutionary selection.

How long would it take for cells with the mutator phenotype to become the dominant clone in a growing tumor? Using the fitness framework introduced in Chapter 14 (Equation 14.1), the evolutionary fitness W of a genotype is defined as:

$$W = \frac{n_{t+1}}{n_t},$$

where n_t is the number of cells in generation, t, and n_{t+1} is the number in the next generation. The excess fold increase after one generation is:

$$E = 1 + \frac{n_{t+1}}{n_t} = 1 + W$$

To calculate the number of generations G required for a clone with this fitness to expand by a factor of 10,000 (i.e., reach population dominance), we use (with base 10 logarithm):

$$(1 + W)^G = 10{,}000 \Rightarrow G = 4/\log(1 + W) \qquad (15.5)$$

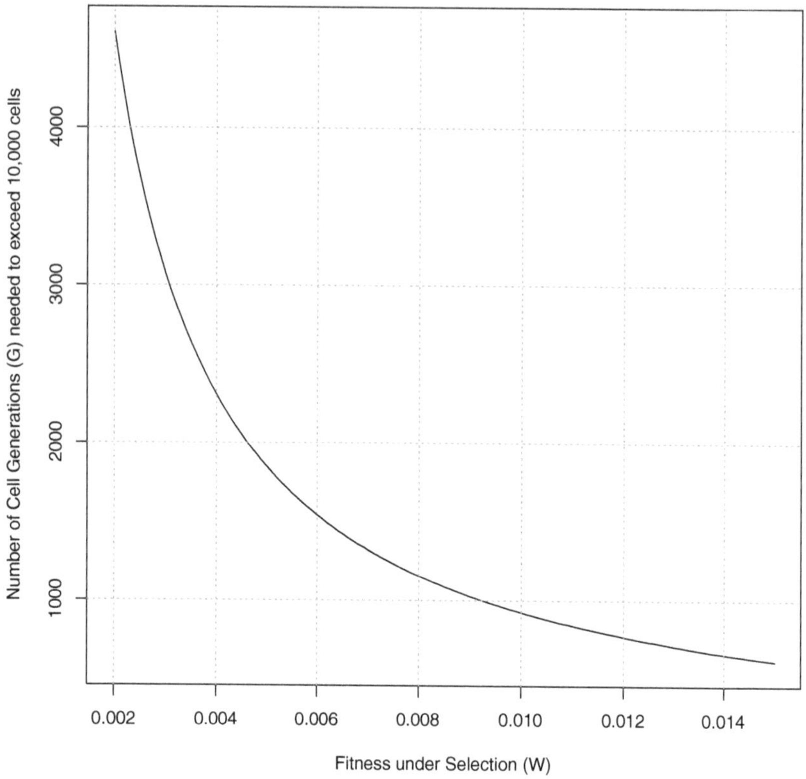

Figure 15.3 The number of generations of growth needed for tumor cells to exceed the threshold of abundance over normal cells (G) as a function of evolutionary fitness under selection (W). The relationship is defined by Equation 15.5 (see text).

Plugging in the values of W, we find that for a modest selective advantage of 1% (i.e., $W = 0.01$), it would take 925 generations to reach population dominance. For a 10% advantage ($W = 0.1$), it would only take ~97 generations (see Fig. 15.3).

These estimates depend on tissue-specific cell division rates. For instance, fast dividing intestinal crypt epithelial cells divide approximately every 3–4 days. Thus, 925 generations would span roughly 465 days — a little over a year — for the mutator subclone to become dominant.

In summary, this model illustrates how even a small selective advantage can lead to the rapid expansion of hypermutating cell populations. This helps explain both the diversity of mutations observed in many tumors (as noted in Part 2, Chapter 2, Section 2.4), and the rapid progression often seen in cancers of rapidly renewing tissues, such as skin, lung, gastrointestinal tract, and certain breast cancers.

Chapter 16

Evolutionary Distance Measures

In this chapter, using algebra and simulation, we explore the idea that all organisms on Earth are related by a common descent. A consequence of that result is that we can develop a metric for evolutionary distance that separates organisms ***on the evolutionary scale****. This chapter explores two such ideas: one uses the relatedness of a pair of proteins and the other uses the relatedness of a pair of DNA sequences to develop metrics for measuring evolutionary distance.*

16.1 All Extant Organisms Have the Same Primordial Ancestor

A striking feature of scientific theories is their ability to explain more than they originally set out to. Newton's law of gravity, developed to explain why apples fall, also predicted the orbits of planets and moons, and the timing of tides. Einstein's general theory of relativity, proposed to explain how mass warps spacetime, unexpectedly explained why light from distant stars appears redshifted.

Karl Popper, one of the most influential philosophers of science, called this quality of a theory its falsifiability — the more predictions a theory makes that can be rigorously tested and potentially proven wrong, the more powerful and scientific it is.

Darwin's theory of evolution by natural selection is a model example of this. Though originally proposed to explain how populations change over time due to selection acting on random variation, it carries profound, numerically testable predictions. One of the strongest of these is that all extant organisms share a common ancestry. In other words, if Darwinian evolution is correct, then all living organisms on Earth should be related by descent from a shared origin.

Let us contrast this with a competing hypothesis: that different organisms arose independently, with no shared genealogies. This is a view sometimes found in creationist or supernatural explanations. Under this alternative hypothesis, there would be no reason to expect systematic patterns of genetic relatedness among species. We will formalize this hypothesis of shared origin with a simple mathematical model to examine its predictive power.

Consider a haploid organism with discrete, non-overlapping generations G_0, G_1, G_2 … G_r…, each consisting of exactly N individuals (i.e., a constant population size). Assume that the organisms reproduce asexually, so that each individual in generation G_{r+1} has exactly one parent in generation G_r. Further assume that the number of offspring per individual is a random integer (0, 1, 2, ...) drawn from a uniform multinomial distribution, meaning that everyone in generation G_{r+1} is descended from its parent independently and uniformly at random from the N individuals in generation G_r. This assumption ensures neutrality — no individual has a reproductive advantage.

Now, consider any two distinct individuals in generation G_r. The probability that they have different parents in the previous generation G_{r-1} is:

$$P_1 = 1 - \frac{1}{N}$$

If you repeat this reasoning for k generations, the probability that these two individuals trace back to different ancestors in generation G_{r-k} is:

$$P(k) = \left(1 - \frac{1}{N}\right)^k \quad (16.1)$$

In other words, this is the probability that coalescence has *not* yet occurred within k generations. Eventually, however, when $k \to \infty$, $P(k) \to 0$ (see Box 16.1).

Box 16.1 The limit in Equation 16.1

Let us write, $x = \left(1 - \frac{1}{N}\right)^k \Rightarrow \text{In}x = k \cdot \text{In}\left(1 - \frac{1}{N}\right)$

Using the approximation, In(1 – ε) ≈ –ε, for small ε, we get:

$$\text{In}\ x \approx -\frac{k}{N} \Rightarrow x \approx e^{-k/N}$$

Now take the limit: $\lim_{k\to\infty} e^{-k/N} = 0$. Thus, $\lim_{k\to\infty}\left(1 - \frac{1}{N}\right)^k = 0$.

For diploid organisms, the probability that two alleles sampled from different individuals do *not* coalesce within *k* generations is:

$$P_{dip}(k) = \left(1 - \frac{1}{2N}\right)^k \tag{16.1a}$$

This simply reflects that diploid organisms have twice as many gene copies per generation, so the chance of coalescence is halved. As illustrated in Figure 16.1, smaller values of *N* lead to faster coalescence (a steeper decline of the probability of coalescence). In diploids, the process takes roughly twice as long as in haploids with the same population size.

For human evolution, the effective population size at key moments — such as the emergence of *Homo sapiens* — may have been no more than a few thousand. Even if there were multiple founding hominid lineages, the genealogies of all present-day humans would converge within roughly 5,000 generations, or about 100,000 years, given a generation time of ~20 years. This mathematical inevitability is the basis for popular ideas of a "Mitochondrial Eve" and a "Y-chromosome Adam" — referring to the most recent individuals from whom all humans today inherit their mitochondrial DNA and Y-chromosomes, respectively. Importantly, this does not mean they were the only humans alive at that time, but rather that all other lineages eventually failed to persist due to the logic explained above.

The same logic applies on a *planetary timescale* for all life forms: even if life arose multiple times independently on early Earth over 3.5 billion years of evolution, it is overwhelmingly likely that all surviving lineages

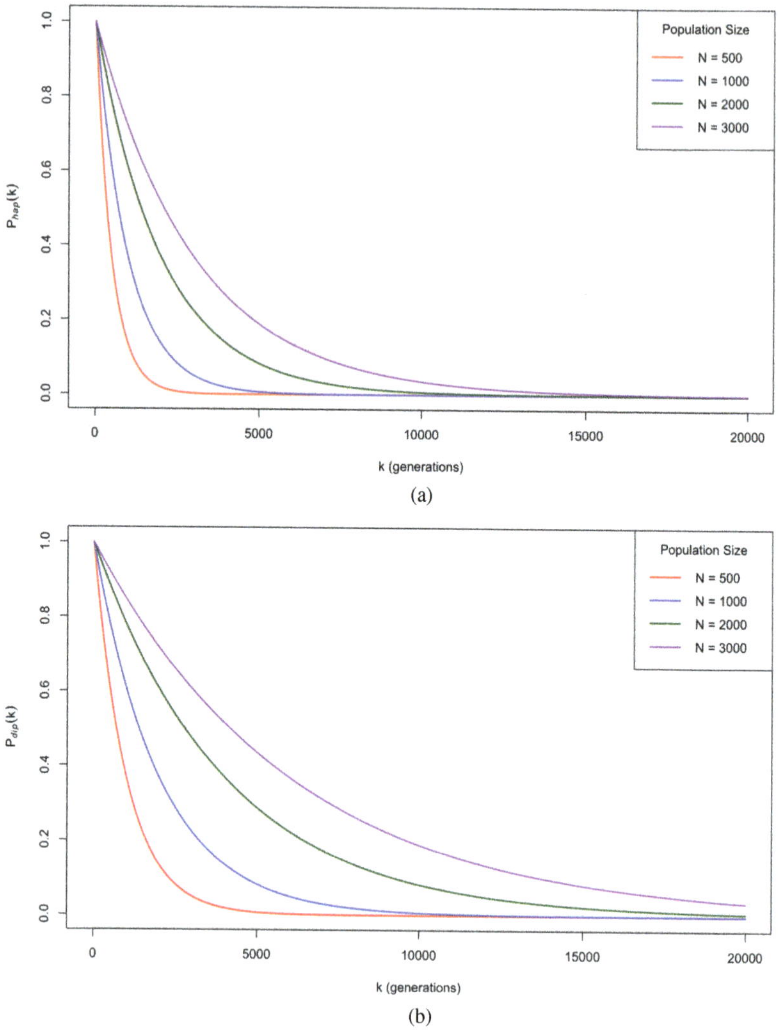

Figure 16.1 The probability that any two separate individuals have distinctly different ancestors plotted as a function of the number of generations. (a) For haploid organisms, Equation 16.1; (b) for diploid organisms, Equation 16.1a, for different values of population sizes (N). For larger population sizes, more generations are necessary for the coalescence of separate ancestries to a single ancestor. Haploid ancestries coalesce faster.

have coalesced to a single common ancestor — the Last Common Universal Ancestor, or LUCA, a cellular entity at the root of the tree of life.

Coalescence theory also applies to genes, not just organisms. The genealogical tree of a particular gene variant (allele) can be traced back to a common ancestral copy. This principle is the foundation of association mapping in medical genetics. However, genealogical convergence of individuals occurs much faster than complete genetic convergence across all loci.w

With the availability of thousands of human genome sequences, it is now possible to test whether certain sequence variants are overrepresented in individuals with a particular disease. If a variant is found significantly more often in cases than in controls, there exists a possibility that it *may be* linked to a causal gene (but not necessarily so).

However, the signal is often weak — particularly for complex diseases caused by many variants of small effect. Population structure, recombination, and incomplete linkage can obscure associations. Developing robust methods to detect such subtle signals remains an active area of research (Larribe, Lessard, and Schork 2002; Morris, Whittaker, and Balding 2002; Rosenberg and Nordborg 2002).

16.2 Evolutionary Distance Can Be Measured from Genealogy

The husband-and-wife team of Steve and Jorja Henikoff at the University of Washington, Seattle, developed a substitution scoring matrix — called the Blocks Substitution Matrix (BLOSUM) — based on observed amino acid replacements in conserved regions of protein families (Henikoff and Henikoff 1992). The alignment between protein sequences that yields the highest total BLOSUM score is chosen as the most likely evolutionary alignment. From this alignment, evolutionary relationships among the proteins can be inferred (Deonier, Tavaré, and Waterman 2005).

Each entry s_{ij} in the BLOSUM matrix represents a log-odds score for substituting amino acid *i* with amino acid *j*, defined as:

$$s_{ij} = \frac{1}{\lambda}\log_2 \frac{p_{ij}}{q_i q_j} \tag{16.2}$$

Here, p_{ij} is the observed probability of substituting amino acid *i* with *j* in homologous sequences; q_i and q_j are the background probabilities of amino acids *i* and *j*, respectively, based on their frequencies in known proteins; λ is a scaling constant used to convert the log-odds scores to integers in practical BLOSUM matrices (see Box 16.2 for details).

This formulation quantifies how much more likely a substitution is to occur in homologous proteins than by random chance. A high value of s_{ij} implies that the observed substitution is unlikely to be random and may reflect either functional conservation or structural constraint. Conversely, a low or negative score suggests either functional divergence or an alignment artifact.

The observed alignment score S_o between two protein sequences is computed as the sum of the individual substitution scores:

$$S_o = \sum_{k=1}^{n} s_{ij}^{(k)} \tag{16.3}$$

where *n* is the number of aligned amino acid positions, and $s_{ij}^{(k)}$ is the score at position *k*.

To normalize this score, the final similarity score *S* is computed using:

$$S = \frac{S_0 - S_r}{S_i - S_r} \tag{16.4}$$

Here, S_r is the expected sore for two randomly generated sequences, and S_i is the score for two identical sequences. Normalization ensures that when $S = 0$, the observed similarity is no better than random, and when $S = 1$, the sequences are identical. Hence, the similarity score *S* ranges between 0 and 1 and provides a standardized measure of evolutionary relatedness between protein sequences.

	Ala	Arg	Asn	Asp	Cys	Gln	Glu	Gly	His	Ile	Leu	Lys	Met	Phe	Pro	Ser	Thr	Trp	Tyr	Val
Ala	4																			
Arg	-1	5																		
Asn	-2	0	6																	
Asp	-2	-2	1	6																
Cys	0	-3	-3	-3	9															
Gln	-1	1	0	0	-3	5														
Glu	-1	0	0	2	-4	2	5													
Gly	0	-2	0	-1	-3	-2	-2	6												
His	-2	0	1	-1	-3	0	0	-2	8											
Ile	-1	-3	-3	-3	-1	-3	-3	-4	-3	4										
Leu	-1	-2	-3	-4	-1	-2	-3	-4	-3	2	4									
Lys	-1	2	0	-1	-3	1	1	-2	-1	-3	-2	5								
Met	-1	-1	-2	-3	-1	0	-2	-3	-2	1	2	-1	5							
Phe	-2	-3	-3	-3	-2	-3	-3	-3	-1	0	0	-3	0	6						
Pro	-1	-2	-2	-1	-3	-1	-1	-2	-2	-3	-3	-1	-2	-4	7					
Ser	1	-1	1	0	-1	0	0	0	-1	-2	-2	0	-1	-2	-1	4				
Thr	0	-1	0	-1	-1	-1	-1	-2	-2	-1	-1	-1	-1	-2	-1	1	5			
Trp	-3	-3	-4	-4	-2	-2	-3	-2	-2	-3	-2	-3	-1	1	-4	-3	-2	11		
Tyr	-2	-2	-2	-3	-2	-1	-2	-3	2	-1	-1	-2	-1	3	-3	-2	-2	2	7	
Val	0	-3	-3	-3	-1	-2	-2	-3	-3	3	1	-2	1	-1	-2	-2	0	-3	-1	4

Figure 16.2 Block Substitution Matrix, or BLOSUM. A BLOSUM can be constructed for families of similar proteins chosen for a particular degree of similarity. In this case, the extent of similarity is 62%. In other words, the substitution matrix was calculated for groups of protein families that were 62% similar. Only one half of the matrix is shown because the other half is identical. See Box 16.2 for how to construct such a matrix.

Box 16.2 BLOSUM score

To see how the BLOSUM similarity score, $S_{ij} = \frac{1}{\lambda}\log\frac{p_{ij}}{q_i q_j}$ works, consider a simplified example using a small set of aligned protein sequences and a hypothetical scenario where we have aligned only four amino acids: A, B, C, and D, and six proteins (Sequences 1–6). We will calculate the penalty scores for a BLOSUM matrix based on the observed substitutions in this hypothetical toy dataset:

Protein Sequence 1: …A-B-C-D-A…
Protein Sequence 2: …A-B-C-B-A…

(Continuied)

Protein Sequence 3: …A-B-C-B-A…
Protein Sequence 4: …A-B-C-B-A…
Protein Sequence 5: …A-B-C-B-A…
Protein Sequence 6: …A-B-C-B-A…

We want to calculate the similarity score for the substitution of amino acid D in column 4 of sequence 1 with amino acid B in sequences 2, 3, 4, and so on. We see that there are six aligned sequences, and each sequence has five residues.

Step 1: Calculate the total number of amino acid pairs contributed by an aligned block to the substitution matrix:

On column 4, the number of all possible pairs of B to itself, $f_{BB} = n(n-1)/2 = 10$, where n is the number of residues; $f_{BD} = 5$ (one D paired with five Bs), and the other pairs are 0.

If there are d rows of proteins in the alignment (i.e., depth = d) and w columns, the block contributes a total of: $w\frac{d(d-1)}{2}$ amino acid pairs to the count. Here, that number $= 5 \cdot \frac{6(6-1)}{2} = 75$.

Step 2: Construct the pair frequency (f_{ij}) matrix:

	A	B	C	D
A	f_{AA}			
B	f_{BA}	f_{AB}		
C	f_{CA}	f_{CB}	f_{CC}	
D	f_{DA}	f_{DB}	f_{DC}	f_{DD}

[In the case of a real protein, there are 20 such rows and 20 columns, but half the entries are identical because an AB pair is the same as a BA pair, and so on. Thus, $f_{AB} = f_{BA}$.]

Step 3: Calculate the observed frequency of substitution (p_{ij}). In our case, there are only 4 amino acids.

$p_{BD} = \frac{f_{BD}}{f_{BD}+f_{BB}} = \frac{5}{5+10} = 0.33.$ $\left[\text{For the real case with 20 amino acids, } p_{ij} = \frac{f_{ij}}{\sum_{i=1}^{20}\sum_{j=1}^{j} f_{ij}} \right]$

Step 4: Calculate q_i and q_j, the probabilities of the occurrences of the amino acids i and j. Here, let us assume each of the four amino acids is equally likely, in which case, $q_i = q_j = 0.25$ because there are only amino acids here. [In the real case, the number of occurrences of each of 20 amino acids can be counted in protein sequence databases and then estimated by dividing by the total.]

Step 5: Calculate BLOSUM substitution score for the alignment, $S_{BD} = \frac{1}{\lambda}\log_2 \frac{p_{ij}}{q_i q_j} = \frac{1}{\lambda}\log_2 \frac{0.33}{0.25\times0.25} = \frac{2.4005}{\lambda} = 2$, where λ is set to 1.20025 to adjust the score to the nearest integer. This example shows how to compute a substitution score for a simplified BLOSUM matrix. In real applications, BLOSUM matrices are based on a larger set of protein sequences and more complex statistical analysis to ensure accuracy and biological relevance.

16.3 Protein Sequence Divergence Is Related to Evolutionary Distance

Clearly, when Darwin lived, no one had any idea of the molecular nature of proteins or their constituent sequence of amino acids. Nonetheless, we will now show that the idea of how closely related two organisms might be over evolution (the evolutionary distance) translates itself to protein sequence similarity.

To relate a metric of evolutionary distance D to the protein sequence similarity score S between two aligned protein sequences, we start with the assumption that amino acid substitutions occur as discrete, random, and independent events — consistent with the Darwinian model of neutral evolution. In this model, each substitution is unaffected by the fitness of the resulting variant. This contrasts with engineered systems, like car engines, where improvements are guided by design and intentional selection. Such a process of random substitution follows a probability distribution known as the Poisson distribution. This distribution is suitable for modeling discrete, random, and independent events, like whether two cars running on freeways would have an accident — the probability of exactly two or three or more cars colliding over a time interval. In an analogous way, the probability of exactly k substitutions occurring on a string of amino acid sequences within an interval of evolutionary time, where one expects an average of μ events, is given by:

$$p_k = \frac{\mu^k e^{-\mu}}{k!}$$

This probability distribution is often used in evolutionary models, including the BLOSUM model. Although the Poisson distribution approximates a normal distribution for $\mu > 10$, in evolutionary contexts μ is typically small (on the order of $10^{-6} - 10^{-9}$ per position per generation), just as the probability of two cars colliding at random is relatively small. The small value of μ reflects the low probability of substitutions per generation per residue position, ensuring that substitutions are spontaneous and independent, adhering to Darwinian principles.

For our purposes, the probability of no substitution occurring at a given position is given by the Poisson distribution as:

$$p_0 = e^{-\mu}$$

The similarity score, *S*, between two aligned protein sequences (as defined by the BLOSUM score in the previous section) is the proportion of residue positions that have not undergone substitution. Therefore, the similarity score is given by the product of the probabilities of no substitution across all positions:

$$S = \prod e^{-\mu}$$

Taking the natural logarithm of both sides, we get:

$$\text{In } S = -\Sigma\ (\mu)$$

Since μ represents the average number of substitutions per residue position per unit of time, the sum of μ over all residue positions represents the total number of substitutions per unit of time. We denote this sum as the evolutionary distance *D*:

$$\Sigma\, \mu = D$$

Substituting this back into the equation, we have:

$$\text{In } S = -D$$

This equation shows that evolutionary distance is inversely related to sequence similarity — the more similar two sequences are, the shorter the time since they diverged.

To express D in terms of a percentage (i.e., as substitutions per 100 residues), we multiply by 100:

$$D = -\ln S \times 100 \tag{16.5}$$

16.4 DNA Sequence Divergence Is Also Related to Evolutionary Distance

As we have seen above, the degree of similarity between two homologous proteins reflects the time elapsed since their common ancestor, with each amino acid substitution acting as a molecular record of that history. However, protein-level comparisons have limitations. Because many amino acid changes are subject to natural selection — either being removed because they are harmful or retained because they are beneficial — they can obscure the underlying molecular clock. To peer deeper into the neutral, clock-like processes of evolution, we must look beyond the proteins themselves to the DNA sequences that encode them.

In particular, substitutions at the third codon position offer a unique window into evolutionary time. These changes often do not alter the encoded amino acid and therefore tend to evolve under minimal selective pressure (see Figure 6.2 in Part I, Chapter 6). By focusing on these silent or synonymous substitutions, we can estimate intrinsic mutation rates that reflect the true tempo of molecular evolution — independent of selection. This approach, pioneered by Motoo Kimura (Kimura 1981), allows us to measure evolutionary divergence with a level of resolution and neutrality that Darwinian postulates — focused on adaptive fitness — cannot achieve alone. What follows is a detailed derivation of how such DNA sequence divergence at the third codon position can be used to compute evolutionary time. The challenge is to find a mechanism by which one can time the rate of evolutionary change that is not strongly affected by selection; otherwise, it will be unclear whether a change in the DNA nucleotide is observed because of selection/adaptation or whether the change is simply there because of random chance.

Kimura proposed a neutral theory of gene evolution by analyzing nucleotide substitutions at the third codon position in protein-coding genes. His reasoning was based on the observation that changes at this position often do not alter the encoded amino acid due to the degeneracy of the genetic code. Specifically, in ~30 out of 64 codons, changing the third base does not change the amino acid; in ~20 codons, the third base can vary between two alternatives with no amino acid change; in the remaining codons, changes at the third base do tend to replace one amino acid with another of similar chemical properties.

By focusing on third-position substitutions, Kimura minimized the confounding effects of natural selection on protein function. These substitutions are therefore considered nearly neutral, allowing them to serve as a proxy for the intrinsic rate of molecular evolution. Note, however, that although most third-position substitutions are approximately neutral, some are still affected by codon usage bias, RNA structure, or other selective pressures.

Kimura also noted a compositional bias in the third codon position. For instance, in mammalian globin genes, the average base composition is: C (40%), G (32%), A (6%), T (22%). To simplify the analysis, the bases are grouped into two classes: Class *a*: A + T; Class *b*: G + C.

Let α be the rate of substitution of class *a* to class *b* ($a \rightarrow b$); likewise, let β be the rate of substitution of class *b* to class *a* ($b \rightarrow a$) (Fig. 16.3). These are *evolutionary substitution rates*, not raw mutation rates. The total substitution rate per site per year is:

$$k = \alpha + \beta$$

Since both sequences in a diverging pair evolve over time *T*, the total expected number of substitutions between them is:

$$K = 2kT = 2(\alpha + \beta)T \quad (16.6)$$

However, we cannot directly measure the rates α and β. So, to estimate *K* empirically, we define:

X: the fraction of aligned sites with $a \rightarrow a$ (within-class *a*) substitutions

Y: the fraction with $b \rightarrow b$ (within-class *b*) substitutions

Z: the fraction of sites with between-class substitutions (i.e., $a \leftrightarrow b$)

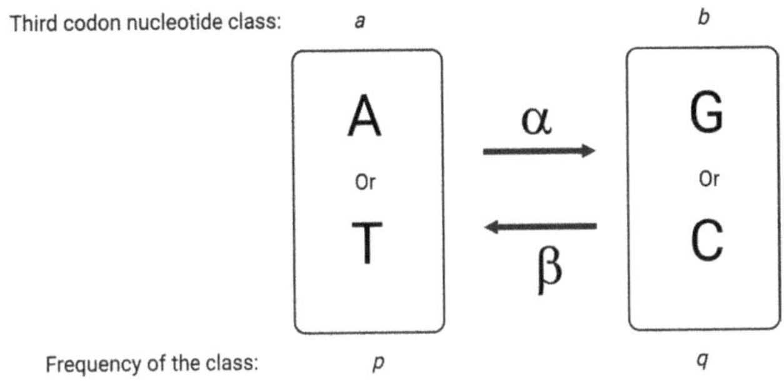

Figure 16.3 Kimura's simplified model of neutral mutation in the third base of codons. The nucleotide classes are *a* and *b*. The probability of the occurrence of a class *a* base in the third position is *p*, that of class *b* base is *q*. The rate of change over evolutionary time of the class *a* base to class *b* is α and β is the rate of the converse change.

By definition, $X + Y + Z = 1$. The goal now is to find a relationship between *K* (or, equivalently, evolutionary time *T*) and the observable quantity *Z*.

Now we are ready to write down three coupled linear differential equations that together satisfy the rate of change of the three classes of substitution pairs per unit of evolutionary time as follows:

$$\frac{dX}{dT} = -2\alpha X + \beta Z \qquad (16.7)$$

$$\frac{dY}{dT} = -2\beta Y + \alpha Z \qquad (16.8)$$

$$\frac{dZ}{dT} = 2\alpha X + 2\beta Y - (\alpha + \beta)Z \qquad (16.9)$$

Let us briefly analyze Equation 16.7 to understand the meaning of these differential equations. Notice that the rate at which the fraction of class $a \rightarrow a$ substitution occurs, i.e., the change in fraction *X* over the time period *T*, occurs in two different ways. One is the change of a base in one of the pair of sequences from class *b* to class *a* by a substitution (i.e., $b \rightarrow a$), resulting in an increase in the class a substitution over time. This contributes half the total class Z fraction because the other half is the reverse, $a \rightarrow b$, in which the class a substitution decreases with time. Thus, we get the βZ term, which represents the gain of within-class a sites due to $b \rightarrow a$ substitutions at between-class ($a \leftrightarrow b$) sites. Similarly, the $-2\alpha X$ arises because each of the two sequences can mutate from $a \rightarrow b$, converting an $a \leftrightarrow a$ pair into an $a \leftrightarrow b$ pair.

The reader can work out the meanings of the other two differential equations in a similar manner.

To solve these equations, we assume equilibrium base composition: the relative proportions of a and b bases remain constant over evolutionary time.

In other words, if p is the frequency of class a and q is the frequency of class b, then $p + q = 1$, because there are only two classes. We get:

$$p = \frac{\beta}{\alpha + \beta} \tag{16.10}$$

So, the substitution rate becomes:

$$K = 2(p\alpha + q\beta)T = 4pq(\alpha + \beta)T$$

By combining Equations 16.6 and 16.10 under the condition $p + q = 1$, and solving the system under this equilibrium condition (Equations 16.7–16.9), we obtain:

$$K = -\,2pq \ln(1 - Z/2pq) \tag{16.11}$$

This gives us a direct formula for the evolutionary divergence K in terms of p and q, the measurable equilibrium frequencies of A+T and G+C, and Z, the fraction of positions with between-class substitutions ($a \leftrightarrow b$). Note that p and q are dimensionless fractions and Z is also a dimensionless proportion, not a rate.

Figure 16.4a plots the values of K against Z for a range of values of the quantity $2pq$. It is important to note that K is relatively insensitive to the values of $2pq$ below ~0.3 for Z in the range 0–0.1 (i.e., the three colored graphs are coincident).

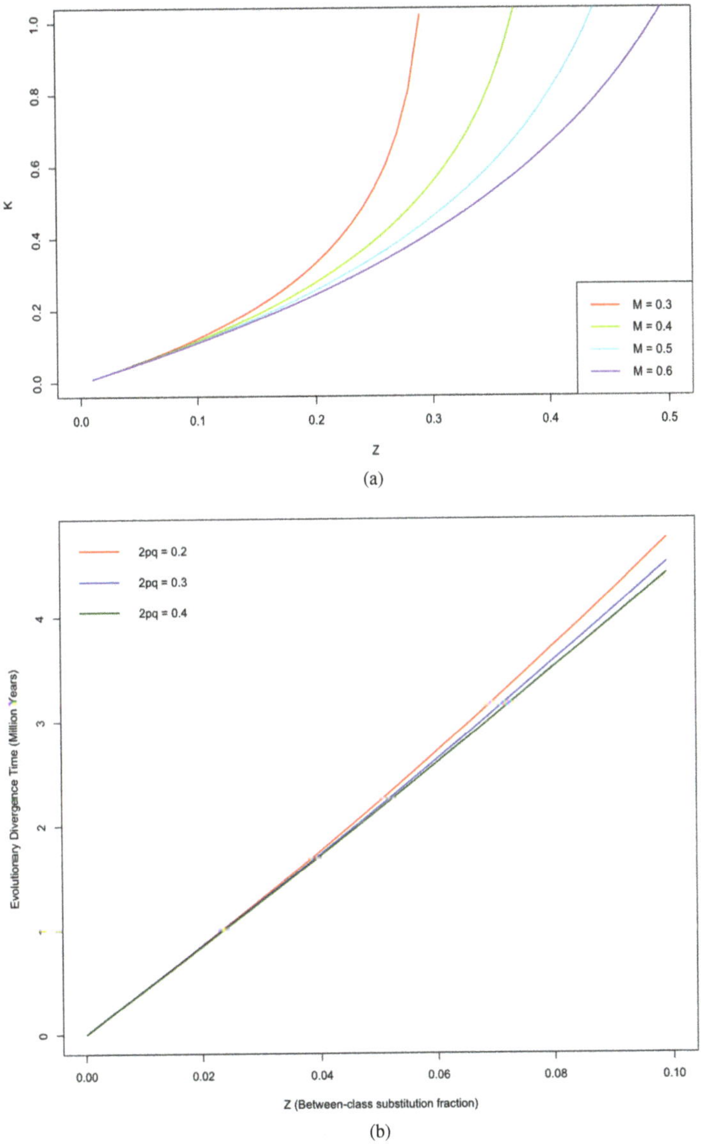

Figure 16.4 Evolutionary divergence estimated from the frequency of third codon base substitutions. (a) A plot of K versus Z (the fraction of class $a \rightarrow b$ or $b \rightarrow a$ substitutions) for M = $2pq$ in the range 0.3–0.6. Note that for $Z < 0.15$, K is unaffected by the value of M. (b) A plot of the evolutionary divergence time T (in years) versus Z, with $k = 1.2 \times 10^{-8}$ per year per nucleotide, where $K = 2kT$. Colors represent different values of M.

Once K is known, we can solve for the evolutionary divergence time T, in units of years, by substituting with $2kT$:

$$T = \frac{K}{2k} = -\frac{-pq}{k} \cdot \text{In}\left(1 - \frac{z}{2pq}\right) \tag{16.12}$$

Within an average neutral evolutionary substitution rate of $k \sim 1.2 \times 10^{-8}$ substitutions per site per year (Kong et al. 2012), we can compute the divergence time in real units. Figure 16.4b provides a plot of the time for evolutionary divergence within the same range of values of the quantity $2pq$ as in Figure 16.4a. This derivation never assumed any dependence of substitution mutations on one another or any extraneous factors influencing mutations, relying solely on intrinsic DNA base substitution frequencies. We can now compare these rates of DNA sequences of organisms against evolutionary divergence ages derived from fossils in geological strata. We have seen in Part II the remarkable congruences between evolutionary times measured by these very different methods.

Chapter 17

Evolutionary Processes in Viral Pandemics

In this chapter, we consider the special case of natural selection pressure on viruses as they spread through a population of organisms, such as humans. We explore the selection pressure on newly arising virus variants. These naturally lead to questions on how viruses spread in a pandemic; hence, we explore a few simple models of virus spread in human populations. These theoretical considerations enable us to appreciate how virus evolution and natural selection affect the progress and management of viral pandemics. We then apply these insights to the last COVID-19 pandemic. These concepts apply to all pandemics, past or future, and Darwinian evolutionary principles are at the heart of this analysis.

17.1 Defining Selection Pressure on Viral Variants in a Pandemic

Let us assume a pandemic begins when a virus spills over into a small group of people who arrive by airplane on a densely populated island. In a few days, the virus begins to infect more and more people, and, as expected, spontaneous mutant variants arise in the virus population.

At this early stage, before any selective pressure is imposed on the viruses, let the number of people infected by the original genotype of the virus be $A = qN$, and those infected by the variant genotype be $V = pN$, where $N = A + V$ is the total number of infected individuals. Here, p and q are the respective frequencies of A and V genotypes, with $p + q = 1$.

As defined earlier, the selection pressure on the variant is given by the selection coefficient (Chapter 14, Equation 14.4):

$$s = s_v - s_a = \frac{W_v - W_a}{\hat{W}} \tag{17.1}$$

where W_a and W_v are the fitnesses of the original and variant virus, respectively, and $\hat{W}$ is the average fitness across the viral population.

The fitness, W, for a particular viral genotype is defined as the change in frequency (or abundance) of individuals infected with that genotype in the next generation. Therefore, the variant V has a selective advantage if $s > 0$ or a disadvantage if $s < 0$.

17.2 The Evolutionary Fitness of a Virus Is Related to Its Rate of Infectious Spread

Let us now define (Dorp et al. 2021b) the absolute fitness of the two viral genotypes based on the exponential rate at which they cause new infections. Let the original virus of genotype A have a growth rate, so its absolute fitness is $W_v = \alpha$. Let the variant genotype V have fitness $W_v = \alpha(1 + s)$. The number of infected individuals at infection-generation t is given by:

$$N_t = N_u e^{\alpha t} \tag{17.2}$$

where N_t is the total number of infected people, and N_u is the number of uninfected people remaining in the population.

We assume $\alpha > 0$ early in the pandemic because most individuals are uninfected and susceptible, leading to an increase in infections across generations.

We define fitness in this way because the original genotype A propagates its infection at a baseline exponential rate α, while the variant genotype V propagates at a slightly altered rate $\alpha(1 + s)$, reflecting its relative advantage or disadvantage.

Next, we derive a relationship between the selection coefficient s and the relative number of individuals infected with each virus after one generation of selection, assuming no migration. This relationship enables estimation of natural selection's impact on the rise of viral variants.

17.3 A Simple Relation between the Selection Coefficient and Virus Genotype Frequencies

Here, we will explore how to estimate the coefficient of natural selection on a variant virus in a pandemic by enumerating the frequency of the variant. Let p^* and q^* denote the frequencies of the variant V and the original A viral genotypes, respectively, after one round of selection. Let A^* and V^* represent the number of individuals infected with the original and the variant virus strains after this round. Given the fitness values defined earlier, the expected number of infections in the next generation is:

$$A^* = W_A \cdot A = \alpha qN \tag{17.3}$$

$$V^* = W_V \cdot V = \alpha(1 + s)pN \tag{17.4}$$

The post-selection genotype frequencies are given by:

$$\begin{aligned} p^* &= \frac{V^*}{A^* + V^*} = \frac{\alpha(1+s)pN}{\alpha qN + \alpha(1+s)pN} \\ &= \frac{\alpha(1+s)pN}{\alpha(1-p)N + \alpha(1+s)pN} = \frac{(1+s)p}{1+sp} \end{aligned} \tag{17.5}$$

$$q^* = \frac{A^*}{A^* + V^*} = \frac{\alpha qN}{\alpha qN + \alpha(1+s)pN} = \frac{1-p}{1+sp} \tag{17.6}$$

Note that the post-selection frequencies are independent of the population size N and the exponential growth rate constant, α, since both factors cancel out in the ratio. We have assumed both $p + q = 1$ and $p^* + q^* = 1$.

Equations 17.5 and 17.6 offer a practical way to estimate the selection coefficient s if the initial and final genotype frequencies (p and p^*) can be measured. These frequencies can be estimated by sequencing viral genomes from infected individuals over a period of time.

It is important to note that this model makes simplifying assumptions. In real-world scenarios, complexities such as multiple introductions of the variant through migration, convergent mutations, population immunity due to prior exposure, incomplete transmission by all infected individuals, and network-based population structure (i.e., who contacts whom) can all influence viral spread. As such, the estimate of s derived by this method should be viewed as an illustrative approximation. For more detailed treatments, readers are referred to works described elsewhere (Dorp et al. 2021b; Otto et al. 2021a).

17.4 Estimating the Fitness of Newly Emerging SARS-CoV-2 Variants in Real Time

How can we estimate the fitness advantage of the emerging SARS-CoV-2 variants during the pandemic? This section presents a simple, real-time calculation made on December 26, 2021. On September 7, 2020, a variant named B.1.1.7 (or Alpha) constituted ~5% of COVID-19 cases in London. By November 9, that number rose to 20%. Using $p = 0.05$ and $p^* = 0.2$, and assuming no migration, the initial selection coefficient estimate is $s \approx 3.75$. Given that ~18 generations of infections occurred over this period (based on an average generation time of ~3.5 days, over 63 days from Sept 7 to Nov 9), the adjusted per-generation selection coefficient over this period is approximately 0.09 — implying a 9% advantage over the previously dominant strain.

Two months later, on January 18, 2021, nearly 100% of Londoners suffering from COVID-19 were infected with this specific variant strain.

More accurate estimates of s from about 5,000 sequenced samples of cases globally, using more refined techniques than those shown above, as well as including effects of migration, were obtained (Dorp et al. 2021a). These were approximately 0.3 for the variant G614G, 0.6 for variant B.1.1.7, 0.25 for variant B.1.351, and –0.99 for variant R.1. Clearly, the variant B.1.1.7 was under significant selective advantage relative to the other three variants or to the prevailing strain, and the variant R.1 was destined for extinction.

Let us now calculate the selective advantage that the newly arising (as of this writing on December 26, 2021) omicron (or B.1.1.529) variant of SARS-CoV-2 had over the prevailing delta variant (or B.1.617.2).

The omicron variant was first detected in the state of California on December 1, 2021. The delta variant constituted the remainder of the new cases, or 4,821 new cases of COVID-19 on that day (plus 1 case of omicron). Therefore, the frequency of the omicron variant on December 1 was 0.0002. By December 22, there were 460 cases in California with this variant. There were 180,600 new COVID-19 cases in total between December 1 and Dec 22 in California. This leads to an estimated frequency of the omicron variant of about 0.0026 in California on December 22.

At the time of this writing (December 26, 2021), somewhat limited data from South Africa, where the omicron variant was first detected, suggested that the incubation period (i.e., the time between two generations of infection) of omicron is approximately 1.5 days, which would have produced about 14 generations of infection between December 1 and 22, 2021.

Using these numbers and ignoring any further migration of infected people into or out of the state, one can estimate, by Equation 17.5, the selection coefficient s per generation for omicron in California as approximately 0.20.

Thus, the omicron variant appears to possess a substantial 20% selective advantage per generation of infection over the previously prevailing delta variant. This estimate, however, is likely somewhat of an overestimate because the migration of multiple people infected with the variant into California could also explain the rate of rise in the omicron numbers over the preceding 21-day period. Nonetheless, we can probably expect this variant to completely saturate new cases in California in another two to three weeks, thereby displacing the delta variant.

Again, caution is advised regarding these estimated values because they are subject to several simplified assumptions mentioned earlier. Despite these important caveats, these examples help illustrate the point that it is possible for variants having strong selective advantage to appear spontaneously in the population, and by enumerating the selection coefficients of these variants, appropriate epidemiological policy measures may, in principle, be taken by public health institutions.

Quite interestingly, however, the above predictions came true within the first few months of 2022: the delta variant was rapidly overshadowed by the omicron variant in the US and then globally. Meanwhile, the author, who had holed himself in a one-bedroom apartment in Quito, Ecuador, to write this book, contracted COVID-19 while venturing to travel to the Galapagos Islands off the coast of Ecuador.

A far more sophisticated analysis reported nearly a year later, with more complete epidemiological data over a longer duration and larger geographic area, which also took into account several additional factors including population migration, delayed reporting of infection, and incomplete reporting, estimated an overall global selection coefficient of 0.69 (range: ~0.3–1.25) for the omicron variant (Dorp et al. 2022).

The author hopes that his readership will be encouraged to engage in such armchair estimates on everything that is observable.

17.5 Pandemic Growth Models

As we discussed earlier, evolution only works on individuals as they are present in a population. We had explored how populations grow (Chapter 13), and how variants arise in proportion to growth rates. But how do viruses that evolve inside human hosts spread and grow? Here, the objects upon which evolution works — namely, the individual virus particles — are rarely measured. What we measure instead are the hosts that are infected and sometimes succumb to the infection or survive and remain immunized, at least for a while. How do we adjust our quantitative reasoning with respect to these additional considerations?

Specifically, we had fleetingly introduced in Equation 17.2 the parameter α, the pandemic growth rate constant, without much explanation. Here, let us briefly examine its basis to complete our understanding of how a pandemic might spread with time.

17.5.1 *A simple pandemic model*

Let y be the fraction of the population who are infected at time t, which means that the fraction $(1 - y)$ is still uninfected but susceptible to future infection. Note that this might not always be true, because some of those uninfected people might be immune (perhaps due to vaccination); we will ignore this possibility. The rate of new infection is therefore the derivative, dy/dt.

Now, this derivative must be proportional to the product of the fraction of infected people and the fraction of still uninfected people:

$$\frac{dy}{dt} \propto y \cdot (1-y) \tag{17.7}$$

$$\Rightarrow \frac{dy}{dt} = \alpha y(1-y) \tag{17.8}$$

where α is a proportionality constant, which in this case is equivalent to the probability of transmission per contact of an infected person with an uninfected person. This parameter depends on both the infectiousness of the virus as well as whether the people are using distancing or other physical measures (e.g., wearing masks of various degrees of effectiveness).

To solve Equation 17.7 for the variable y, we integrate on both sides, assuming the initial value of y at $t = 0$ was 0:

$$\int_0^y \frac{1}{y(1-y)} dy = \alpha \int_0^t dt$$

We get:

$$\text{In}\left(\frac{y}{1-y}\right) = \alpha t$$

Or:

$$\frac{y}{1-y} = e^{\alpha t} \tag{17.9}$$

If N is the total number of people, and N_t the number of people infected at time t, then $y = N_t/N$, and $1 - y = (N - N_t)/N$. Substituting and rearranging, we get:

$$N_t = (N - N_t)e^{\alpha t} = N_u e^{\alpha t} \tag{17.10}$$

This is identical to the exponential growth equation we had assumed in Equation 17.2.

17.5.2 *Flattening the pandemic curve*

Now let us re-examine Equation 17.8. For solving this equation by integration, we had assumed that the value of y at $t = 0$ was zero. But this is obviously unreasonable because if it really were zero, there would be no infection, as the virus couldn't have originated spontaneously in the population, though we might approximate it as zero in a very large population to derive Equation 17.10. We really do not know what the initial value of y was, though we agree that it is a small number relative to the population. Thus, the integral is no longer a definite integral:

$$\int \frac{1}{y(1-y)} \cdot dy = \alpha \int dt$$

Or $\quad \text{In} \frac{y}{1-y} = \alpha \cdot t + C$, where C is a constant.

Therefore, $$\frac{y}{1-y} = e^{\alpha t + C} \tag{17.11}$$

Now, let us define $k = e^C$, where k is another constant.
Thus, $e^{\alpha t + C} = e^{\alpha t} \cdot e^C = k \cdot e^{\alpha t}$
Substituting, we get:

$$\frac{y}{1-y} = k \cdot e^{\alpha t}$$

Solving for y, we get: $$y = \frac{ke^{\alpha t}}{1 + ke^{\alpha t}} \tag{17.12}$$

If we now ask at what rate the fraction of infected people changes with time, we simply take the derivative of y with respect to time (using the quotient rule and simplifying):

$$\frac{dy}{dt} = \frac{d}{dt}\left(\frac{ke^{\alpha t}}{1 + ke^{\alpha t}}\right) = \frac{k\alpha e^{\alpha t}}{\left(1 + ke^{\alpha t}\right)^2} \tag{17.13}$$

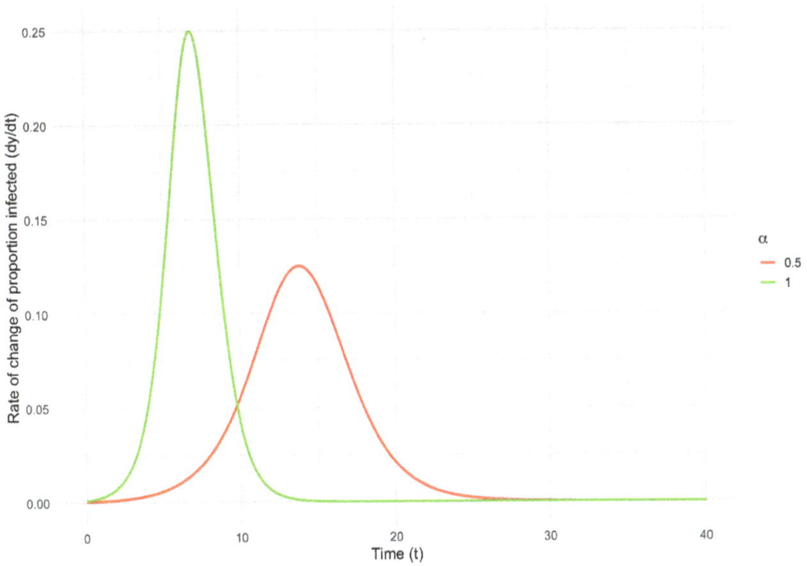

Figure 17.1 Flattening the curve of pandemic rate of rise. A plot of Equation 17.13, where the y-axis represents the derivative of y with respect to time. Green curve, $\alpha = 1.0$, red curve, $\alpha = 0.5$; $k = 0.001$ for both. With respect to the green curve, the behavior of the red curve is what public health experts desire — to vie for more time and resources so as to optimize response against a pandemic.

In Fig. 17.1 we plot Equation 17.13 for two arbitrary values of the infection growth rate parameter α (for a chosen value of k). An examination of how the curve changes will convince the reader that this simple equation explains the trend of a pandemic spread in relation to the parameter α. Specifically, a reduction in the value of α has the effect of "flattening" the rate of infection. Note how the areas under the curves (i.e., the total number of people eventually infected) remain unchanged, but the peak rate of infection is delayed. Thus, masking or distancing, which can reduce the "contact" incidences between a pair of people — one of whom is infected and the other yet uninfected — should enable more time for the public health infrastructure of a country to better cope with critical care and lessen the rate of impact of a pandemic on the local economy.[1]

[1] As previously alluded, this logistic model assumes a closed population where recovered individuals do not re-enter the susceptible pool, and immunity is ignored.

17.5.3 *More complex pandemic models*

A real pandemic may have more complexities than we allowed so far. For example, not all individuals in a population may be randomly in contact with one another, an assumption that we implicitly made in our formulation in Section 17.1 above. Therefore, we will need to consider the probability of transmission per contact and the rate of contact per unit time, which can be modeled as two coupled differential equations (the so-called Susceptible-Infectious or SI model).

Even more complex models assume that the population does not remain fixed during the time but grows, that there is a birth rate and a mortality rate, and a recovery rate from infection such that the recovered fraction becomes resistant to further infection (Susceptible-Infectious-Recovered or SIR model). A simplified version of this model was used to describe the dynamics of the appearance of new variants, and how such a model can be applied to guide policy for vaccination (Otto et al. 2021b).

The reader may consult these more sophisticated models in previously published work (Chang and Kaplan 2021; Giordano et al. 2020; "The SIR Model for Spread of Disease — The Differential Equation Model | Mathematical Association of America" n.d.).

Chapter 18

Origins and Evolution of Morality

The ability to reengineer our own genes, even to extinguish a species, by genetic engineering raises many ethical and moral questions. When and why should we manipulate the human genome? What are the moral bounds that should guide our policy? Do we have the moral right to deliberately extinguish a species? To what risk do we put our descendant generations by extinguishing biodiversity on Earth, the only place we call home? These are complex questions, and this book is about raising, not answering, these questions. There is, however, a notion that the Darwinian idea of the "struggle for existence" allows a tacit justification for purely rational and selfish behavior by humans, which somehow must maximize individual adaptation. In this chapter, we provide thoughts to the contrary. We explore a fundamental question: what are the origins of morality? Can Darwinian evolution, at least in theory, form the basis of human morality? To begin to answer this question, we provide a glimpse of current theoretical insights on how natural selection can, in principle, lead to the origin of moral behavior by compromising purely individual gains. We attempt to provide a philosophical framework for the existence of a continuum of animal and human morality. In this view of life, human morality becomes an evolvable trait that we must examine periodically to navigate a complex technological world we had never experienced before, but within which we find ourselves inexorably enmeshed.

***

In "*Faust*," Johann Wolfgang von Goethe (1749–1832) explores profound philosophical themes, notably the human desire for knowledge and the pursuit of meaning in life. The protagonist, Faust, seeks fulfillment through the acquisition of knowledge and experience but faces moral dilemmas and the consequences of his actions (Goethe, Johann Wolfgang von and Translator, George Madison Priest 1941):

> *"...Two souls alas! are dwelling in my breast;*
> *And each is fain to leave its brother.*
> *The one, fast clinging, to the world adheres*
> *With clutching organs, in love's sturdy lust;*
> *The other strongly lifts itself from dust*
> *To yonder high, ancestral spheres..."*

Goethe's *Faust* delves into the complexities of human nature, the quest for enlightenment, and the eternal struggle between good and evil. Ultimately, the work serves as a reflection on the human condition and the pursuit of a meaningful existence. Must we, like Faust, make a deal with the supposed "evils of Darwinism"?

The advent of Darwinism in its time immediately provoked moral outrage, because the two central tenets of evolution as proposed by Darwin implied, for many, a world without compassion — a dystopian world guided by the selfish outcome of higher reproductive fitness. Authors such as Herbert Spencer, writing in *The Man Versus the State* (1884), applied the Darwinian concept of "survival of the fittest" to human societies and argued against government intervention in social welfare, suggesting that competition is the natural driver of human progress. Ayn Rand's works, particularly *Atlas Shrugged* (1957) and *The Fountainhead* (1943), reflect elements akin to social Darwinism, though she did not explicitly use Darwinian theory or biological evolution as a framework. Instead, Rand's philosophy of Objectivism promotes the idea that selfishness (or rational self-interest) is a virtue and that individuals should act in accordance with their own personal interests rather than for the collective good. In her worldview, the most capable individuals (whom she refers to as "producers" or "creators") rise to success and power, while those who rely on altruism and collectivism are portrayed as parasitic.

Today, there is sufficient evidence that this simplistic notion of "social Darwinism" is incorrect. Below we discuss some evidence in favor of the idea that Darwinian evolution has, in fact, selected traits in humans that form the very foundations of morality and cooperation. First, we consider the evidence that traits underlying basic principles of human morality do indeed exist in other animals. However, the evidence of the existence of a phenomenon is not evidence of the mechanism by which it is formed. Therefore, we must further examine whether some properties of morality can at all evolve in a Darwinian world. If the answer is favorable to such a possibility, then we are free to assume that there might be a biological basis for human morality — that we are a moral animal. As a community, we can then reject the ideals proposing purely selfish competition as the end-all for human survival and flourishing — as dogmas unsupported by Darwinian evolutionary theory.

18.1 Do Non-Human Animals Possess Notions of Morality?

Morality is thought to reflect the innate fairness of the human mind. It is probably not the case that a sense of morality, at least of fairness, is unique to humans (Hauser 2006). Dogs are surprised, display hurt, when abused by owners; prides of lionesses are known to exhibit intolerance to violent behavior by other members. Are these perhaps elements or building blocks of purely human morality? If true, does morality evolve?

The word "empathy" is often used to suggest the cognitive state that allows an individual to sense the cognitive state of another being (de Waal and Preston 2017). It does not, however, immediately follow that empathy is necessarily benign or altruistic, although it is generally used in that sense: the cognitive state of a police interrogator might be quite the opposite yet in accord with that definition. It is quite possible, though it might not be socially acceptable, to define empathy as positive, neutral, or negative depending on the outcome of the respective cognitive states. Nonetheless, we will only consider empathy in its positive domain.

By the general definition of empathy, as defined above, it immediately suggests that an individual can alter their own cognitive state of perception (exhibited by an output of action) in response to observing how a set of conditions impacts the cognitive states of other individuals.

For example, great apes can perceive the potential need for a tool by another member of the same tribe and hand out that tool to that individual (Krupenye et al. 2016), in an apparent effort to be helpful. Intimately related to positive empathy is an innate sense of fairness, from which, it is argued, the system of justice in human societies has emerged and has been institutionalized (Rawls 1985; 1971; Sen 2006) — the idea of a measure of justice that can be ranked as degrees of fairness.

Non-human primates display an innate sense of fairness when subjected to trickery: they decline to participate in earning a reward for an effort if they have seen another member of their tribe obtain a better payoff for the same or a smaller effort, or, more strongly, for no effort at all (Brosnan and de Waal 2003).

18.2 Evolution of Morality by Group Selection

Clearly, emotions such as jealousy, competitive aggression, and unfair action are common in children and emerge spontaneously. And yet, children are quick to spot a lack of fairness and are known to develop a moral code by the time they are three years old (Blake, McAuliffe, and Warneken 2014).

Is our innate sense of fairness biological in origin and therefore subject to the same natural selection as many of the molecular traits of cells and living beings? If so, is there a genetic basis for morality, which there must be if morality is of biological origin?

Some think that morality has its roots in biology, specifically through "group" or "kin" selection (Wilson 2005). This "group selection" idea means that individuals with behaviors that help them get along in a group might have been favored over time. Paradoxically, ants live in big colonies, and one would think they seem to get along just fine without any moral code — at least by our standards — some even bite. Indeed, ants of various kinds, including termites and honeybees, have been carefully documented to exhibit various degrees of extremely altruistic behavior, even of self-sacrifice, such as rescuing colony-mates from imperiled conditions with no evidence of coercion (Miler and Turza 2021; Ratnieks and Helanterä 2009).

It will be contentious to equate such examples of insect altruism with morality. If morality, as is generally accepted in human societies, does come from biology, it is probably linked to having more complex brains than insects, and goes beyond simply altruistic behavior. And if it is true that morality has a biological basis, we should be able to see signs of it in our brain activity even when we are not interacting with others in a group. It might have evolved to make socializing easier, but it is still there even when we are flying solo. Indeed, neurological imaging of human brain does support the mapping of empathic response to certain cortical regions (Mathur et al. 2010), and the neural basis of empathy in mammals has been under intense studies (de Waal and Preston 2017).

Rats, those social creatures of sewers and dusty holes underground, do exhibit the evidence of empathic behavior that passes muster with human standards (Cox and Reichel 2020). Rats who had previously experienced struggling in a pool of water "rescue" other struggling rats from the pool by pulling a chain (which they had learned to pull to affect the water level in a separate learning experience). They repeatedly and readily rescued their own previous cage-mates even though following the rescue they were prevented from social interaction. They didn't just save their buddies from the same cage merely once; they kept rescuing them even when they were kept apart after the heroic act. The rats were at first reluctant to rescue stranger rats but after some initial hesitation eventually they rescued stranger rats as readily as they rescued their former buddies (Cox and Reichel 2020).

Empathy, therefore, is not unique to humans. Is it possible that such a trait, manifested in the relatively simple brains of rats, exhibits higher sophistication in more complex human brains and therefore is under selective evolutionary pressure?

The idea that empathic behavior — the neurological basis of altruism, can be a product of natural selection directly contradicts a naïve view of Darwinism — the caricature of Darwinian evolution as selfish competition, a view favored by many Nazis in the mid-20th century.

18.3 Evolution of Fairness

By contrast, can fairness emerge spontaneously in an evolving population? Does fairness only incur a cost to its practitioners, as alleged in certain political contexts, or can fair play, in principle, do better than purely selfish behavior? Is fairness contingent upon Darwinian evolution?

To answer these questions, we need to turn from behavioral observations to mathematical models that simulate evolution in populations, to ask: can cooperation and fairness emerge in a Darwinian framework? This is an important endeavor despite the commonsense examples of apparently empathetic behavior observed in animals, because there is always the possibility that what we observe is influenced by our own anthropomorphic biases. Mathematical demonstration, on the other hand, can allow us to overcome these biases, so long as we carefully construct our definitions and set our boundaries with precision.

Game theory, a branch of mathematics and economics, analyzes strategic interactions between rational decision-makers. It provides a framework for understanding and predicting behaviors in situations where the outcomes of one participant's choices depend on the choices of others. Game theory has diverse applications, ranging from economics and political science to biology and computer science. By studying strategic interactions, it helps in making informed decisions in competitive scenarios, designing optimal strategies, and understanding the dynamics of negotiations. Game theory offers valuable insights into human behavior and decision-making, making it a powerful tool in various fields to model, analyze, and improve strategic interactions. Many problems in economics and political strategy can be cast in terms of game theory and analyzed for an optimum solution. Consulting companies use proprietary game theory-based computer software to predict optimum strategies for their clients.

An example of a conventional game theory problem is the famous Prisoner's Dilemma, where two prisoners are held apart. Now the problem that faces each of the two prisoners is whether to betray the fellow prisoner and gain a concession from the authorities, or to not betray and therefore suffer a harsher punishment. Each knows that if the other betrays, then one will get the full punishment, whereas the betrayer will get a lesser sentence. What will the prisoners do? Betray the prison-mate? Or will they not betray each other and each, therefore, risk a hard punishment?

The Prisoner's Dilemma game becomes more interesting if the game is played sequentially, such that after each prisoner's move, the other prisoner responds with their own move, thus either retaliating and punishing the other party for a selfish move, or rewarding them for a cooperative move. In this form, it is called an Iterated Prisoner's Dilemma game.

A mathematical analysis of the Iterative Prisoner's Dilemma has shown that purely selfish behavior provides an advantage (Press and Dyson 2012). The strategy that one prisoner can engage is to unilaterally set the other prisoner's payoff, or else to extort that other prisoner. Mathematics showed that this strategy should become dominant upon repeated playing. Further work on the same Iterative Prisoner's Dilemma, however, has shown that while cooperation will evolve to maximize individual benefits given certain strategies of rewards and punishment, the cooperation will collapse if the very strategy of cooperation can coevolve with the reward (Stewart and Plotkin 2014). If cooperation is mathematically improbable, it might also be improbable in a biological world.

And yet a baboon, seeing a lion, screams its warning call; its tribe is alerted; the predator's attention falls on the screaming baboon. How could such cooperative behavior evolve if evolution always favors selfish behavior? The warning baboons should be eliminated from the population soon enough if that were the case. Thus, the apparent mathematical demonstration directly contradicts observation.

Before we jump to the conclusion that cooperative behavior cannot naturally evolve, we must remember that the mathematical solution of the Iterated Prisoner's Dilemma concerned two players, played iteratively. That is not the playground of natural evolution — natural evolution must be considered within a population. It has been shown that a strategy that unilaterally sets the other's payoff in an Iterated Prisoner's Dilemma game itself proliferates among the "progeny" of the evolving entities that inherit that strategy — thus they compete among each other and lead to extinction of the strategy itself in competition with other competing strategies (Adami and Hintze 2013). When multiple pairs of players are introduced into the game, reminiscent of more realistic natural evolving systems, cooperative behavior can survive with stability; especially in small groups of players, cooperative strategies tend to dominate (Shen et al. 2022). Fig. 18.1 shows simulations of the iterative prisoner's dilemma game played by multiple players. When the number of players is large, then defection generally wins (Fig. 18.1A). However, when the number of players is small, even over a small number of generations, cooperation emerges (Fig. 18.1B).

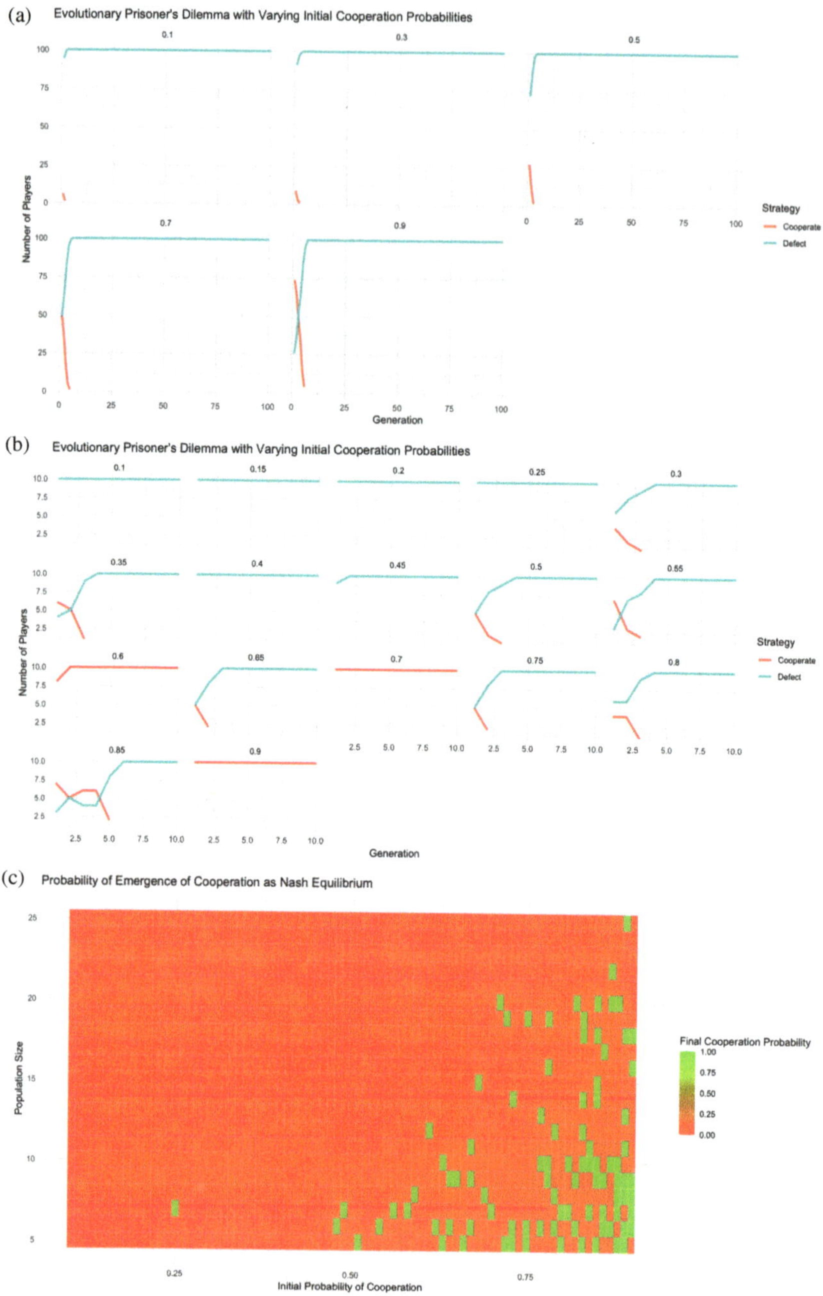

Figure 18.1 (*Continued*)

Evolutionary game theory constitutes an extension of traditional game theory, incorporating evolutionary principles into the analysis of strategic interactions within a population (Maynard-Smith 1986). Unlike conventional game theory, this framework introduces the dimension of natural selection, elucidating how strategies employed by individuals undergo changes over time because of evolutionary processes. In essence, evolutionary game theory provides a theoretical framework to comprehend the dynamics of strategic decision-making in the context of biological evolution. It examines how various strategies for survival and reproduction may evolve and persist in populations, shedding light on the interplay between game-theoretic considerations and the forces of natural selection. This interdisciplinary approach facilitates a nuanced understanding of the adaptive strategies that emerge within populations, offering insights into the intricate relationship between strategic interactions and evolutionary dynamics.

When an evolutionary game of the Prisoner's dilemma is played over a small population, even over a relatively small number of generations, cooperation frequently emerges as the optimum strategy in equilibrium (Fig. 18.1C). If cooperative behavior can emerge as an optimum strategy, then it may evolve in nature.

The case of baboons warning while risking themselves, or the behavior of a mother protecting her children from danger at her personal peril, can be understood in terms of "group selection" — natural selection in favor of strategies for successfully passing the gene pool to the next generation, even to the extinction of itself. Thus, what really evolves is the

Figure 18.1 (Figure on facing page) Defection or cooperation in an evolutionary game of prisoner's dilemma. Pairs of prisoners iteratively choose to either defect or cooperate with each other when they are kept separate. Both prisoners get penalty if they cooperate with each other and do not defect; if one defects and the other does not, the defector gets a reward while the other gets an even worse penalty; if both defect then each gets a relatively small payoff. They play iteratively. (a) Defection is the overwhelming choice if the number of pairs of prisoners is large (green lines). (b) For a small number of pairs (10), cooperation emerges (red lines): 3 out of 17. (c) The emergencc of cooperation at equilibrium as a function of the population size and the initial probability of cooperation in an *evolutionary* game of iterative prisoner's dilemma played over 20 generations (x-axis). Cooperation at equilibrium: green.

strategy of gene pool survival and not merely the species that contains that gene pool.

Now we will consider a slightly different case: can the quality of fairness, as opposed to a shorter-term incentive for reward (i.e., greed), emerge naturally in a population? Can wisdom evolve in a population?

18.3.1 *Fair play is the only way to maximize personal benefit in a resource-limited competition*

In a recent cooperative game model ("divide a lottery"), fairness emerged as an optimal strategy even among self-interested agents competing for uncertain resources (Kim and Lee 2023). When fair players dominated, mutual benefit was maximized — demonstrating that fairness is not just morally desirable but evolutionarily advantageous. Let us examine this game in more detail.

Imagine a pot of money whose value (v) is not known to a group of players. To simplify the discussion, let's assume that the value is a fraction that varies between 0 and 1 ($0 \leq v \leq 1$). You could imagine it as a fraction of a sum of money, say, \$1000. The pot might be empty, or it may contain any fraction of \$1000, or the entire amount. Players, unaware of a pot's value v (a fraction between 0 and 1), sequentially guess fractions x (by the first player) and y (by the second player), such that $x + y \leq v$, each player receives their guessed amount. Otherwise, neither wins anything. Too high a call by the second player has a higher chance of yielding nothing at all for both, because there is a chance of overbidding that could destroy the second player's own chance of a payoff in the first round, and, of course, the first player would also lose.

Two types of players participate: rational players (R), aiming to maximize their own payoff, and fair players (F), seeking equitable outcomes. For rational players, the second player maximizes payoff using

$$y_R = y(1 - x - y),$$

with the optimal strategy being $\frac{1-x}{2}$ (Box 18.1).

For example, if the first player guesses $x = 0.75$ (75% of the pot containing, say, \$800, i.e., \$600), the second player optimally guesses $y = 0.125$ (or \$100, if the pot contained \$800) to avoid overbidding. Over sequential rounds, equilibrium strategies (x_{eq}, y_{eq}) converge to (Kim and Lee 2023):

$$x_{eq} = \frac{1}{2}, \text{ and } y_{eq} = y_R | x_{eq} = \frac{1}{4}.$$

Box 18.1

To find a value of y that maximizes the value of the function, $y_R = y(1 - x - y)$ where $0 \le x \le 1$ and $(0 \le y \le 1)$ we can proceed by taking the derivative of y_R with respect to y, then setting it to zero and solving for y.

Given, $y_R = y(1 - x - y)$

First expand and simplify: $y_R = y - xy - y^2$

Now take the derivative and set it equal to zero: $\frac{d_{yR}}{d_y} = 1 - x - 2y = 0$

Solving, $y = \frac{1-x}{2}$

Check that it is indeed the maximum by taking the second derivative: $\frac{d^2 y_R}{dy^2} = -2$, which is negative. Hence, $\frac{1-x}{2}$ is the maximum value of y.

However, when a rational player competes against a fair player — or two fair players compete — the dynamics shift (Kim and Lee 2023). Fair players prioritize eliminating disparities in payoffs. This approach often maximizes benefits for both participants, particularly when both adopt fair strategies. For example, fair play ensures balanced outcomes even when the exact value of the resource is unknown.

These findings underscore that fairness, rather than pure selfish rationality, is the optimal strategy in cooperative games involving finite but uncertain resources (Kim and Lee 2023). Fairness promotes mutual benefit, making it a superior evolutionary strategy.

Winning (or losing) a fraction of a sum of money is akin to gaining (or losing) a fitness value (the fraction of reproductive success), and multiple players (akin to individuals in a population) compete with one another. Thus, this game emulates a Darwinian idea of evolution by competition in finite-resource environments. Therefore, if the payoff is equivalent to the evolutionary fitness under a finite-resource environment, then fairness emerges as the best strategy in this evolutionary game, where players compete for their resources to optimize their own shares.

Here is the crux of the significance of the above results: this mathematical contingency suggests that a biological brain, complex enough to remember past outcomes of similar competitions (through nurture by grandmothers, for example) or by personal experience in the real world, knowing the reputation of the participants, and given limited resources, should evolve a biological propensity for fairness — simply because it optimizes fitness. If true, might there be genes involved in shaping the human brain to determine this "property of fairness"? Or is it entirely the effect of nurture in human societies that remembers the past and passes on strands of wisdom that allow us to compromise? The answer is likely both, because non-human mammals such as rats tend to show examples of compassion to fellow individuals — properties that are likely explained by genes — and because people embedded deeply in a civil society tend to behave in a fair manner depending on their personal experiences and nurture. The latter thought emphasizes the role of societal imprinting — nurture through social embedding — in perpetuating the quality of fairness for achieving and maintaining our evolutionary fitness. Indeed, experimental observations in non-human primates provide strong evidence in favor of the evolution of cognitive properties that tend to discourage partner dissatisfaction and its potentially negative impact on future cooperation (Brosnan and de Waal 2014). Thus, the human sense of fairness might have evolved from this cognitive adaptation that encourages cooperation at a future time — certainly an advantageous trait on which natural selection can work.

18.3.2 *Genetics of human morality?*

In a rather provocatively titled study (Graham, Haidt, and Nosek 2009), it has been claimed that there are five neuro-behavioral "intuitions" that govern moral choices in humans. These are the intuitive tendencies for (1) caring, (2) expectation of fairness, (3) kinship to the group in which a person belongs, (4) respect for authority, and, finally, (5) giving value to the idea of purity/sanctity. According to extensive studies on human subjects by these authors and others, moral choices by humans might be mapped on this five-dimensional vector-space, where individuals have their own intuitive positions. Interestingly, people tend to cluster into two main groups — some intuitively value caring (1) and fairness (2) more than the other three regimes of intuitive choice; the second cluster gives roughly the same weight to all five regimes (Kivikangas et al. 2021). If indeed human morality has a foundation in intuitive behavior, it must have a biological basis rooted in the structure and development of the human brain; hence, it must have been under evolutionary selection.

How might natural selection have influenced the evolution of human morality?

According to most anthropologists, modern humans, *Homo sapiens*, exhibit remarkable evidence of "creativity" relative to other hominids (Hill, Barton, and Hurtado 2009; Tattersall 2008; Tattersall 2009), from whom we have diverged about 500,000 years ago. However, the creative burst of modern humans reached a near explosive rate barely 50,000 years ago. This relatively rapid evolution of human cognition remains a mystery. If creativity is a defining characteristic of cognition in modern humans, then what exactly is creativity?

A succinct definition of creativity is the use of "imagination and original ideas in achieving valued goals" (Zwir et al. 2022). In a resource-limited environment, a "valued goal" is to optimize (not necessarily, maximize) personal benefit (or adaptation), which we have just seen (Section 18.2.1) to be critically dependent on the cognitive idea of "fairness". What then might be the biological basis of creativity that likely would have used, among other psychological traits not discussed here, "fairness" as an adaptive mechanism?

A series of recent studies involving several different ethnic groups have revealed approximately 900 genes that are statistically associated with a constellation of measurable properties of human creativity (Zwir et al. 2022). Out of these, only 267 genes are uniquely human, not present in either chimpanzees or neanderthals (both of whose genome sequences are known), and 509 genes are common among all three. A remarkable and highly significant pattern emerged when we asked what these genes encode: besides protein-coding genes among this set of ~900 genes, the proportion of genes that encoded no proteins but only RNA constituted about 8% of the genes carried in the chimpanzee genome, about 15% in Neanderthal, and about 27–29% in the modern human genome. Even more strikingly, among these non-protein-coding genes, there was zero gene (within this set of ~900 genes) that could code for long noncoding RNAs (lncRNAs)[1] in chimpanzees, whereas the Neanderthal and human gene sets (of this set of ~900 genes) encoded lncRNAs of various kinds. In fact, 251 out of the 267 genes unique to *Homo sapiens* encoded noncoding RNAs, most of these lncRNAs. Over the past ten years, it has become clear that lncRNAs control the expression of hundreds of genes through diverse mechanisms. One would therefore assume that, in modern humans, this novel set of regulatory genes is playing important functions. Could some of these uniquely human functions related to cognitive sophistication, expressions of creativity, lead to wisdom--the ability to see the place of the individual and the consequences of individual human action far into spatio-temporal distances? Is not wisdom the kernel of morality?

Now the real excitement! These 267 uniquely human lncRNA-encoding genes are highly expressed in human brain regions called amygdala, hippocampus, cingulate cortex, thalamus and the cerebellum — the very human brain regions that light up a bonfire of neuronal electrical activities when humans engage in creative thoughts (Zwir et al. 2022). Evidently, a uniquely human genetic network of uniquely human cognition defines this neuronal network. The authors rightly wonder:

> *"What mechanism promoted the emergence of the genetic network for creative self-awareness in behaviorally modern human beings? The brains of Sapiens are unique in having a system for self-awareness... These most recently evolved regions of the brain are the final association areas in which information is integrated and evaluated...self-transcendent thinking involves the ability to perceive oneself as a local aspect of a larger spatio-temporal whole, which permits thinking that is free and creative...as expressed in art, science, spirituality, and narrative syntactical language.*

---

[1] lncRNAs are generally longer than ~200 nucleotides in length, can be much larger, and their gene sequences are often highly conserved (even more than most protein-coding genes) among related species. Many of these lncRNA molecules are known to coordinately regulate the expression of hundreds of other genes.

...The newly emergent creativity may have provided selective advantages to behaviorally modern humans beyond its purely cognitive advantages. Physiologically, it is associated with enhanced memory functions, health, and well-being... Living longer and healthier lives may have allowed behaviorally modern Homo sapiens to disperse rapidly and widely around the world, and it may also have helped individuals support their children, grandchildren, and others in interconnected social communities, thereby possibly leading to positive selection for traits such as creativity, innovativeness, prosociality, and wisdom..."

(Zwir et al. 2022).

18.4 Darwinism Is Consistent with Inclusivity

Having established, at least in theory, that evolution can indeed produce conditions for fairness and altruism to evolve, and having cited results that demonstrate the presence of empathy in non-human animals, we are now unburdened by the false belief that social Darwinism must reject altruism in favor of economic progress. Thus, unfettered, our concerns for the potential evil unleashed by unregulated genetic engineering for human progress — curing diseases, environmental engineering, genetically engineered organisms — are indeed well-grounded in evolutionary theory. And the opposite — pure selfish rationalism — is not.

We may thus take pleasure and comfort in the thought that Darwinism is not, after all, a theory of dispassionate rational men fighting for survival in a dystopian world. Evolution does not move toward morality with inevitability; selfishness, tribalism, violence, and exploitation can also emerge under natural selection. The point is simply that Darwinian evolution does not uniquely favor such traits. Under appropriate ecological, social, and cognitive conditions, natural selection can also favor cooperation, fairness, empathy, creativity, and forms of moral behavior. Darwinism is consistent with the existence of the very traits of morality, sense of fairness and justice, empathy, love, passion, even poetry. These traits likely have been selected in human social evolution by fashioning genes and their interactions, which act through neuronal networks of the human brain, and thus have ensured our survival as the dominant species of mammals on Earth. We are only beginning to appreciate that these valued human traits might be the consequences of genes whose functions are shaped by natural selection. Our future survival as a species may depend on how our sense of fairness and cooperation, in the broadest sense, play out in the global context.

18.5 A Parting Thought

In the beginning of this book, I claimed that the reader does not need to believe in the truth or falsity of Darwinian evolution, but only that the reader assumed, for the moment, that it was true — and then followed the thread of reasoning. We have tried to do this throughout. We have seen how this assumption led to numerous interesting and unexpected connections, some mathematically reasoned and consistent with observations. We have witnessed the powers of this assumption in curing previously incurable diseases and in producing biotechnological innovations worth trillions of dollars. We have also seen controversies and conflicting ideas among scientists on how things really work. I leave the reader with the thought that what she has experienced is known as the scientific method, where all beliefs — except a belief in the method of science — are suspended such that the reasoned consequences of our assumptions can be tested in the crucible of our experimentations. This belief in the scientific method is all that is necessary. Whether or not Darwinian evolution is true or false is irrelevant. What matters is that it is a useful concept of profound power for understanding the biological universe.

***

Three men sit around a campfire by the side of an observatory in a deserted landscape in our parallel universe. The night is clear and crisp. Bertrand Russell was one of the most influential philosophers of the past 200 years, whose investigations into the foundations of mathematics laid, indirectly, the foundations of theoretical computer science; Albert Einstein needs no introduction; Carl Sagan, an astronomer, was probably the most prolific popularizer of science in the past century.

Russell: Gentlemen, we are at an inflection point in human history. Morality, once rooted in religion, faces the challenges of biotechnology, and artificial intelligence. Can humanity adapt to these new realities?

Sagan: Sir Russell, I am optimistic. Morality, like our understanding of the universe, evolves. We have expanded our circles of empathy — first to kin, then to tribes, nations, and now, perhaps, to the entire Earth. Science, especially biotechnology and medicine, forces us to consider moral dilemmas on a planetary scale.

Einstein: True, but empathy alone is insufficient. Reason must guide our morality. Science is a tool. Biotechnology can save or destroy, depending on how wisely we use it.

Russell: But Professor Einstein, how do we ensure wisdom? History is rife with examples of technological progress outpacing moral development. Take the atomic bomb — your own work when you were only twenty six years old enabled it. Its consequences remain a moral quagmire.

Einstein: A burden I carry, yes. But it illustrates my point. The problem is not science itself but the failure to temper its application with foresight and humility. The same applies to gene editing, and artificial intelligence.

Sagan: There lies the paradox. As we reveal the secrets of life, we play the role of cosmic gardeners by engineering life itself. Yet we are but a young species, prone to error, shortsightedness, and hubris. We make mistakes, large and small.

Russell: Cosmic gardeners, Sagan? An optimistic metaphor. I might prefer cosmic adolescents — reckless and unaware of our own ignorance. The climate crisis alone demonstrates our inability to plan beyond the next generation.

Sagan: Perhaps. But morality, a property of the biological brain, evolves, like life itself. The existential threats we face — climate change, pandemics, inequality, artificial intelligence robbing us of our freedom — could put sufficient selection pressure to evolve a new, more inclusive ethical framework. Consider the idea of a "communal brain", highly connected through social media, which prioritizes our shared humanity and the biosphere.

Einstein: A noble vision, but ideals require enforcement. How do we ensure that the communal brain's ethical instincts triumph over narrow self-interest? Nations and corporations often act as though morality is negotiable, bending to power and profit.

Russell: Education, Professor Einstein. The key is to teach critical thinking, nourish empathy, and provide the will to foresee the consequences of our actions. If we fail to educate, we fail to evolve morally.

Sagan: Education, yes. AI can instruct, it can solve queries that you can frame. But science can inspire awe, a sense of wonder, not just utility. This sense of wonder is the mother of creativity. If people grasp the interconnectedness of things — the pale blue dot in the vast universe — they may act with greater care.

Einstein: Wonder is indeed a teacher. Perhaps the greatest danger lies not in our technology but in forgetting our humanity.

Russell: Then let us hope that humanity chooses to evolve morally, not just technologically. By combining reason with compassion.

Einstein: With humility and wonder.

References

Abram, C. L., and S. A. Courtneidge. 2000. "Src Family Tyrosine Kinases and Growth Factor Signaling." *Experimental Cell Research* 254(1): 1–13. https://doi.org/10.1006/excr.1999.4732.

Abzhanov, Arhat, Meredith Protas, B. Rosemary Grant, Peter R. Grant, and Clifford J. Tabin. 2004. "Bmp4 and Morphological Variation of Beaks in Darwin's Finches." *Science (New York, N.Y.)* 305(5689): 1462–65. https://doi.org/10.1126/science.1098095.

Adami, Christoph, and Arend Hintze. 2013. "Evolutionary Instability of Zero-Determinant Strategies Demonstrates That Winning Is Not Everything." *Nature Communications* 4(August): 2193. https://doi.org/10.1038/ncomms3193.

Adhikari, Badri. 2020. "A Fully Open-Source Framework for Deep Learning Protein Real-Valued Distances." *Scientific Reports* 10(1): 13374. https://doi.org/10.1038/s41598-020-70181-0.

Akhoundi, Maryam, Mahsa Mohammadi, Seyedeh Saeideh Sahraei, Mohsen Sheykhhasan, and Nashmin Fayazi. 2021. "CAR T Cell Therapy as a Promising Approach in Cancer Immunotherapy: Challenges and Opportunities." *Cellular Oncology* 44(3): 495–523. https://doi.org/10.1007/s13402-021-00593-1.

Alharbi, Naif Khalaf, Eriko Padron-Regalado, Craig P. Thompson, Alexandra Kupke, Daniel Wells, Megan A. Sloan, Keith Grehan, et al. 2017. "ChAdOx1 and MVA Based Vaccine Candidates against MERS-CoV Elicit Neutralising Antibodies and Cellular Immune Responses in Mice." *Vaccine* 35(30): 3780–88. https://doi.org/10.1016/j.vaccine.2017.05.032.

Aliota, Matthew T., Stephen A. Peinado, Ivan Dario Velez, and Jorge E. Osorio. 2016. "The wMel Strain of Wolbachia Reduces Transmission of Zika Virus by Aedes Aegypti." *Scientific Reports* 6(July): 28792. https://doi.org/10.1038/srep28792.

Anandakrishnan, Ramu, Robin T. Varghese, Nicholas A. Kinney, and Harold R. Garner. 2019. "Estimating the Number of Genetic Mutations (Hits) Required for Carcinogenesis Based on the Distribution of Somatic Mutations." *PLOS Computational Biology* 15(3): e1006881. https://doi.org/10.1371/journal.pcbi.1006881.

Armitage, P., and R. Doll. 1954a. "The Age Distribution of Cancer and a Multi-Stage Theory of Carcinogenesis." *British Journal of Cancer* 8(1): 1–12. https://doi.org/10.1038/bjc.1954.1.

Armitage, P., and R. Doll. 1954b. "The Age Distribution of Cancer and a Multi-Stage Theory of Carcinogenesis." *British Journal of Cancer* 8(1): 1–12. https://www.ncbi.nlm.nih.gov/pmc/articles/PMC2007940/.

Arnold, Frances H. 1988. "Protein Design for Non-Aqueous Solvents." *Protein Engineering, Design and Selection* 2(1): 21–25. https://doi.org/10.1093/protein/2.1.21.

Association, British Medical. 1992. "Medicine Betrayed: The Participation of Doctors in Human Rights Abuses." Zed Books/Bloomsbury Academic (London) 1992. https://philpapers.org/rec/ASSMBT.

Bailey, Susan F., Nicolas Rodrigue, and Rees Kassen. 2015. "The Effect of Selection Environment on the Probability of Parallel Evolution." *Molecular Biology and Evolution* 32(6): 1436–48. https://doi.org/10.1093/molbev/msv033.

Baltimore, David. 1970. "Viral RNA-Dependent DNA Polymerase: RNA-Dependent DNA Polymerase in Virions of RNA Tumour Viruses." *Nature* 226(5252): 1209–11. https://doi.org/10.1038/2261209a0.

Bartel, David P. 2018. "Metazoan MicroRNAs." *Cell* 173(1): 20–51. https://doi.org/10.1016/j.cell.2018.03.006.

Bate, N., C. G. Savva, P. C. E. Moody, E. A. Brown, S. E. Evans, J. K. Ball, J. W. R. Schwabe, J. E. Sale, N. P. J. Brindle. 2022. "In vitro evolution predicts emerging SARS-CoV-2 mutations with high affinity for ACE2 and cross-species binding." *PLOS Pathogens* 18, e1010733(2022). https://doi.org/10.1371/journal.ppat.1010733

Battistuzzi, Fabia U., Andreia Feijao, and S. Blair Hedges. 2004. "A Genomic Timescale of Prokaryote Evolution: Insights into the Origin of Methanogenesis, Phototrophy, and the Colonization of Land." *BMC Evolutionary Biology* 4(1): 44. https://doi.org/10.1186/1471-2148-4-44.

Baylin, S. B., M. Esteller, M. R. Rountree, K. E. Bachman, K. Schuebel, and J. G. Herman. 2001. "Aberrant Patterns of DNA Methylation, Chromatin Formation and Gene Expression in Cancer." *Human Molecular Genetics* 10(7): 687–92. https://doi.org/10.1093/hmg/10.7.687.

Beck, Alain, Thierry Wurch, Christian Bailly, and Nathalie Corvaia. 2010. "Strategies and Challenges for the next Generation of Therapeutic Antibodies." *Nature Reviews Immunology* 10(5): 345–52. https://doi.org/10.1038/nri2747.

Beckwith, J. R., and E. R. Signer. 1966. "Transposition of the Lac Region of *Escherichia Coli*. I. Inversion of the Lac Operon and Transduction of Lac by Phi80." *Journal of Molecular Biology* 19(2): 254–65. https://doi.org/10.1016/s0022-2836(66)80003-7.

Bertrand Russell. 1940. *An Inquiry into Meaning and Truth*. New York: Routledge. https://www.routledge.com/An-Inquiry-into-Meaning-and-Truth/Russell/p/book/9780203714331.

Beutler, E. 1993. "Gaucher Disease as a Paradigm of Current Issues Regarding Single Gene Mutations of Humans." *Proceedings of the National Academy of Sciences* 90(12): 5384–90. https://doi.org/10.1073/pnas.90.12.5384.

Blackburn, Elizabeth H., Carol W. Greider, and Jack W. Szostak. 2006. "Telomeres and Telomerase: The Path from Maize, Tetrahymena and Yeast to Human Cancer and Aging." *Nature Medicine* 12(10): 1133–38. https://doi.org/10.1038/nm1006-1133.

Blackburn, E. H., and J. W. Szostak. 1984. "The Molecular Structure of Centromeres and TELomeres." *Annual Review of Biochemistry* 53(1): 163–94. https://doi.org/10.1146/annurev.bi.53.070184.001115.

Blagih, Julianna, Michael D. Buck, and Karen H. Vousden. 2020. "P53, Cancer and the Immune Response." *Journal of Cell Science* 133(5): jcs237453. https://doi.org/10.1242/jcs.237453.

Blake, Peter R., Katherine McAuliffe, and Felix Warneken. 2014. "The Developmental Origins of Fairness: The Knowledge–Behavior Gap." *Trends in Cognitive Sciences* 18(11): 559–61. https://doi.org/10.1016/j.tics.2014.08.003.

Blurton Jones, Nicholas G., Kristen Hawkes, and James F. O'Connell. 2002. "Antiquity of Postreproductive Life: Are There Modern Impacts on Hunter-Gatherer Postreproductive Life Spans?" *American Journal of Human Biology* 14(2): 184–205. https://doi.org/10.1002/ajhb.10038.

Bolotin, Alexander, Benoit Quinquis, Alexei Sorokin, and S. Dusko Ehrlich. 2005. "Clustered Regularly Interspaced Short Palindrome Repeats (CRISPRs) Have Spacers of Extrachromosomal Origin." *Microbiology (Reading, England)* 151(Pt 8): 2551–61. https://doi.org/10.1099/mic.0.28048-0.

Bradley CC, Wang C, Gordon AJE, Wen AX, Luna PN, Cooke MB, Kohrn BF, Kennedy SR, Avadhanula V, Piedra PA, Lichtarge O, Shaw CA, Ronca SE, Herman C. Targeted accurate RNA consensus sequencing (tARC-seq) reveals mechanisms of replication error affecting SARS-CoV-2 divergence. *Nat Microbiol.* 2024 May;9(5):1382-1392. doi: 10.1038/s41564-024-01655-4

Briney, Bryan, Anne Inderbitzin, Collin Joyce, and Dennis R. Burton. 2019. "Commonality despite Exceptional Diversity in the Baseline Human Antibody Repertoire." *Nature* 566(7744): 393–97. https://doi.org/10.1038/s41586-019-0879-y.

Brosnan, Sarah F., and Frans B. M. de Waal. 2003. "Monkeys Reject Unequal Pay." *Nature* 425(6955): 297–99. https://doi.org/10.1038/nature01963.

Brosnan, Sarah F., and Frans B. M. de Waal. 2014. "Evolution of Responses to (Un)Fairness." *Science (New York, N.Y.)* 346(6207): 1251776. https://doi.org/10.1126/science.1251776.

Brown, Kyle M., Marna S. Costanzo, Wenxin Xu, Scott Roy, Elena R. Lozovsky, and Daniel L. Hartl. 2010. "Compensatory Mutations Restore Fitness during the Evolution of Dihydrofolate Reductase." *Molecular Biology and Evolution* 27(12): 2682–90. https://doi.org/10.1093/molbev/msq160.

Brown, M. T., and J. A. Cooper. 1996. "Regulation, Substrates and Functions of Src." *Biochimica Et Biophysica Acta* 1287(2–3): 121–49. https://doi.org/10.1016/0304-419x(96)00003-0.

Buchanan, Adam H., Kandamurugu Manickam, Michelle N. Meyer, Jennifer K. Wagner, Miranda L. G. Hallquist, Janet L. Williams, Alanna Kulchak Rahm, et al. 2018. "Early Cancer Diagnoses through BRCA1/2 Screening of Unselected Adult Biobank Participants." *Genetics in Medicine: Official Journal of the American College of Medical Genetics* 20(5): 554–58. https://doi.org/10.1038/gim.2017.145.

Butts, Brittany, Hanfeng Huang, William T. Hu, Patrick Gavin Kehoe, James Scott Miners, Danielle D. Verble, Henrik Zetterberg, et al. 2024. "sPDGFRβ and Neuroinflammation Are Associated with AD Biomarkers and Differ by Race: The ASCEND Study." *Alzheimer's & Dementia: The Journal of the Alzheimer's Association* 20(2): 1175–89. https://doi.org/10.1002/alz.13457.

Cairns, J. 1975. "Mutation Selection and the Natural History of Cancer." *Nature* 255(5505): 197–200. https://doi.org/10.1038/255197a0.

Cairns, John. 2002. "Somatic Stem Cells and the Kinetics of Mutagenesis and Carcinogenesis." *Proceedings of the National Academy of Sciences* 99(16): 10567–70. https://doi.org/10.1073/pnas.162369899.

Cairns, John. 2006. "Cancer and the Immortal Strand Hypothesis." *Genetics* 174(3): 1069–72. https://doi.org/10.1534/genetics.104.66886.

Callaway, Ewen. 2020. "'It Will Change Everything': DeepMind's AI Makes Gigantic Leap in Solving Protein Structures." *Nature* 588(7837): 203–4. https://doi.org/10.1038/d41586-020-03348-4.

Capecchi, Mario R. 2005. "Gene Targeting in Mice: Functional Analysis of the Mammalian Genome for the Twenty-First Century." *Nature Reviews. Genetics* 6(6): 507–12. https://doi.org/10.1038/nrg1619.

Carroll, Dana. 2011. "Genome Engineering with Zinc-Finger Nucleases." *Genetics* 188(4): 773–82. https://doi.org/10.1534/genetics.111.131433.

Casado-Pelaez, Marta, Alberto Bueno-Costa, and Manel Esteller. 2022. "Single Cell Cancer Epigenetics." *Trends in Cancer* 8(10): 820–38. https://doi.org/10.1016/j.trecan.2022.06.005.

Chai, Heidi. 2008. "Tumor Suppressor (TS) Genes and the Two-Hit Hypothesis | Learn Science at Scitable." 2008. http://www.nature.com/scitable/topicpage/tumor-suppressor-ts-genes-and-the-two-887.

Chala, Bayissa, and Feyissa Hamde. 2021. "Emerging and Re-Emerging Vector-Borne Infectious Diseases and the Challenges for Control: A Review." *Frontiers in Public Health* 9: 715759. https://www.frontiersin.org/articles/10.3389/fpubh.2021.715759.

Chan, Andrew C., and Paul J. Carter. 2010. "Therapeutic Antibodies for Autoimmunity and Inflammation." *Nature Reviews Immunology* 10(5): 301–16. https://doi.org/10.1038/nri2761.

Chang, Joseph T., and Edward H. Kaplan. 2021. "Modeling Local Coronavirus Outbreaks." *European Journal of Operational Research* (August). https://doi.org/10.1016/j.ejor.2021.07.049. Vol. 304: pp. 57–68.

Chen, K, and F. H. Arnold. 1993. "Tuning the Activity of an Enzyme for Unusual Environments: Sequential Random Mutagenesis of Subtilisin E for Catalysis in Dimethylformamide." *Proceedings of the National Academy of Sciences* 90(12): 5618–22. https://doi.org/10.1073/pnas.90.12.5618.

Chen, K. Q., and F. H. Arnold. 1991. "Enzyme engineering for nonaqueous solvents: random mutagenesis to enhance activity of subtilisin E in polar organic media." *Biotechnology (NY)* 9, 1073–1077(1991). https://doi.org/10.1038/nbt1191-1073.

Chen, Xiaorui, Yang Luo, Qing Zhu, Jingzi Zhang, Huan Huang, Yansheng Kan, Dian Li, et al. 2024. "Small Extracellular Vesicles from Young Plasma Reverse Age-Related Functional Declines by Improving Mitochondrial Energy Metabolism." *Nature Aging* 4(6): 814–38. https://doi.org/10.1038/s43587-024-00612-4.

Cheng, Hua, R. Dustin Schaeffer, Yuxing Liao, Lisa N. Kinch, Jimin Pei, Shuoyong Shi, Bong-Hyun Kim, and Nick V. Grishin. 2014. "ECOD: An Evolutionary Classification of Protein Domains." *PLoS Computational Biology* 10(12): e1003926. https://doi.org/10.1371/journal.pcbi.1003926.

Cheng, Junyun, Gaole Lin, Tianhao Wang, Yunzhu Wang, Wenbo Guo, Jie Liao, Penghui Yang, et al. 2023. "Massively Parallel CRISPR-Based Genetic Perturbation Screening at Single-Cell Resolution." *Advanced Science* 10(4): 2204484. https://doi.org/10.1002/advs.202204484.

Cheong, Yoon Ling, Wasi A. Nazni, Han Lim Lee, Ahmad NoorAfizah, Ibrahim C. MohdKhairuddin, Ghazali M. R. Kamarul, Nasir M. N. Nizam, et al. 2023. "Spatial Distribution and Long-Term Persistence of Wolbachia-Infected Aedes Aegypti in the Mentari Court, Malaysia." *Insects* 14(4): 373. https://doi.org/10.3390/insects14040373.

Chiurazzi, M., A. Ray, J. F. Viret, R. Perera, X. H. Wang, A. M. Lloyd, and E. R. Signer. 1996. "Enhancement of Somatic Intrachromosomal Homologous Recombination in Arabidopsis by the HO Endonuclease." *Plant Cell* 8(11): 2057–66. https://doi.org/10.1105/tpc.8.11.2057.

Cho, Seung Woo, Sojung Kim, Jong Min Kim, and Jin-Soo Kim. 2013. "Targeted Genome Engineering in Human Cells with the Cas9 RNA-Guided Endonuclease." *Nature Biotechnology* 31(3): 230–32. https://doi.org/10.1038/nbt.2507.

Chowdhury, Ratul, Nazim Bouatta, Surojit Biswas, Christina Floristean, Anant Kharkar, Koushik Roy, Charlotte Rochereau, et al. 2022. "Single-Sequence Protein Structure Prediction Using a Language Model and Deep Learning." *Nature Biotechnology* 40(11): 1617–23. https://doi.org/10.1038/s41587-022-01432-w.

Clevers, Hans. 2005. "Stem Cells, Asymmetric Division and Cancer." *Nature Genetics* 37(10): 1027–28. https://doi.org/10.1038/ng1005-1027.

Coller, Barry S. 2019. "Ethics of Human Genome Editing." *Annual Review of Medicine* 70(1): 289–305. https://doi.org/10.1146/annurev-med-112717-094629.

Collins, Francis, and David Galas. 1993. "A New Five-Year Plan for the U.S. Human Genome Project." *Science* 262(5130): 43–46. https://doi.org/10.1126/science.8211127.

Conboy, Irina M., Michael J. Conboy, and Justin Rebo. 2015. "Systemic Problems: A Perspective on Stem Cell Aging and Rejuvenation." *Aging (Albany NY)* 7(10): 754–65. http://www.ncbi.nlm.nih.gov/pmc/articles/PMC4637204/.

Conboy, Michael J., Ariela O. Karasov, and Thomas A. Rando. 2007. "High Incidence of Non-Random Template Strand Segregation and Asymmetric Fate Determination in Dividing Stem Cells and Their Progeny." Edited by Margaret A Goodell. *PLoS Biology* 5(5): e102. https://doi.org/10.1371/journal.pbio.0050102.

Cong, Le, F. Ann Ran, David Cox, Shuailiang Lin, Robert Barretto, Naomi Habib, Patrick D. Hsu, et al. 2013. "Multiplex Genome Engineering Using CRISPR/Cas Systems." *Science (New York, N.Y.)* 339(6121): 819–23. https://doi.org/10.1126/science.1231143.

Corman, Victor Max, Ndapewa Laudika Ithete, Leigh Rosanne Richards, M. Corrie Schoeman, Wolfgang Preiser, Christian Drosten, and Jan Felix Drexler. 2014. "Rooting the Phylogenetic Tree of Middle East Respiratory Syndrome Coronavirus by Characterization of a Conspecific Virus from an African Bat." *Journal of Virology*, 88(19): 11297. https://doi.org/10.1128/JVI.01498-14.

Couzin, Jennifer. 2002. "Small RNAs Make Big Splash." *Science* 298(5602): 2296–97. https://doi.org/10.1126/science.298.5602.2296.

Cox, Stewart S., and Carmela M. Reichel. 2020. "Rats Display Empathic Behavior Independent of the Opportunity for Social Interaction." *Neuropsychopharmacology* 45(7): 1097–1104. https://doi.org/10.1038/s41386-019-0572-8.

Crawford, L. V., and E. M. Crawford. 1961. "The Properties of Rous Sarcoma Virus Purified by Density Gradient Centrifugation." *Virology* 13(2): 227–32. https://doi.org/10.1016/0042-6822(61)90056-3.

Crick, F. H. C. 1968. "The Origin of the Genetic Code." *Journal of Molecular Biology* 38(3): 367–79. https://doi.org/10.1016/0022-2836(68)90392-6.

Cyranoski, David. 2019. "The CRISPR-Baby Scandal: What's next for Human Gene-Editing." *Nature* 566(7745): 440–42. https://doi.org/10.1038/d41586-019-00673-1.

Dawkins, Richard. 2015. *The Blind Watchmaker: Why the Evidence of Evolution Reveals a Universe without Design*. Reissue ed. New York: W. W. Norton & Company.

De Meyer, Tim, Tim Nawrot, Sofie Bekaert, Marc L. De Buyzere, Ernst R. Rietzschel, and Vicente Andrés. 2018. "Telomere Length as Cardiovascular Aging Biomarker: JACC Review Topic of the Week." *Journal of the American College of Cardiology* 72(7): 805–13. https://doi.org/10.1016/j.jacc.2018.06.014.

Deonier, Richard C., Simon Tavaré, and Michael S. Waterman. 2005. *Computational Genome Analysis: An Introduction*. New York: Springer.

Diamond, Jared. 1987. "The Worst Mistake in the History of the Human Race." *Discover Magazine*, May, 64–66. https://www.discovermagazine.com/the-worst-mistake-in-the-history-of-the-human-race-12157

Dobzhansky, Theodosius. 1973. "Nothing in Biology Makes Sense except in the Light of Evolution." *The American Biology Teacher* 35(3): 125–29. https://doi.org/10.2307/4444260.

Dorp, Christiaan H. van, Emma E. Goldberg, Nick Hengartner, Ruian Ke, and Ethan O. Romero-Severson. 2021. "Estimating the Strength of Selection for New SARS-CoV-2 Variants." *Nature Communications* 12(1): 7239. https://doi.org/10.1038/s41467-021-27369-3.

Dorp, Christiaan van, Emma Goldberg, Ruian Ke, Nick Hengartner, and Ethan Romero-Severson. 2022. "Global Estimates of the Fitness Advantage of SARS-CoV-2 Variant Omicron." *Virus Evolution* 8(2): veac089. https://doi.org/10.1093/ve/veac089.

Dronamraju, Krishna R. 1987. "On Some Aspects of the Life and Work of John Burdon Sanderson Haldane, F.R.S., in India." *Notes and Records of the Royal Society of London* 41(2): 211–37. https://www.jstor.org/stable/531546.

Duerr, Ralf, Dacia Dimartino, Christian Marier, Paul Zappile, Guiqing Wang, Fritz François, Mila B. Ortigoza, et al. 2023. "Selective Adaptation of SARS-CoV-2 Omicron under Booster Vaccine Pressure: A Multicentre Observational Study." *eBioMedicine* 97(October): 104843. https://doi.org/10.1016/j.ebiom.2023.104843.

Duesberg, Peter H., and Peter K. Vogt. 1970. "Differences between the Ribonucleic Acids of Transforming and Nontransforming Avian Tumor Viruses." *Proceedings of the National Academy of Sciences* 67(4): 1673–80. https://doi.org/10.1073/pnas.67.4.1673.

Dugar, Gaurav, Ryan T. Leenay, Sara K. Eisenbart, Thorsten Bischler, Belinda U. Aul, Chase L. Beisel, and Cynthia M. Sharma. 2018. "CRISPR RNA-Dependent Binding and Cleavage of Endogenous RNAs by the Campylobacter Jejuni Cas9." *Molecular Cell* 69(5): 893–905.e7. https://doi.org/10.1016/j.molcel.2018.01.032.

Earnest, Mark. 2020. "On Becoming a Plague Doctor." *New England Journal of Medicine* 383(10): e64. https://doi.org/10.1056/NEJMp2011418.

Ellington, Andrew D., and Jack W. Szostak. 1990. "In Vitro Selection of RNA Molecules That Bind Specific Ligands." *Nature* 346(6287): 818–22. https://doi.org/10.1038/346818a0.

Elliott, Arthur R. 1911. "A Case of Myositis Ossificans Progressiva." *Journal of the American Medical Association* LVII (11) 57(11): 873–77. https://doi.org/10.1001/jama.1911.04260090095006.

Emert, Benjamin L., Christopher J. Cote, Eduardo A. Torre, Ian P. Dardani, Connie L. Jiang, Naveen Jain, Sydney M. Shaffer, and Arjun Raj. 2021. "Variability within Rare Cell States Enables Multiple Paths toward Drug Resistance." *Nature Biotechnology* 39(7): 865–76. https://doi.org/10.1038/s41587-021-00837-3.

Enriquez-Navas, Pedro M., Jonathan W. Wojtkowiak, and Robert A. Gatenby. 2015. "Application of Evolutionary Principles to Cancer Therapy." *Cancer Research* 75(22): 4675–80. https://doi.org/10.1158/0008-5472.CAN-15-1337.

Esteller, Manel. 2002. "CpG Island Hypermethylation and Tumor Suppressor Genes: A Booming Present, a Brighter Future." *Oncogene* 21(35): 5427–40. https://doi.org/10.1038/sj.onc.1205600.

Esvelt, Kevin M, Andrea L Smidler, Flaminia Catteruccia, and George M Church. 2014. "Concerning RNA-Guided Gene Drives for the Alteration of Wild Populations." *eLife* 3 (July): e03401. https://doi.org/10.7554/eLife.03401.

Finch, Caleb E. 2010. “Evolution of the Human Lifespan and Diseases of Aging: Roles of Infection, Inflammation, and Nutrition.” *Proceedings of the National Academy of Sciences* 107(suppl_1): 1718–24. https://doi.org/10.1073/pnas.0909606106.

Finkel, Toren, and Nikki J. Holbrook. 2000. “Oxidants, Oxidative Stress and the Biology of Ageing.” *Nature* 408(6809): 239–47. https://doi.org/10.1038/35041687.

Fisher, J. C. 1958. “Multiple-Mutation Theory of Carcinogenesis.” *Nature* 181(4609): 651–52. https://doi.org/10.1038/181651b0.

Fittall, Matthew W., and Peter Van Loo. 2019. “Translating Insights into Tumor Evolution to Clinical Practice: Promises and Challenges.” *Genome Medicine* 11(1): 20. https://doi.org/10.1186/s13073-019-0632-z.

Folger, K., K. Thomas, and M. R. Capecchi. 1984. “Analysis of Homologous Recombination in Cultured Mammalian Cells.” *Cold Spring Harbor Symposia on Quantitative Biology* 49:123–38. https://doi.org/10.1101/sqb.1984.049.01.016.

Folger, K. R., K. Thomas, and M. R. Capecchi. 1985. “Nonreciprocal Exchanges of Information between DNA Duplexes Coinjected into Mammalian Cell Nuclei.” *Molecular and Cellular Biology* 5(1): 59–69. https://doi.org/10.1128/mcb.5.1.59-69.1985.

Foretz, Marc, Bruno Guigas, Luc Bertrand, Michael Pollak, and Benoit Viollet. 2014. “Metformin: From Mechanisms of Action to Therapies.” *Cell Metabolism* 20(6): 953–66. https://doi.org/10.1016/j.cmet.2014.09.018.

Foulkes, William D., Bartha Maria Knoppers, and Clare Turnbull. 2016. “Population Genetic Testing for Cancer Susceptibility: Founder Mutations to Genomes.” *Nature Reviews Clinical Oncology* 13(1): 41–54. https://doi.org/10.1038/nrclinonc.2015.173.

Fraenkel, Aviezri S. 1993. “Complexity of Protein Folding.” *Bulletin of Mathematical Biology* 55(6): 1199–1210. https://doi.org/10.1016/S0092-8240(05)80170-3.

Frangoul, Haydar, David Altshuler, M. Domenica Cappellini, Yi-Shan Chen, Jennifer Domm, Brenda K. Eustace, Juergen Foell, et al. 2021. “CRISPR-Cas9 Gene Editing for Sickle Cell Disease and β-Thalassemia.” *New England Journal of Medicine* 384(3): 252–60. https://doi.org/10.1056/NEJMoa2031054.

Frosina, Guido. 2010. “The Bright and the Dark Sides of DNA Repair in Stem Cells.” *Journal of Biomedicine and Biotechnology* 2010:845396. https://doi.org/10.1155/2010/845396.

Fulop, Tamas, Anis Larbi, Gilles Dupuis, Aurélie Le Page, Eric H. Frost, Alan A. Cohen, Jacek M. Witkowski, and Claudio Franceschi. 2018. "Immunosenescence and Inflamm-Aging as Two Sides of the Same Coin: Friends or Foes?" *Frontiers in Immunology* 8(January):1960. https://doi.org/10.3389/fimmu.2017.01960.

Gantz, Valentino M., and Ethan Bier. 2015. "The Mutagenic Chain Reaction: A Method for Converting Heterozygous to Homozygous Mutations." *Science (New York, N.Y.)* 348(6233): 442. https://doi.org/10.1126/science.aaa5945.

Gatenby, Robert A., and Joel S. Brown. 2020. "Integrating Evolutionary Dynamics into Cancer Therapy." *Nature Reviews Clinical Oncology* 17(11): 675–86. https://doi.org/10.1038/s41571-020-0411-1.

Gerstung, Moritz, Clemency Jolly, Ignaty Leshchiner, Stefan C. Dentro, Santiago Gonzalez, Daniel Rosebrock, Thomas J. Mitchell, et al. 2020. "The Evolutionary History of 2,658 Cancers." *Nature* 578(7793): 122–28. https://doi.org/10.1038/s41586-019-1907-7.

Gilbert, Walter. 1986. "Origin of Life: The RNA World." *Nature* 319(6055): 618–618. https://doi.org/10.1038/319618a0.

Gilbert, Walter. 1991. "Towards a Paradigm Shift in Biology." *Nature* 349(6305): 99–99. https://doi.org/10.1038/349099a0.

Giordano, Giulia, Franco Blanchini, Raffaele Bruno, Patrizio Colaneri, Alessandro Di Filippo, Angela Di Matteo, and Marta Colaneri. 2020. "Modelling the COVID-19 Epidemic and Implementation of Population-Wide Interventions in Italy." *Nature Medicine* 26(6): 855–60. https://doi.org/10.1038/s41591-020-0883-7.

Goethe, Johann Wolfgang von and Translator, George Madison Priest. 1941. *Faust*. Translated and Edited from the Original German by George Madison Priest. Parts One & Two. Princeton: Alfred A. Knopf.

Gostimskaya, Irina. 2022. "CRISPR–Cas9: A History of Its Discovery and Ethical Considerations of Its Use in Genome Editing." *Biochemistry. Biokhimiia* 87(8): 777–88. https://doi.org/10.1134/S0006297922080090.

Gould, Stephen Jay. 1989. *Wonderful Life*. BOMC Edition. W. W. New York: Norton & Company, Inc.

Goyal, Yogesh, Gianna T. Busch, Maalavika Pillai, Jingxin Li, Ryan H. Boe, Emanuelle I. Grody, Manoj Chelvanambi, et al. 2023. "Diverse Clonal Fates Emerge upon Drug Treatment of Homogeneous Cancer Cells." *Nature* 620(7974): 651–59. https://doi.org/10.1038/s41586-023-06342-8.

Graham, Jesse, Jonathan Haidt, and Brian A. Nosek. 2009. "Liberals and Conservatives Rely on Different Sets of Moral Foundations." *Journal of*

Personality and Social Psychology 96(5): 1029–46. https://doi.org/10.1037/a0015141.

Graham, Rachel L., and Ralph S. Baric. 2010. "Recombination, Reservoirs, and the Modular Spike: Mechanisms of Coronavirus Cross-Species Transmission." *Journal of Virology* 84(7): 3134–46. https://doi.org/10.1128/JVI.01394-09.

Gregg, R. G., and O. Smithies. 1986. "Targeted Modification of Human Chromosomal Genes." *Cold Spring Harbor Symposia on Quantitative Biology* 51(Pt 2): 1093–99. https://doi.org/10.1101/sqb.1986.051.01.127.

Greider, C. W., and E. H. Blackburn. 1985. "Identification of a Specific Telomere Terminal Transferase Activity in Tetrahymena Extracts." *Cell* 43(2 Pt 1): 405–13. https://doi.org/10.1016/0092-8674(85)90170-9.

Greider, Carol W., and Elizabeth H. Blackburn. 1987. "The Telomere Terminal Transferase of Tetrahymena Is a Ribonucleoprotein Enzyme with Two Kinds of Primer Specificity." *Cell* 51(6): 887–98. https://doi.org/10.1016/0092-8674(87)90576-9.

Gunadi, Mohamad Saifudin Hakim, Hendra Wibawa, Marcellus, Ika Trisnawati, Endah Supriyati, Afiahayati, et al. 2021. "Association between Prognostic Factors and the Outcomes of Patients Infected with SARS-CoV-2 Harboring Multiple Spike Protein Mutations." *Scientific Reports* 11(1): 21352. https://doi.org/10.1038/s41598-021-00459-4.

Guo, Fang, Sanjib Das, Barbara M. Mueller, Carlos F. Barbas, Richard A. Lerner, and Subhash C. Sinha. 2006. "Breaking the One Antibody — One Target Axiom." *Proceedings of the National Academy of Sciences of the United States of America* 103(29): 11009–14. https://doi.org/10.1073/pnas.0603822103.

Gurven, Michael, and Hillard Kaplan. 2007. "Longevity Among Hunter-Gatherers: A Cross-Cultural Examination." *Population and Development Review* 33(2): 321–65. https://doi.org/10.1111/j.1728-4457.2007.00171.x.

Gutzmer, Ralf, Daniil Stroyakovskiy, Helen Gogas, Caroline Robert, Karl Lewis, Svetlana Protsenko, Rodrigo P. Pereira, et al. 2020. "Atezolizumab, Vemurafenib, and Cobimetinib as First-Line Treatment for Unresectable Advanced BRAFV600 Mutation-Positive Melanoma (IMspire150): Primary Analysis of the Randomised, Double-Blind, Placebo-Controlled, Phase 3 Trial." *The Lancet* 395(10240): 1835–44. https://doi.org/10.1016/S0140-6736(20)30934-X.

Haldane, J. B. S. 1924. "A Mathematical Theory of Artificial and Natural Selection." *Transactions of the Cambridge Philosophical Society* 23: 19–41.

Haldane, J. B. S. 1957. "The Cost of Natural Selection." *Journal of Genetics* 55(3): 511. https://doi.org/10.1007/BF02984069.

Haldane, J. B. S. 1941. *The Causes of Evolution. Princeton Science Library*. London: Allen & Unwin.

Hall, David W., Matthew Agan, and Sara C. Pope. 2010. "Fitness Epistasis among 6 Biosynthetic Loci in the Budding Yeast Saccharomyces Cerevisiae." *The Journal of Heredity* 101(Suppl 1): S75–84. https://doi.org/10.1093/jhered/esq007.

Hall, Michael J., Julia E. Reid, Lynn A. Burbidge, Dmitry Pruss, Amie M. Deffenbaugh, Cynthia Frye, Richard J. Wenstrup, Brian E. Ward, Thomas A. Scholl, and Walter W. Noll. 2009. "BRCA1 and BRCA2 Mutations in Women of Different Ethnicities Undergoing Testing for Hereditary Breast-Ovarian Cancer." *Cancer* 115(10): 2222–33. https://doi.org/10.1002/cncr.24200.

Hanahan, Douglas, and Robert A. Weinberg. 2011. "Hallmarks of Cancer: The next Generation." *Cell* 144(5): 646–74. https://doi.org/10.1016/j.cell.2011.02.013.

Hanson, Carl Veith, Yasuhiro Nishiyama, and Sudhir Paul. 2005. "Catalytic Antibodies and Their Applications." *Current Opinion in Biotechnology* 16(6): 631–36. https://doi.org/10.1016/j.copbio.2005.10.003.

Hariri, Yuval Noah. 2015. *Sapiens: A Brief History of Humankind (Hardcover) | Penguin Bookshop*. New York: Harper, Penguin. https://www.penguinbookshop.com/book/9780062316097.

Harish, Ajith, and Gustavo Caetano-Anollés. 2012. "Ribosomal History Reveals Origins of Modern Protein Synthesis." *PLOS One* 7(3): e32776. https://doi.org/10.1371/journal.pone.0032776.

Haura, Eric B. 2006. "SRC and STAT Pathways." *Journal of Thoracic Oncology* 1(5): 403–5. https://doi.org/10.1016/S1556-0864(15)31601-4.

Hauser, Marc. 2006. *Moral Minds: How Nature Designed Our Universal Sense of Right and Wrong*. New York: Ecco/HarperCollins Publishers.

Hemingway, Ernest. 1933. *Winner Take Nothing*. New York, London: Charles Scribner and Sons.

Henikoff, S., and J. G. Henikoff. 1992. "Amino Acid Substitution Matrices from Protein Blocks." *Proceedings of the National Academy of Sciences of the United States of America* 89(22): 10915–19. https://www.ncbi.nlm.nih.gov/pmc/articles/PMC50453/.

Hill, Kim, Michael Barton, and A. Magdalena Hurtado. 2009. "The Emergence of Human Uniqueness: Characters Underlying Behavioral Modernity." *Evolutionary Anthropology: Issues, News, and Reviews* 18(5): 187–200. https://doi.org/10.1002/evan.20224.

Hill, M., and Jana Hillova. 1972. "Virus Recovery in Chicken Cells Tested with Rous Sarcoma Cell DNA." *Nature New Biology* 237(71): 35–39. https://doi.org/10.1038/newbio237035a0.

Holliday, Robin. 1964. "A Mechanism for Gene Conversion in Fungi." *Genetics Research* 5(2): 282–304. https://doi.org/10.1017/S0016672300001233.

Huang, Shi, Niina Haiminen, Anna-Paola Carrieri, Rebecca Hu, Lingjing Jiang, Laxmi Parida, Baylee Russell, et al. 2020. "Human Skin, Oral, and Gut Microbiomes Predict Chronological Age." *mSystems* 5(1): e00630–19. https://doi.org/10.1128/msystems.00630-19.

Huebner, Robert J., and George J. Todaro. 1969. "Oncogenes of RNA Tumor Viruses as Determinants of Cancer." *Proceedings of the National Academy of Sciences* 64(3): 1087–94. https://doi.org/10.1073/pnas.64.3.1087.

Ilan, Yaron, Deborah Elstein, and Ari Zimran. 2009. "Glucocerebroside: An Evolutionary Advantage for Patients with Gaucher Disease and a New Immunomodulatory Agent." *Immunology & Cell Biology* 87(7): 514–24. https://doi.org/10.1038/icb.2009.42.

Ishino, Y., H. Shinagawa, K. Makino, M. Amemura, and A. Nakata. 1987. "Nucleotide Sequence of the Iap Gene, Responsible for Alkaline Phosphatase Isozyme Conversion in *Escherichia Coli*, and Identification of the Gene Product." *Journal of Bacteriology* 169(12): 5429–33. https://www.ncbi.nlm.nih.gov/pmc/articles/PMC213968/.

Jacot, Damien, Gilbert Greub, Katia Jaton, and Onya Opota. 2020. "Viral Load of SARS-CoV-2 across Patients and Compared to Other Respiratory Viruses." *Microbes and Infection* 22(10): 617–21. https://doi.org/10.1016/j.micinf.2020.08.004.

Jansen, Ruud., Jan. D. A. van Embden, Wim Gaastra, and Leo. M. Schouls. 2002. "Identification of Genes That Are Associated with DNA Repeats in Prokaryotes." *Molecular Microbiology* 43(6): 1565–75. https://doi.org/10.1046/j.1365-2958.2002.02839.x.

Ji, Jingran, Jacqueline V. Aredo, Andrew Piper-Vallillo, Laura Huppert, Julia K. Rotow, Hatim Husain, Susan Stewart, et al. 2023. "Osimertinib in NSCLC With Atypical EGFR-Activating Mutations: A Retrospective Multicenter Study." *JTO Clinical and Research Reports* 4(3): 100459. https://doi.org/10.1016/j.jtocrr.2022.100459.

Jinek, Martin, Alexandra East, Aaron Cheng, Steven Lin, Enbo Ma, and Jennifer Doudna. 2013. "RNA-Programmed Genome Editing in Human Cells." *eLife* 2(January):e00471. https://doi.org/10.7554/eLife.00471.

Jinek, Martin, Krzysztof Chylinski, Ines Fonfara, Michael Hauer, Jennifer A. Doudna, and Emmanuelle Charpentier. 2012. "A Programmable Dual RNA-Guided DNA Endonuclease in Adaptive Bacterial Immunity." *Science* 337(6096): 816–21. https://doi.org/10.1126/science.1225829.

Jones, P. A., and P. W. Laird. 1999. "Cancer Epigenetics Comes of Age." *Nature Genetics* 21(2): 163–67. https://doi.org/10.1038/5947.

Joung, J. Keith, and Jeffry D. Sander. 2013. "TALENs: A Widely Applicable Technology for Targeted Genome Editing." *Nature Reviews. Molecular Cell Biology* 14(1): 49–55. https://doi.org/10.1038/nrm3486.

Karikó, Katalin, Michael Buckstein, Houping Ni, and Drew Weissman. 2005. "Suppression of RNA Recognition by Toll-like Receptors: The Impact of Nucleoside Modification and the Evolutionary Origin of RNA." *Immunity* 23(2): 165–75. https://doi.org/10.1016/j.immuni.2005.06.008.

Karimzadeh, Sedighe, Raj Bhopal, and Huy Nguyen Tien. 2021. "Review of Infective Dose, Routes of Transmission and Outcome of COVID-19 Caused by the SARS-COV-2: Comparison with Other Respiratory Viruses." *Epidemiology & Infection* 149: e96. https://doi.org/10.1017/S0950268821000790.

Karlstrom, Amelie, Guofu Zhong, Christoph Rader, Nicholas A. Larsen, Andreas Heine, Roberta Fuller, Benjamin List, et al. 2000. "Using Antibody Catalysis to Study the Outcome of Multiple Evolutionary Trials of a Chemical Task." *Proceedings of the National Academy of Sciences of the United States of America* 97(8): 3878–83. https://www.ncbi.nlm.nih.gov/pmc/articles/PMC18110/.

Kelley, Brian. 2020. "Developing Therapeutic Monoclonal Antibodies at Pandemic Pace." *Nature Biotechnology* 38(5): 540–45. https://doi.org/10.1038/s41587-020-0512-5.

Kendal, Wayne S., and Philip Frost. 1988. "Pitfalls and Practice of Luria-Delbrück Fluctuation Analysis: A Review." *Cancer Research* 48(5): 1060–65. http://cancerres.aacrjournals.org/content/48/5/1060.

Kent, Bethany N., and Seth R. Bordenstein. 2010. "Phage WO of Wolbachia: Lambda of the Endosymbiont World." *Trends in Microbiology* 18(4): 173–81. https://doi.org/10.1016/j.tim.2009.12.011.

Khan, Aisha I., Duy M. Dinh, Dominique Schneider, Richard E. Lenski, and Tim F. Cooper. 2011. "Negative Epistasis between Beneficial Mutations in an Evolving Bacterial Population." *Science (New York, N.Y.)* 332(6034): 1193–96. https://doi.org/10.1126/science.1203801.

Kidwell, M. G., and D. R. Lisch. 2001. "Perspective: Transposable Elements, Parasitic DNA, and Genome Evolution." *Evolution; International Journal of Organic Evolution* 55(1): 1–24. https://doi.org/10.1111/j.0014-3820.2001.tb01268.x.

Kim, Jeong-Yoo, and Kyu-Min Lee. 2023. "Evolution of Fairness in the Divide-a-Lottery Game." *Scientific Reports* 13(1): 7048. https://doi.org/10.1038/s41598-023-34131-w.

Kim, Y. G., J. Cha, and S. Chandrasegaran. 1996. "Hybrid Restriction Enzymes: Zinc Finger Fusions to Fok I Cleavage Domain." *Proceedings of the National Academy of Sciences of the United States of America* 93(3): 1156–60. https://www.ncbi.nlm.nih.gov/pmc/articles/PMC40048/.

Kimura, M. 1981. "Estimation of Evolutionary Distances between Homologous Nucleotide Sequences." *Proceedings of the National Academy of Sciences of the United States of America* 78(1): 454–58. https://doi.org/10.1073/pnas.78.1.454.

Kirkwood, Thomas B. L. 2015. "Deciphering Death: A Commentary on Gompertz (1825) 'On the Nature of the Function Expressive of the Law of Human Mortality, and on a New Mode of Determining the Value of Life Contingencies.'" *Philosophical Transactions of the Royal Society B: Biological Sciences* 370(1666): 20140379. https://doi.org/10.1098/rstb.2014.0379.

Kivikangas, J. Matias, Belén Fernández-Castilla, Simo Järvelä, Niklas Ravaja, and Jan-Erik Lönnqvist. 2021. "Moral Foundations and Political Orientation: Systematic Review and Meta-Analysis." *Psychological Bulletin* 147(1): 55–94. https://doi.org/10.1037/bul0000308.

Köhler, G., and C. Milstein. 1975. "Continuous Cultures of Fused Cells Secreting Antibody of Predefined Specificity." *Nature* 256(5517): 495–97. https://doi.org/10.1038/256495a0.

Kong, Augustine, Michael L. Frigge, Gisli Masson, Soren Besenbacher, Patrick Sulem, Gisli Magnusson, Sigurjon A. Gudjonsson, et al. 2012. "Rate of de Novo Mutations and the Importance of Father's Age to Disease Risk." *Nature* 488(7412): 471–75. https://doi.org/10.1038/nature11396.

Kontomanolis, Emmanuel N., Antonios Koutras, Athanasios Syllaios, Dimitrios Schizas, Aikaterini Mastoraki, Nikolaos Garmpis, Michail Diakosavvas, et al. 2020. "Role of Oncogenes and Tumor-Suppressor Genes in Carcinogenesis: A Review." *Anticancer Research* 40(11): 6009–15. https://doi.org/10.21873/anticanres.14622.

Krupenye, Christopher, Fumihiro Kano, Satoshi Hirata, Josep Call, and Michael Tomasello. 2016. "Great Apes Anticipate That Other Individuals Will Act According to False Beliefs." *Science* 354(6308): 110–14. https://doi.org/10.1126/science.aaf8110.

Kukidome, Daisuke, Takeshi Nishikawa, Kazuhiro Sonoda, Koujiro Imoto, Kazuo Fujisawa, Miyuki Yano, Hiroyuki Motoshima, Tetsuya Taguchi, Takeshi Matsumura, and Eiichi Araki. 2006. "Activation of AMP-Activated Protein Kinase Reduces Hyperglycemia-Induced Mitochondrial Reactive Oxygen Species Production and Promotes Mitochondrial Biogenesis in

Human Umbilical Vein Endothelial Cells." *Diabetes* 55(1): 120–27. https://doi.org/10.2337/diabetes.55.01.06.db05-0943.

Kurian, Allison W., Paul Abrahamse, Allison Furgal, Kevin C. Ward, Ann S. Hamilton, Rachel Hodan, Rachel Tocco, et al. 2023. "Germline Genetic Testing After Cancer Diagnosis." *JAMA* 330(1): 43–51. https://doi.org/10.1001/jama.2023.9526.

Kuroki, Toshio, and Yoshinori Murakami. 1989. "Random Segregation of DNA Strands in Epidermal Basal Cells." *Japanese Journal of Cancer Research: Gann* 80(7): 637–42. https://doi.org/10.1111/j.1349-7006.1989.tb01690.x.

Lan, Jun, Jiwan Ge, Jinfang Yu, Sisi Shan, Huan Zhou, Shilong Fan, Qi Zhang, et al. 2020. "Structure of the SARS-CoV-2 Spike Receptor-Binding Domain Bound to the ACE2 Receptor." *Nature* 581(7807): 215–20. https://doi.org/10.1038/s41586-020-2180-5.

Larribe, Fabrice, Sabin Lessard, and Nicholas J. Schork. 2002. "Gene Mapping via the Ancestral Recombination Graph." *Theoretical Population Biology* 62(2): 215–29. https://doi.org/10.1006/tpbi.2002.1601.

Lau, Billy T., Dmitri Pavlichin, Anna C. Hooker, Alison Almeda, Giwon Shin, Jiamin Chen, Malaya K. Sahoo, et al. 2021. "Profiling SARS-CoV-2 Mutation Fingerprints That Range from the Viral Pangenome to Individual Infection Quasispecies." *Genome Medicine* 13(1): 62. https://doi.org/10.1186/s13073-021-00882-2.

Leavy, Olive. 2010. "Therapeutic Antibodies: Past, Present and Future." *Nature Reviews Immunology* 10(5): 297. https://doi.org/10.1038/nri2763.

Leavy, Olive. 2016. "The Birth of Monoclonal Antibodies." *Nature Immunology* 17(1): S13–S13. https://doi.org/10.1038/ni.3608.

Legraverend, Catherine, Marion Escobar, and Philippe Jay. 2010. "'The Immortal DNA Strand': Difficult to Digest?" *Cell Stem Cell* 6(4): 298–99. https://doi.org/10.1016/j.stem.2010.03.005.

Leiphrakpam, Premila D., Senthil J. Rajappa, Mridula Krishnan, Rishi Batra, Shilpa S. Murthy, and Chandrakanth Are. 2023. "Colorectal Cancer: Review of Signaling Pathways and Associated Therapeutic Strategies." *Journal of Surgical Oncology* 127(8): 1277–95. https://doi.org/10.1002/jso.27295.

Lenharo, Mariana. 2023. "Massive Mosquito Factory in Brazil Aims to Halt Dengue." *Nature* 616(7958): 637–38. https://doi.org/10.1038/d41586-023-01266-9.

Lerman, Caryn, and Alexandra E. Shields. 2004. "Genetic Testing for Cancer Susceptibility: The Promise and the Pitfalls." *Nature Reviews Cancer* 4(3): 235–41. https://doi.org/10.1038/nrc1301.

Levine, Arnold J. 2020. "P53: 800 Million Years of Evolution and 40 Years of Discovery." *Nature Reviews. Cancer* 20(8): 471–80. https://doi.org/10.1038/s41568-020-0262-1.

Levinthal, Cyrus. 1954. "Recombination in Phage T2: Its Relationship to Heterozygosis and Growth." *Genetics* 39(2): 169–84. https://doi.org/10.1093/genetics/39.2.169.

Lewnard, Joseph A., Vennis Hong, Jeniffer S. Kim, Sally F. Shaw, Bruno Lewin, Harpreet Takhar, Marc Lipsitch, and Sara Y. Tartof. 2023. "Increased Vaccine Sensitivity of an Emerging SARS-CoV-2 Variant." *Nature Communications* 14(1): 3854. https://doi.org/10.1038/s41467-023-39567-2.

Li, Benshang, Samuel W. Brady, Xiaotu Ma, Shuhong Shen, Yingchi Zhang, Yongjin Li, Karol Szlachta, et al. 2020. "Therapy-Induced Mutations Drive the Genomic Landscape of Relapsed Acute Lymphoblastic Leukemia." *Blood* 135(1): 41–55. https://doi.org/10.1182/blood.2019002220.

Li, Wendong, Zhengli Shi, Meng Yu, Wuze Ren, Craig Smith, Jonathan H. Epstein, Hanzhong Wang, et al. 2005. "Bats Are Natural Reservoirs of SARS-Like Coronaviruses." *Science*, 310: 676–9 October. https://doi.org/10.1126/science.1118391.

Li, Xia, Chentao Li, Wanying Zhang, Yanan Wang, Pengxu Qian, and He Huang. 2023. "Inflammation and Aging: Signaling Pathways and Intervention Therapies." *Signal Transduction and Targeted Therapy* 8(1): 1–29. https://doi.org/10.1038/s41392-023-01502-8.

Lin, Zeming, Halil Akin, Roshan Rao, Brian Hie, Zhongkai Zhu, Wenting Lu, Nikita Smetanin, et al. 2023. "Evolutionary-Scale Prediction of Atomic-Level Protein Structure with a Language Model." *Science* 379(6637): 1123–30. https://doi.org/10.1126/science.ade2574.

Liu, Yilun, and Stephen C. West. 2004. "Happy Hollidays: 40th Anniversary of the Holliday Junction." *Nature Reviews. Molecular Cell Biology* 5(11): 937–44. https://doi.org/10.1038/nrm1502.

Lozovsky, Elena R., Thanat Chookajorn, Kyle M. Brown, Mallika Imwong, Philip J. Shaw, Sumalee Kamchonwongpaisan, Daniel E. Neafsey, Daniel M. Weinreich, and Daniel L. Hartl. 2009. "Stepwise Acquisition of Pyrimethamine Resistance in the Malaria Parasite." *Proceedings of the National Academy of Sciences of the United States of America* 106(29): 12025–30. https://doi.org/10.1073/pnas.0905922106.

Luria, S. E., and M. Delbrück. 1943. "Mutations of Bacteria from Virus Sensitivity to Virus Resistance." *Genetics* 28(6): 491–511. https://doi.org/10.1093/genetics/28.6.491.

Lynce, Filipa, Karen Lisa Smith, Julie Stein, Tiffani DeMarco, Yiru Wang, Hongkun Wang, Melissa Fries, Beth N. Peshkin, and Claudine Isaacs. 2015. "Deleterious BRCA1/2 Mutations in an Urban Population of Black Women." *Breast Cancer Research and Treatment* 153(1): 201–9. https://doi.org/10.1007/s10549-015-3527-8.

Mai, Phuong L., Susan Thomas Vadaparampil, Nancy Breen, Timothy S. McNeel, Louise Wideroff, and Barry I. Graubard. 2014. "Awareness of Cancer Susceptibility Genetic Testing." *American Journal of Preventive Medicine* 46(5): 440–48. https://doi.org/10.1016/j.amepre.2014.01.002.

Makarova, Kira S., L. Aravind, Nick V. Grishin, Igor B. Rogozin, and Eugene V. Koonin. 2002. "A DNA Repair System Specific for Thermophilic Archaea and Bacteria Predicted by Genomic Context Analysis." *Nucleic Acids Research* 30(2): 482–96. https://www.ncbi.nlm.nih.gov/pmc/articles/PMC99818/.

Makarova, Kira S., Nick V. Grishin, Svetlana A. Shabalina, Yuri I. Wolf, and Eugene V. Koonin. 2006. "A Putative RNA-Interference-Based Immune System in Prokaryotes: Computational Analysis of the Predicted Enzymatic Machinery, Functional Analogies with Eukaryotic RNAi, and Hypothetical Mechanisms of Action." *Biology Direct* 1(March):7. https://doi.org/10.1186/1745-6150-1-7.

Mali, Prashant, Luhan Yang, Kevin M. Esvelt, John Aach, Marc Guell, James E. DiCarlo, Julie E. Norville, and George M. Church. 2013. "RNA-Guided Human Genome Engineering via Cas9." *Science (New York, N.Y.)* 339(6121): 823–26. https://doi.org/10.1126/science.1232033.

Market, Eleonora, and F. Nina Papavasiliou. 2003. "V(D)J Recombination and the Evolution of the Adaptive Immune System." *PLOS Biology* 1(1): e16. https://doi.org/10.1371/journal.pbio.0000016.

Mathur, Vani A., Tokiko Harada, Trixie Lipke, and Joan Y. Chiao. 2010. "Neural Basis of Extraordinary Empathy and Altruistic Motivation." *NeuroImage* 51(4): 1468–75. https://doi.org/10.1016/j.neuroimage.2010.03.025.

Maynard Smith, John. 1970. "Natural Selection and the Concept of a Protein Space." *Nature* 225(5232): 563–64. https://doi.org/10.1038/225563a0.

Maynard-Smith, John. 1986. "Evolutionary Game Theory." *Physica D: Nonlinear Phenomena*, Proceedings of the Fifth Annual International Conference, 22(1): 43–49. https://doi.org/10.1016/0167-2789(86)90232-0.

McNeil, Donald G. Jr. 2019. "Scientists Were Hunting for the Next Ebola. Now the U.S. Has Cut Off Their Funding." *The New York Times*, October 25, 2019, sec. Health. https://www.nytimes.com/2019/10/25/health/predict-usaid-viruses.html.

Mergener, Svenja, and Samuel Peña-Llopis. 2022. "A New Perspective on Immune Evasion: Escaping Immune Surveillance by Inactivating Tumor Suppressors." *Signal Transduction and Targeted Therapy* 7(1): 1–2. https://doi.org/10.1038/s41392-022-00875-6.

Merok, Joshua R., Janice A. Lansita, James R. Tunstead, and James L. Sherley. 2002. "Cosegregation of Chromosomes Containing Immortal DNA Strands in Cells That Cycle with Asymmetric Stem Cell Kinetics." *Cancer Research* 62(23): 6791–95.

Meselson, M. S., and C. M. Radding. 1975. "A General Model for Genetic Recombination." *Proceedings of the National Academy of Sciences* 72(1): 358–61. https://doi.org/10.1073/pnas.72.1.358.

Miler, Krzysztof, and Filip Turza. 2021. "'O Sister, Where Art Thou?' — A Review on Rescue of Imperiled Individuals in Ants." *Biology* 10(11): 1079. https://doi.org/10.3390/biology10111079.

Mohamad Anuar, Nur Najmi, Nur Syahidah Nor Hisam, Sze Ling Liew, and Azizah Ugusman. 2020. "Clinical Review: Navitoclax as a Pro-Apoptotic and Anti-Fibrotic Agent." *Frontiers in Pharmacology* 11: 564108. https://www.frontiersin.org/articles/10.3389/fphar.2020.564108.

Mojica, F. J., C. Díez-Villaseñor, E. Soria, and G. Juez. 2000. "Biological Significance of a Family of Regularly Spaced Repeats in the Genomes of Archaea, Bacteria and Mitochondria." *Molecular Microbiology* 36(1): 244–46. https://doi.org/10.1046/j.1365-2958.2000.01838.x.

Mojica, Francisco J. M., César Díez-Villaseñor, Jesús García-Martínez, and Elena Soria. 2005. "Intervening Sequences of Regularly Spaced Prokaryotic Repeats Derive from Foreign Genetic Elements." *Journal of Molecular Evolution* 60(2): 174–82. https://doi.org/10.1007/s00239-004-0046-3.

"Monoclonal Antibodies Market Size & Share Report, 2030." n.d. Accessed May 25, 2024. https://www.grandviewresearch.com/industry-analysis/monoclonal-antibodies-market.

Morris, A. P., J. C. Whittaker, and D. J. Balding. 2002. "Fine-Scale Mapping of Disease Loci via Shattered Coalescent Modeling of Genealogies." *The American Journal of Human Genetics* 70(3): 686–707. https://doi.org/10.1086/339271.

Morrison, Sean J., and Judith Kimble. 2006. "Asymmetric and Symmetric Stem-Cell Divisions in Development and Cancer." *Nature* 441(7097): 1068–74. https://doi.org/10.1038/nature04956.

Nakagawa, Kazuhiko, Edward B. Garon, Takashi Seto, Makoto Nishio, Santiago Ponce Aix, Luis Paz-Ares, Chao-Hua Chiu, et al. 2019. "Ramucirumab plus Erlotinib in Patients with Untreated, EGFR-Mutated, Advanced

Non-Small-Cell Lung Cancer (RELAY): A Randomised, Double-Blind, Placebo-Controlled, Phase 3 Trial." *The Lancet Oncology* 20(12): 1655–69. https://doi.org/10.1016/S1470-2045(19)30634-5.

Nidhi, Sweta, Uttpal Anand, Patrik Oleksak, Pooja Tripathi, Jonathan A. Lal, George Thomas, Kamil Kuca, and Vijay Tripathi. 2021. "Novel CRISPR-Cas Systems: An Updated Review of the Current Achievements, Applications, and Future Research Perspectives." *International Journal of Molecular Sciences* 22(7): 3327. https://doi.org/10.3390/ijms22073327.

Nordling, C. O. 1953. "A New Theory on Cancer-Inducing Mechanism." *British Journal of Cancer* 7(1): 68–72. https://doi.org/10.1038/bjc.1953.8.

Nowell, P. C. 1976. "The Clonal Evolution of Tumor Cell Populations." *Science (New York, N.Y.)* 194(4260): 23–28. https://doi.org/10.1126/science.959840.

Ohno, Susumu, Ulrich Wolf, and Niels B. Atkin. 1968. "Evolution from Fish to Mammals by Gene Duplication." *Hereditas* 59(1): 169–87. https://doi.org/10.1111/j.1601-5223.1968.tb02169.x.

Olovnikov, A. M. 1973. "A Theory of Marginotomy: The Incomplete Copying of Template Margin in Enzymic Synthesis of Polynucleotides and Biological Significance of the Phenomenon." *Journal of Theoretical Biology* 41(1): 181–90. https://doi.org/10.1016/0022-5193(73)90198-7.

Onken, Brian, Christine A. Sedore, Anna L. Coleman-Hulbert, David Hall, Erik Johnson, Eleanor Grace Jones, Stephen A. Banse, et al. 2022. "Metformin Treatment of Diverse Caenorhabditis Species Reveals the Importance of Genetic Background in Longevity and Healthspan Extension Outcomes." *Aging Cell* 21(1): e13488. https://doi.org/10.1111/acel.13488.

Orr, H. Allen. 2005. "The Probability of Parallel Evolution." *Evolution* 59(1): 216–20. https://doi.org/10.1111/j.0014-3820.2005.tb00907.x.

Orr-Weaver, T. L., J. W. Szostak, and R. J. Rothstein. 1981. "Yeast Transformation: A Model System for the Study of Recombination." *Proceedings of the National Academy of Sciences of the United States of America* 78(10): 6354–58. https://doi.org/10.1073/pnas.78.10.6354.

Ostrov, Nili, Akos Nyerges, Anush Chiappino-Pepe, Alexandra Rudolph, Maximilien Baas-Thomas, and George M. Church. 2020. "Synthetic Genomes with Altered Genetic Codes." *Current Opinion in Systems Biology*, Systems Immunology & Host-Pathogen Interaction (2020), 24(December): 32–40. https://doi.org/10.1016/j.coisb.2020.09.007.

Otto, Sarah P., Troy Day, Julien Arino, Caroline Colijn, Jonathan Dushoff, Michael Li, Samir Mechai, et al. 2021a. "The Origins and Potential Future of SARS-CoV-2 Variants of Concern in the Evolving COVID-19 Pandemic." *Current Biology: CB* 31(14): R918–29. https://doi.org/10.1016/j.cub.2021.06.049.

"Pan-Cancer Analysis of Whole Genomes." 2020. *Nature* 578(7793): 82–93. https://doi.org/10.1038/s41586-020-1969-6.

Park, Sungshim L., Iona Cheng, and Christopher A. Haiman. 2018. "Genome-Wide Association Studies of Cancer in Diverse Populations." *Cancer Epidemiology, Biomarkers & Prevention: A Publication of the American Association for Cancer Research, Cosponsored by the American Society of Preventive Oncology* 27(4): 405–17. https://doi.org/10.1158/1055-9965.EPI-17-0169.

Pederson, Thoru. 2004. "RNA Interference and mRNA Silencing, 2004: How Far Will They Reach?" *Molecular Biology of the Cell* 15(2): 407–10. https://doi.org/10.1091/mbc.E03-10-0726.

Pernicova, Ida, and Márta Korbonits. 2014. "Metformin — Mode of Action and Clinical Implications for Diabetes and Cancer." *Nature Reviews. Endocrinology* 10(3): 143–56. https://doi.org/10.1038/nrendo.2013.256.

Pinto, Sofia B., Thais I. S. Riback, Gabriel Sylvestre, Guilherme Costa, Julia Peixoto, Fernando B. S. Dias, Stephanie K. Tanamas, et al. 2021. "Effectiveness of Wolbachia-Infected Mosquito Deployments in Reducing the Incidence of Dengue and Other Aedes-Borne Diseases in Niterói, Brazil: A Quasi-Experimental Study." *PLoS Neglected Tropical Diseases* 15(7): e0009556. https://doi.org/10.1371/journal.pntd.0009556.

Potten, Christopher S., Gary Owen, and Dawn Booth. 2002. "Intestinal Stem Cells Protect Their Genome by Selective Segregation of Template DNA Strands." *Journal of Cell Science* 115(11): 2381–88. https://doi.org/10.1242/jcs.115.11.2381.

Pourcel, C., G. Salvignol, and G. Vergnaud. 2005. "CRISPR Elements in Yersinia Pestis Acquire New Repeats by Preferential Uptake of Bacteriophage DNA, and Provide Additional Tools for Evolutionary Studies." *Microbiology (Reading, England)* 151(Pt 3): 653–63. https://doi.org/10.1099/mic.0.27437-0.

Press, William H., and Freeman J. Dyson. 2012. "Iterated Prisoner's Dilemma Contains Strategies That Dominate Any Evolutionary Opponent." *Proceedings of the National Academy of Sciences of the United States of America* 109(26): 10409–13. https://doi.org/10.1073/pnas.1206569109.

Puchta, H., B. Dujon, and B. Hohn. 1996. "Two Different but Related Mechanisms Are Used in Plants for the Repair of Genomic Double-Strand Breaks by Homologous Recombination." *Proceedings of the National Academy of Sciences of the United States of America* 93(10): 5055–60. https://doi.org/10.1073/pnas.93.10.5055.

Pūras, Dainius. 2017. "Human Rights and the Practice of Medicine." *Public Health Reviews* 38(1): 9. https://doi.org/10.1186/s40985-017-0054-7.

Rader, Christoph, Subhash C. Sinha, Mikhail Popkov, Richard A. Lerner, and Carlos F. Barbas. 2003. "Chemically Programmed Monoclonal Antibodies for Cancer Therapy: Adaptor Immunotherapy Based on a Covalent Antibody Catalyst." *Proceedings of the National Academy of Sciences of the United States of America* 100(9): 5396–5400. https://doi.org/10.1073/pnas.0931308100.

Ramadan, Nour, and Houssam Shaib. 2019. "Middle East Respiratory Syndrome Coronavirus (MERS-CoV): A Review." *Germs* 9(1): 35–42. https://doi.org/10.18683/germs.2019.1155.

Rambhatla, Lakshmi, Sumati Ram-Mohan, Jennifer J. Cheng, and James L. Sherley. 2005. "Immortal DNA Strand Cosegregation Requires P53/IMPDH-Dependent Asymmetric Self-Renewal Associated with Adult Stem Cells." *Cancer Research* 65(8): 3155–61. https://doi.org/10.1158/0008-5472.CAN-04-3161.

Ratnieks, Francis L. W., and Heikki Helanterä. 2009. "The Evolution of Extreme Altruism and Inequality in Insect Societies." *Philosophical Transactions of the Royal Society B: Biological Sciences* 364(1533): 3169–79. https://doi.org/10.1098/rstb.2009.0129.

Rawls, John. 1971. *A Theory of Justice: Original Edition*. Harvard University Press. https://doi.org/10.2307/j.ctvjf9z6v.

Rawls, John1985. "Justice as Fairness: Political Not Metaphysical." *Philosophy & Public Affairs* 14(3): 223–51. https://www.jstor.org/stable/2265349.

Ray, A., I. Siddiqi, A. L. Kolodkin, and F. W. Stahl. 1988. "Intra-Chromosomal Gene Conversion Induced by a DNA Double-Strand Break in Saccharomyces Cerevisiae." *Journal of Molecular Biology* 201(2): 247–60. http://www.ncbi.nlm.nih.gov/entrez/query.fcgi?cmd=Retrieve&db=PubMed&dopt=Citation&list_uids=3047399.

Roa, Benjamin B., Alfred A. Boyd, Kelly Volcik, and C. Sue Richards. 1996. "Ashkenazi Jewish Population Frequencies for Common Mutations in BRCA1 and BRCA2." *Nature Genetics* 14(2): 185–87. https://doi.org/10.1038/ng1096-185.

Robson, Arthur J., and Hillard S. Kaplan. 2003. "The Evolution of Human Life Expectancy and Intelligence in Hunter-Gatherer Economies." *American Economic Review* 93(1): 150–69. https://doi.org/10.1257/000282803321455205.

Rodriguez Moore, Gabriela, Isabel Melo-Escobar, David Stegner, and Oliver Bracko. 2024. "One Immune Cell to Bind Them All: Platelet Contribution to Neurodegenerative Disease." *Molecular Neurodegeneration* 19(1): 65. https://doi.org/10.1186/s13024-024-00754-4.

Rosenberg, Noah A., and Magnus Nordborg. 2002. "Genealogical Trees, Coalescent Theory and the Analysis of Genetic Polymorphisms." *Nature Reviews Genetics* 3(5): 380–90. https://doi.org/10.1038/nrg795.

Roth, Richard B., and Leona D. Samson. 2002. "3-Methyladenine DNA Glycosylase-Deficient Aag Null Mice Display Unexpected Bone Marrow Alkylation Resistance1." *Cancer Research* 62(3): 656–60.

Rous, Peyton. 1967. "The Challenge to Man of the Neoplastic Cell." *Science* 157(3784): 24–28. https://doi.org/10.1126/science.157.3784.24.

Rudin, N., and J. E. Haber. 1988. "Efficient Repair of HO-Induced Chromosomal Breaks in Saccharomyces Cerevisiae by Recombination between Flanking Homologous Sequences." *Molecular and Cellular Biology* 8(9): 3918–28. https://doi.org/10.1128/mcb.8.9.3918-3928.1988.

Sabir, Jamal S. M., Tommy T.-Y. Lam, Mohamed M. M. Ahmed, Lifeng Li, Yongyi Shen, Salah E. M. Abo-Aba, Muhammd I. Qureshi, et al. 2016. "Co-Circulation of Three Camel Coronavirus Species and Recombination of MERS-CoVs in Saudi Arabia." *Science* 351: 81–4, January. https://doi.org/10.1126/science.aac8608.

Scheidel, Walter. 2001. "Roman Age Structure: Evidence and Models." *The Journal of Roman Studies* 91:1–26. https://doi.org/10.2307/3184767.

Scherer, S., and R. W. Davis. 1979. "Replacement of Chromosome Segments with Altered DNA Sequences Constructed in Vitro." *Proceedings of the National Academy of Sciences of the United States of America* 76(10): 4951–55. https://doi.org/10.1073/pnas.76.10.4951.

Schrago, C. G., B. Mello, and A. E. R. Soares. 2013. "Combining Fossil and Molecular Data to Date the Diversification of New World Primates." *Journal of Evolutionary Biology* 26(11): 2438–46. https://doi.org/10.1111/jeb.12237.

Schroer, Adam B., Patrick B. Ventura, Juliana Sucharov, Rhea Misra, M. K. Kirsten Chui, Gregor Bieri, Alana M. Horowitz, et al. 2023. "Platelet Factors Attenuate Inflammation and Rescue Cognition in Ageing." *Nature* 620(7976): 1071–79. https://doi.org/10.1038/s41586-023-06436-3.

Schubart, Christoph, Robert Stöhr, Lars Tögel, Florian Fuchs, Horia Sirbu, Gerhard Seitz, Ruth Seggewiss-Bernhardt, et al. 2021. "MET Amplification in Non-Small Cell Lung Cancer (NSCLC) — A Consecutive Evaluation Using Next-Generation Sequencing (NGS) in a Real-World Setting." *Cancers* 13(19): 5023. https://doi.org/10.3390/cancers13195023.

Sen, Amartya. 2006. "What Do We Want from a Theory of Justice?" *The Journal of Philosophy* 103(5): 215–38. https://www.jstor.org/stable/20619936.

Senior, Andrew W., Richard Evans, John Jumper, James Kirkpatrick, Laurent Sifre, Tim Green, Chongli Qin, et al. 2020. "Improved Protein Structure Prediction Using Potentials from Deep Learning." *Nature* 577(7792): 706–10. https://doi.org/10.1038/s41586-019-1923-7.

Shaffer, Sydney M., Margaret C. Dunagin, Stefan R. Torborg, Eduardo A. Torre, Benjamin Emert, Clemens Krepler, Marilda Beqiri, et al. 2017. "Rare Cell Variability and Drug-Induced Reprogramming as a Mode of Cancer Drug Resistance." *Nature* 546(7658): 431–35. https://doi.org/10.1038/nature22794.

Shalitin, C., and F. W. Stahl. 1965. "Additional Evidence for Two Kinds of Heterozygotes in Phage T4." *Proceedings of the National Academy of Sciences* 54(5): 1340–41. https://doi.org/10.1073/pnas.54.5.1340.

Shen, Aizhong, Zilin Gao, Xiang Gao, and Dan Cui. 2022. "The Evolutionary Extortion Game of Multiple Groups in Hypernetworks." *Scientific Reports* 12(December):20953. https://doi.org/10.1038/s41598-022-25294-z.

Sherr, Charles J. 2004. "Principles of Tumor Suppression." *Cell* 116(2): 235–46. https://doi.org/10.1016/S0092-8674(03)01075-4.

Sicheri, Frank, Ismail Moarefi, and John Kuriyan. 1997. "Crystal Structure of the Src Family Tyrosine Kinase Hck." *Nature* 385(6617): 602–9. https://doi.org/10.1038/385602a0.

Siegel, Rebecca L., Kimberly D. Miller, Hannah E. Fuchs, and Ahmedin Jemal. 2022. "Cancer Statistics, 2022." *CA: A Cancer Journal for Clinicians* 72(1): 7–33. https://doi.org/10.3322/caac.21708.

Siena, S., A. Sartore-Bianchi, S. Marsoni, H. I. Hurwitz, S. J. McCall, F. Penault-Llorca, S. Srock, A. Bardelli, and L. Trusolino. 2018. "Targeting the Human Epidermal Growth Factor Receptor 2(HER2) Oncogene in Colorectal Cancer." *Annals of Oncology* 29(5): 1108–19. https://doi.org/10.1093/annonc/mdy100.

Sinervo, B. 2005. "Evodevo: Darwin's Finch Beaks, Bmp4, and the Developmental Origins of Novelty." *Heredity* 94(2): 141–42. https://doi.org/10.1038/sj.hdy.6800618.

Slatkin, Montgomery. 2004. "A Population-Genetic Test of Founder Effects and Implications for Ashkenazi Jewish Diseases." *The American Journal of Human Genetics* 75(2): 282–93. https://doi.org/10.1086/423146.

Smithies, O. 1986. "Direct Alteration of a Gene in the Human Genome." *Journal of Inherited Metabolic Disease* 9 Suppl 1:92–97. https://doi.org/10.1007/BF01800862.

Smithies, O., and P. A. Powers. 1986. "Gene Conversions and Their Relation to Homologous Chromosome Pairing." *Philosophical Transactions of the Royal*

Society of London. Series B, Biological Sciences 312(1154): 291–302. https://doi.org/10.1098/rstb.1986.0008.

Song, Jiangning, Hao Tan, Mingjun Wang, Geoffrey I. Webb, and Tatsuya Akutsu. 2012. "TANGLE: Two-Level Support Vector Regression Approach for Protein Backbone Torsion Angle Prediction from Primary Sequences." *PLOS ONE* 7(2): e30361. https://doi.org/10.1371/journal.pone.0030361.

Spector, D. H, H. E. Varmus, and J. M. Bishop. 1978. "Nucleotide Sequences Related to the Transforming Gene of Avian Sarcoma Virus Are Present in DNA of Uninfected Vertebrates." *Proceedings of the National Academy of Sciences of the United States of America* 75(9): 4102–6. https://www.ncbi.nlm.nih.gov/pmc/articles/PMC336059/.

Stehelin, D., H. E. Varmus, J. M. Bishop, and P. K. Vogt. 1976. "DNA Related to the Transforming Gene(s) of Avian Sarcoma Viruses Is Present in Normal Avian DNA." *Nature* 260(5547): 170–73. https://doi.org/10.1038/260170a0.

Stewart, Alexander J., and Joshua B. Plotkin. 2014. "Collapse of Cooperation in Evolving Games." *Proceedings of the National Academy of Sciences of the United States of America* 111(49): 17558–63. https://doi.org/10.1073/pnas.1408618111.

Streisinger, George, Joyce Emrich, and Mary Morgan Stahl. 1967. "Chromosome Structure in Phage T4, Iii. Terminal Redundancy and Length Determination*." *Proceedings of the National Academy of Sciences* 57(2): 292–95. https://doi.org/10.1073/pnas.57.2.292.

Suda, Kenichi, Ryoichi Onozato, Yasushi Yatabe, and Tetsuya Mitsudomi. 2009. "EGFR T790M Mutation: A Double Role in Lung Cancer Cell Survival?" *Journal of Thoracic Oncology: Official Publication of the International Association for the Study of Lung Cancer* 4(1): 1–4. https://doi.org/10.1097/JTO.0b013e3181913c9f.

Sung, Hyuna, Jacques Ferlay, Rebecca L. Siegel, Mathieu Laversanne, Isabelle Soerjomataram, Ahmedin Jemal, and Freddie Bray. 2021. "Global Cancer Statistics 2020: GLOBOCAN Estimates of Incidence and Mortality Worldwide for 36 Cancers in 185 Countries." *CA: A Cancer Journal for Clinicians* 71(3): 209–49. https://doi.org/10.3322/caac.21660.

Sydney M. Shaffer, Benjamin L. Emert, Raúl A. Reyes Hueros, Christopher Cote, Guillaume Harmange, Dylan L. Schaff, Ann E. Sizemore, Rohit Gupte, Eduardo Torre, Abhyudai Singh, Danielle S. Bassett, Arjun Raj, Memory Sequencing Reveals Heritable Single-Cell Gene Expression Programs Associated with Distinct Cellular Behaviors, *Cell* 182, 947-959.e17 (2020), https://doi.org/10.1016/j.cell.2020.07.003. (https://www.sciencedirect.com/science/article/pii/S0092867420308680)

Szabo, C. I., and M. C. King. 1995. "Inherited Breast and Ovarian Cancer." *Human Molecular Genetics* 4(Suppl): 1811–17. https://doi.org/10.1093/hmg/4.suppl_1.1811.

Szostak, J. W., and E. H. Blackburn. 1982. "Cloning Yeast Telomeres on Linear Plasmid Vectors." *Cell* 29(1): 245–55. https://doi.org/10.1016/0092-8674(82)90109-x.

Szostak, J. W., T. L. Orr-Weaver, R. J. Rothstein, and F. W. Stahl. 1983. "The Double-Strand-Break Repair Model for Recombination." *Cell* 33(1): 25–35. https://doi.org/10.1016/0092-8674(83)90331-8.

Tattersall, Ian. 2008. "An Evolutionary Framework for the Acquisition of Symbolic Cognition by Homo Sapiens." *Comparative Cognition & Behavior Reviews* 3(January):99–114. https://doi.org/10.3819/ccbr.2008.30006.

Tattersall, Ian2009. "Human Origins: Out of Africa." *Proceedings of the National Academy of Sciences of the United States of America* 106(38): 16018–21. https://doi.org/10.1073/pnas.0903207106.

Temin, Howard M., and Satoshi Mizutani. 1970. "Viral RNA-Dependent DNA Polymerase: RNA-Dependent DNA Polymerase in Virions of Rous Sarcoma Virus." *Nature* 226(5252): 1211–13. https://doi.org/10.1038/2261211a0.

Tenaillon, Olivier, Jeffrey E. Barrick, Noah Ribeck, Daniel E. Deatherage, Jeffrey L. Blanchard, Aurko Dasgupta, Gabriel C. Wu, et al. 2016. "Tempo and Mode of Genome Evolution in a 50,000-Generation Experiment." *Nature* 536(7615): 165–70. https://doi.org/10.1038/nature18959.

Teoh, Phaik Ju, and Wee Joo Chng. 2021. "CAR T-Cell Therapy in Multiple Myeloma: More Room for Improvement." *Blood Cancer Journal* 11(4): 1–18. https://doi.org/10.1038/s41408-021-00469-5.

Teterina, Anastasia A., Anna L. Coleman-Hulbert, Stephen A. Banse, John H. Willis, Viviana I. Perez, Gordon J. Lithgow, Monica Driscoll, and Patrick C. Phillips. 2022. "Genetic Diversity Estimates for the Caenorhabditis Intervention Testing Program Screening Panel." *microPublication Biology* 10.17912 2022. https://doi.org/10.17912/micropub.biology.000518.

"The SIR Model for Spread of Disease — The Differential Equation Model | Mathematical Association of America." n.d. Accessed December 26, 2021. https://www.maa.org/press/periodicals/loci/joma/the-sir-model-for-spread-of-disease-the-differential-equation-model.

The Telomeres Mendelian Randomization Collaboration. 2017. "Association Between Telomere Length and Risk of Cancer and Non-Neoplastic Diseases: A Mendelian Randomization Study." *JAMA Oncology* 3(5): 636–51. https://doi.org/10.1001/jamaoncol.2016.5945.

Thomas, K. R., and M. R. Capecchi. 1986a. "Introduction of Homologous DNA Sequences into Mammalian Cells Induces Mutations in the Cognate Gene." *Nature* 324(6092): 34–38. https://doi.org/10.1038/324034a0.

Thomas, K. R., and M. R. Capecchi1986b. "Targeting of Genes to Specific Sites in the Mammalian Genome." *Cold Spring Harbor Symposia on Quantitative Biology* 51(Pt 2):1101–13. https://doi.org/10.1101/sqb.1986.051.01.128.

Thomas, K. R., K. R. Folger, and M. R. Capecchi. 1986. "High Frequency Targeting of Genes to Specific Sites in the Mammalian Genome." *Cell* 44(3): 419–28. https://doi.org/10.1016/0092-8674(86)90463-0.

Tomlinson, I. P. M., M. R. Novelli, and W. F. Bodmer. 1996. "The Mutation Rate and Cancer." *Proceedings of the National Academy of Sciences of the United States of America* 93(25): 14800–803. https://www.ncbi.nlm.nih.gov/pmc/articles/PMC26216/.

Torri, Alessandro, Johannes Jaeger, Thomas Pradeu, and Maria-Carla Saleh. 2022. "The Origin of RNA Interference: Adaptive or Neutral Evolution?" *PLOS Biology* 20(6): e3001715. https://doi.org/10.1371/journal.pbio.3001715.

Torre et al. (2021) full citation: Torre, E.A., Arai, E., Bayatpour, S. et al. Genetic screening for single-cell variability modulators driving therapy resistance. *Nat Genet* 53, 76–85 (2021). https://doi-org.ccl.idm.oclc.org/10.1038/s41588-020-00749-z

Torrisi, Mirko, Gianluca Pollastri, and Quan Le. 2020. "Deep Learning Methods in Protein Structure Prediction." *Computational and Structural Biotechnology Journal* 18(January):1301–10. https://doi.org/10.1016/j.csbj.2019.12.011.

Tsimberidou, Apostolia M., Elena Fountzilas, Mina Nikanjam, and Razelle Kurzrock. "Review of Precision Cancer Medicine: Evolution of the Treatment Paradigm." *Cancer Treatment Reviews* 86(June):102019. https://doi.org/10.1016/j.ctrv.2020.102019.

Tuerk, Craig, and Larry Gold. 1990. "Systematic Evolution of Ligands by Exponential Enrichment: RNA Ligands to Bacteriophage T4 DNA Polymerase." *Science* 249(4968): 505–10. https://doi.org/10.1126/science.2200121.

Uecker, Hildegard, and Joachim Hermisson. 2016. "The Role of Recombination in Evolutionary Rescue." *Genetics* 202(2): 721–32. https://doi.org/10.1534/genetics.115.180299.

Upasani, Vinit, Izabela Rodenhuis-Zybert, and Tineke Cantaert. 2021. "Antibody-Independent Functions of B Cells during Viral Infections." *PLOS Pathogens* 17(7): e1009708. https://doi.org/10.1371/journal.ppat.1009708.

Uribe, Mary Luz, Ilaria Marrocco, and Yosef Yarden. 2021. "EGFR in Cancer: Signaling Mechanisms, Drugs, and Acquired Resistance." *Cancers* 13(11): 2748. https://doi.org/10.3390/cancers13112748.

Varmus, Harold E. 1990. "Retroviruses and Oncogenes I (Nobel Lecture)." *Angewandte Chemie International Edition in English* 29(7): 707–15. https://doi.org/10.1002/anie.199007073.

Verosloff, Matthew S., Sarah J. Shapiro, Elizabeth M. Hawkins, Emel Alpay, Deepika Verma, Emma G. Stanfield, Lior Kreindler, et al. 2022. "CRISPR-Cas Enzymes: The Toolkit Revolutionizing Diagnostics." *Biotechnology Journal* 17(7): 2100304. https://doi.org/10.1002/biot.202100304.

Virani, Shami, Justin A. Colacino, Jung H. Kim, and Laura S. Rozek. 2012. "Cancer Epigenetics: A Brief Review." *ILAR Journal* 53(3–4): 359–69. https://doi.org/10.1093/ilar.53.3-4.359.

Waal, Frans B. M. de, and Stephanie D. Preston. 2017. "Mammalian Empathy: Behavioural Manifestations and Neural Basis." *Nature Reviews Neuroscience* 18(8): 498–509. https://doi.org/10.1038/nrn.2017.72.

Wang, Haifeng, Marie La Russa, and Lei S. Qi. 2016. "CRISPR/Cas9 in Genome Editing and Beyond." *Annual Review of Biochemistry* 85(1): 227–64. https://doi.org/10.1146/annurev-biochem-060815-014607.

Wang, Jiguang, Emanuela Cazzato, Erik Ladewig, Veronique Frattini, Daniel I. S. Rosenbloom, Sakellarios Zairis, Francesco Abate, et al. 2016. "Clonal Evolution of Glioblastoma under Therapy." *Nature Genetics* 48(7): 768–76. https://doi.org/10.1038/ng.3590.

Wang, Joy Y., Patrick Pausch, and Jennifer A. Doudna. 2022. "Structural Biology of CRISPR-Cas Immunity and Genome Editing Enzymes." *Nature Reviews Microbiology* 20(11): 641–56. https://doi.org/10.1038/s41579-022-00739-4.

Wang, Yanqun, Daxi Wang, Lu Zhang, Wanying Sun, Zhaoyong Zhang, Weijun Chen, Airu Zhu, et al. 2021. "Intra-Host Variation and Evolutionary Dynamics of SARS-CoV-2 Populations in COVID-19 Patients." *Genome Medicine* 13(1): 30. https://doi.org/10.1186/s13073-021-00847-5.

Watson, J. D. 1972. "Origin of Concatemeric T7DNA." *Nature New Biology* 239(94): 197–201. https://doi.org/10.1038/newbio239197a0.

Watson, James D., Berry, Andrew, and Davies, Kevin. 2017. *DNA: The Story of the Genetic Revolution*. Knopf Doubleday Publishing Group, New York (NY).

Weinreich, Daniel M., Nigel F. Delaney, Mark A. Depristo, and Daniel L. Hartl. 2006. "Darwinian Evolution Can Follow Only Very Few Mutational Paths to Fitter Proteins." *Science (New York, N.Y.)* 312(5770): 111–14. https://doi.org/10.1126/science.1123539.

Werner, Benjamin, and Andrea Sottoriva. 2018. "Variation of Mutational Burden in Healthy Human Tissues Suggests Non-Random Strand Segregation and Allows Measuring Somatic Mutation Rates." Edited by Natalia L. Komarova. *PLOS Computational Biology* 14(6): e1006233. https://doi.org/10.1371/journal.pcbi.1006233.

Werren, John H., Laura Baldo, and Michael E. Clark. 2008. "Wolbachia: Master Manipulators of Invertebrate Biology." *Nature Reviews Microbiology* 6(10): 741–51. https://doi.org/10.1038/nrmicro1969.

Wichman, Holly A., and Celeste J. Brown. 2010. "Experimental Evolution of Viruses: Microviridae as a Model System." *Philosophical Transactions of the Royal Society B: Biological Sciences* 365(1552): 2495–2501. https://doi.org/10.1098/rstb.2010.0053.

Wilson, Edward O. 2005. "Kin Selection as the Key to Altruism: Its Rise and Fall." *Social Research* 72(1): 159–66. https://www.jstor.org/stable/40972006.

Wu, Fan, Su Zhao, Bin Yu, Yan-Mei Chen, Wen Wang, Zhi-Gang Song, Yi Hu, et al. 2020. "A New Coronavirus Associated with Human Respiratory Disease in China." *Nature* 579(7798): 265–69. https://doi.org/10.1038/s41586-020-2008-3.

Xu, Lei, Jun Wang, Yulin Liu, Liangfu Xie, Bin Su, Danlei Mou, Longteng Wang, et al. 2019. "CRISPR-Edited Stem Cells in a Patient with HIV and Acute Lymphocytic Leukemia." *New England Journal of Medicine* 381(13): 1240–47. https://doi.org/10.1056/NEJMoa1817426.

Xu, Mary Jue, Daniel E. Johnson, and Jennifer R. Grandis. 2017. "EGFR-Targeted Therapies in the Post-Genomic Era." *Cancer Metastasis Reviews* 36(3): 463–73. https://doi.org/10.1007/s10555-017-9687-8.

Zhang, Bohan, David E. Lee, Alexandre Trapp, Alexander Tyshkovskiy, Akc T. Lu, Akshay Bareja, Csaba Kerepesi, et al. 2023. "Multi-Omic Rejuvenation and Lifespan Extension on Exposure to Youthful Circulation." *Nature Aging* 3(8): 948–64. https://doi.org/10.1038/s43587-023-00451-9.

Zhang, Jie, Shailaja Kalyankrishna, Marie Wislez, Nishan Thilaganathan, Babita Saigal, Wei Wei, Long Ma, Ignacio I. Wistuba, Faye M. Johnson, and Jonathan M. Kurie. 2007. "Src-Family Kinases Are Activated in Non-Small Cell Lung Cancer and Promote the Survival of Epidermal Growth Factor Receptor-Dependent Cell Lines." *The American Journal of Pathology* 170(1): 366–76. https://doi.org/10.2353/ajpath.2007.060706.

Zou, Roger S., Alberto Marin-Gonzalez, Yang Liu, Hans B. Liu, Leo Shen, Rachel K. Dveirin, Jay X. J. Luo, Reza Kalhor, and Taekjip Ha. 2022. "Massively Parallel Genomic Perturbations with Multi Target CRISPR

Interrogates Cas9 Activity and DNA Repair at Endogenous Sites." *Nature Cell Biology* 24(9): 1433–44. https://doi.org/10.1038/s41556-022-00975-z.
Zwanzig, R, A Szabo, and B Bagchi. 1992. "Levinthal's Paradox." *Proceedings of the National Academy of Sciences of the United States of America* 89(1): 20–22. https://www.ncbi.nlm.nih.gov/pmc/articles/PMC48166/.
Zwir, I., C. Del-Val, M. Hintsanen, K. M. Cloninger, R. Romero-Zaliz, A. Mesa, J. Arnedo, et al. 2022. "Evolution of Genetic Networks for Human Creativity." *Molecular Psychiatry* 27(1): 354–76. https://doi.org/10.1038/s41380-021-01097-y.